Lecture Notes in Computer Science 877

Edited by G. Goos, J. Hartmanis and J. van Leeuwen

Advisory Board: W. Brauer   D. Gries   J. Stoer

Leonard M. Adleman
Ming-Deh Huang (Eds.)

Algorithmic Number Theory

First International Symposium, ANTS-I
Ithaca, NY, USA, May 6-9, 1994
Proceedings

Springer-Verlag

Berlin Heidelberg New York
London Paris Tokyo
Hong Kong Barcelona
Budapest

Series Editors

Gerhard Goos
Universität Karlsruhe
Vincenz-Priessnitz-Straße 3, D-76128 Karlsruhe, Germany

Juris Hartmanis
Department of Computer Science, Cornell University
4130 Upson Hall, Ithaka, NY 14853, USA

Jan van Leeuwen
Department of Computer Science, Utrecht University
Padualaan 14, 3584 CH Utrecht, The Netherlands

Volume Editors

Leonard M. Adleman
Ming-Deh Huang
Department of Computer Science, University of Southern California
Los Angeles, CA 90089, USA

CR Subject Classification (1991): I.1, F.2.2, G.2, E.3-4, J.2

1991 Mathematics Subject Classification: 11Yxx, 11T71, 68P25, 68Q40, 68Q25,
68Q20, 12Y05, 94A60

ISBN 3-540-58691-1 Springer-Verlag Berlin Heidelberg New York

CIP data applied for

© Springer-Verlag Berlin Heidelberg 1994
Printed in Germany

Typesetting: Camera-ready by author
SPIN: 10479227   45/3140-543210 - Printed on acid-free paper

Preface

ANTS-I, the first Algorithmic Number Thoery Symposium was held at Cornell University on May 6-9, 1994. It was intended to be the first in a long series of international conferences on the algorithmic, computational and complexity theoretic aspects of number theory. It appeared to be rather successful and attracted over one hundred participants, primarily from Europe and North America. It will be followed in the summer of 1996 by ANTS-II in Bordeaux, France under the guidance of Professor Henri Cohen.

ANTS-I was supported in part by the United States Army Research Office through the Army Center of Excellence for Symbolic Methods in Algorithmic Mathematics (ACSyAM), Mathematical Sciences Institute of Cornell University.

ANTS-I was organized by Professors Leonard Adleman and Moss Sweedler. Professor Sweedler is director of ACSyAM and along with Valerie Kaine, ACSyAM conference coordinator, deserves great credit for providing a lovely forum for our conference. In addition, Kristin G. Sweedler generously provided our ANTS-I logo.

The programming committee was headed by Professors Leonard Adleman and Ming-Deh Huang. The other memebers of the programming committee were: Eric Bach, Ernie Brickell, Johannes Buchmann, Henri Cohen, Arjen Lenstra and Nelson Stephens. We thank them for their fine efforts in assuring a high quality symposium.

Leonard M. Adleman
Ming-Deh Huang
University of Southern California

Table of Contents

On the difficulty of finding reliable witnesses

W. R. Alford, Andrew Granville, and Carl Pomerance

Department of Mathematics, The University of Georgia, Athens, GA 30602, USA

Abstract. For an odd composite number n, let $w(n)$ denote the least witness for n; that is, the least positive number w for which n is not a strong pseudoprime to the base w. It is widely conjectured, but not proved, that $w(n) > 3$ for infinitely many n. We show the stronger result that $w(n) > (\log n)^{1/(3 \log \log \log n)}$ for infinitely many n. We also show that there are finite sets of odd composites which do not have a *reliable witness*, namely a common witness for all of the numbers in the set.

Introduction

Fermat's 'little' theorem asserts that

$$a^{n-1} \equiv 1 \bmod n, \tag{1}$$

whenever n is a prime that does not divide a. If (1) holds for a composite integer n then we call n a '*pseudoprime to base a*'. If a composite number n is a pseudoprime to every base a, for which $(a, n) = 1$, then we call n a '*Carmichael number*'. One can identify Carmichael numbers fairly easily by using

Korselt's criterion: A composite number n is a Carmichael number if and only if n is squarefree and $p - 1$ divides $n - 1$ for every prime p dividing n.

The smallest such number, 561 ($= 3 \times 11 \times 17$), was found by Carmichael in 1910. Recently we proved that there are infinitely many Carmichael numbers; in fact, that there are more than $x^{2/7}$ Carmichael numbers up to x, once x is sufficiently large (see [AGP]).

If n is neither prime nor a Carmichael number, then there are more than $n/2$ integers a in $[1, n-1]$ for which the congruence (1) does not hold. Thus if we pick an integer a at random in the interval $[1, n-1]$ then there is a better than even chance that (1) will fail and so we will have a proof that n is composite. If we repeat this 'test' say 100 times, then there is only a minuscule chance that we will fail to recognize such an integer n as composite (and, in fact, we expect to obtain such a 'witness' a in no more than two such tests). This algorithm is very efficient because one can determine powers modulo n extremely rapidly.

The only way that this test will recognize a Carmichael number n as composite is if some a is chosen in $[1, n-1]$ with $(a, n) > 1$. In section 2 we will show that there are infinitely many Carmichael numbers with no "small" prime factors[1]. So we cannot skirt this difficulty by instructing our algorithm to just

---

[1] specifically, we will show that for each fixed k, there are infinitely many Carmichael numbers n such that the probability that $(a, n) > 1$, for a random integer a in $[1, n-1]$, is $< 1/\log^k n$

look out for a finite list of exceptional integers. However this difficulty can be neatly resolved by replacing the 'Fermat test' based on (1) by a slightly stronger test: For any given odd integer $n > 1$, let t be the largest odd factor of $n - 1$, so we can write $n - 1 = 2^u t$ for some positive integer u. If n is a prime which does not divide a, then

Either $a^t \equiv 1 \bmod n$, **Or** $a^{2^i t} \equiv -1 \bmod n$ for some $i \in \{0, 1, \ldots, u - 1\}$. (2)

If this is true when n is an odd composite then we call n a 'strong pseudoprime to base a'. In the mid-70's Selfridge used (2) to rapidly identify composite numbers, which works whether or not they are Carmichael numbers.

An integer a is called a 'witness'[2] for n if n does not divide a and if (2) fails[3]. Selfridge had observed that there are always a lot of witnesses for any odd composite integer n; more precisely, Monier [M] and Rabin [R] independently proved that at least three-quarters of the integers a in the interval $[1, n - 1]$ are witnesses for n. Imitating the procedure above, we note that now if we select an integer a at random in the interval $[1, n - 1]$ then there is a far better than even chance that it will be a witness for n; and so we can be almost certain that we identify any composite number in just a few such tests[4].

So maybe we can use (2) to test whether a number is prime? Indeed, Miller proved, assuming the truth of an appropriate generalization of the Riemann Hypothesis (GRH), that if n is composite then there must be some 'small' value of a for which (2) fails, thus giving a 'polynomial time' deterministic primality test. Let $w(n)$ denote the least positive witness for n. Following work of Miller, Ankeny, Weinberger, Oesterlé and Bach (see [B]), we now know that

If the GRH is true then $w(n) < 2 \log^2 n$. (3)

We are concerned in this paper with determining how large $w(n)$ can get. It is known that 2 is a witness for most odd composite numbers (see [E] and [P]). However it is also known that there are infinitely many strong pseudoprimes base 2, so that the least witness is then at least 3. Specific examples have been found in which $w(n)$ is fairly large: For instance $w(3215031751) = 11$ ([PSW]) and $w(341550071728321) = 23$ ([J])[5]. [Ar] provides an extraordinary example of a 337-digit odd composite, whose least prime witness exceeds 200.

Prior to this paper it had not been proved that $w(n) > 3$ for infinitely many n, even though it has long been expected that $w(n)$ can get arbitrarily large. Here we prove this and more:

---

[2] to the fact that n is composite

[3] that is, n is *not* a strong pseudoprime to base a. Perhaps bases a to which n is a strong pseudoprime should be referred to as 'alibis', though the usual terminology is 'liars'.

[4] Actually Lehmer [Leh] and Solovay and Strassen [SS] noted that one can obtain such a surefire compositeness test using a procedure intermediate in strength between (1) and (2).

[5] These numbers are, in fact, 'champions', in that $w(n)$ is smaller for all smaller n.

Theorem 1. *There are infinitely many Carmichael numbers n with least witness larger than $(\log n)^{1/(3 \log \log \log n)}$. In fact, there are at least $x^{1/(35 \log \log \log x)}$ such n up to x, once x is sufficiently large.*

In section 3 we will argue that the maximal order of $w(n)$ is presumably $c \log n \log \log \log n$, for some constant $c > 0$, though there are many obstacles to turning our 'argument' into a proof[6]. However under the assumption of a suitable uniform version of the prime k-tuplets conjecture we are able to show that the maximal order of $w(n)$ is at least $\alpha \log n$ for some constant $\alpha > 0$. (A set of linear forms $\{a_i x + b_i, \ 1 \leq i \leq k\}$ is called 'admissible' if $1 \leq b_i < a_i$ for each i, and for every prime p, there exists an integer n_p such that p does not divide any of the $a_i n_p + b_i$. Hardy and Littlewood's 'prime k-tuplets conjecture' [HL] contends that for any admissible set of linear forms, there are infinitely many integers n for which each $a_i n + b_i$ is prime.)

Uniform prime k-tuplets conjecture. *For each integer $k \geq 1$, there exist constants $A_k, \gamma_k > 0$ such that for any 'admissible' set of linear forms $\{a_i x + b_i, \ 1 \leq i \leq k\}$ there exists an integer $n \leq \gamma_k (a_1 a_2 \ldots a_k)^{A_k}$ such that each $a_i n + b_i$ is prime.*

Such a result is known for $k = 1$ (Linnik's Theorem) and even with $A_1 = 5.5$ (see [HB], and [C] for a related theorem); and it is widely believed that the above uniform version of the prime k-tuplets conjecture is true. In section 3 we prove the following result.

Theorem 2. *Suppose that the 'Uniform prime triplets conjecture' is true (that is for $k = 3$). There exists a constant $\alpha > 0$ such that there are infinitely many Carmichael numbers n whose least witness is larger than $\alpha \log n$. Moreover there are at least x^β such n up to x, once x is sufficiently large, for some constant $\beta > 0$.*

Lenstra [Len] asked whether, for any given finite set of odd, composite numbers, there exists an integer w, perhaps very large, which serves as a witness for every number in the set (we will call w a 'reliable witness'). In particular, we would like to have a reliable witness for the set of odd composites up to x. Unfortunately we will prove that there cannot be a reliable witness once x is sufficiently large[7]. We shall actually prove that one needs quite a few numbers to correctly identify all of the odd, composite numbers up to x:

Theorem 3. *For all sufficiently large numbers x and for any set $\mathcal{W}$ of at most $(\log x)^{1/(3 \log \log \log x)}$ integers, there are more than $x^{1/(35 \log \log \log x)}$ Carmichael numbers $n \leq x$ with no witness in the set $\mathcal{W}$.*

---

[6] See also [BH].

[7] Two interesting computational problems come to mind: to find the smallest integer x for which there is no reliable witness for all of the odd composites up to x, and to find the smallest set of odd composites without a reliable witness.

Theorem 1 is a corollary of Theorem 3. If $\mathcal{W}$ is not so large then we can obtain larger sets of Carmichael numbers which have no witnesses in $\mathcal{W}$.

Theorem 4. *For any fixed δ, $0 < \delta < 1$, there exists a constant $c_\delta > 0$, such that for any set $\mathcal{W}$ of $\leq e^{c_\delta (\log \log x)^{(1-\delta)}}$ integers, there are more than $x^{3\delta/25}$ Carmichael numbers $n \leq x$ with no witness in the set $\mathcal{W}$.*

Besides determining $w(n)$, it is also of interest to determine the size of the smallest *'reliable set'* $\mathcal{W}$ of witnesses; this is a set $\mathcal{W}$ with the property that every composite integer up to x has a witness in $\mathcal{W}$. Theorem 3 implies that any such set contains more than $(\log x)^{1/(3 \log \log \log x)}$ numbers. We might wish to restrict the members of $\mathcal{W}$ to themselves be $\leq x$. By (3) we know that if the GRH is true then there is such a set of size $< 2 \log^2 x$. Adleman [A] and Dixon [D, Exercise 12] have shown how to get such a set of size $O(\log x)$ unconditionally (we shall also prove this in Proposition 3.1). We will further argue, in section 3, that it seems unlikely that there is a reliable set of witnesses of size $o(\log x)$.

On the other hand we are not sure what to conjecture about the size of the smallest set of reliable witnesses for the odd composites up to x, where we make no restriction on the size of the witnesses.

Although our arguments construct Carmichael numbers as obstructions to efficient primality testing, we point out that they are well known to be very easy to factor. Indeed, if $n = 2^u t + 1$ is a Carmichael number and w is a witness for n that is coprime to n, then $w^{2^i t}$ is a non-trivial square root of 1 mod n for some i, $0 \leq i \leq u - 1$, and so $w^{2^i t} - 1$ has a non-trivial gcd with n. See [BBCGP] for more in this vein.

Style and notation: Most of the proofs given involve modifications of the proofs in [AGP]; for brevity's sake we suppress details that remain exactly the same, referring the reader to [AGP]; though we have tried to make our explanations here as self-contained as possible. Throughout the paper there are inexplicit constants 'c_j', as well as 'for sufficiently large' hypotheses; these can be made explicit with considerable extra work.

1 Tools

We begin with a simple characterization of strong pseudoprimes which is stated without proof in [PSW]. For any pair of coprime integers a and n with $n > 0$, let $\ell_a(n)$ denote the order[8] of a modulo n .

Proposition 1.1. *Let n be a positive, odd composite integer. Then n is a strong pseudoprime to base a if and only if $a^{n-1} \equiv 1 \bmod n$ and there exists an integer k such that, for every prime factor p of n, 2^k divides $\ell_a(p)$ but 2^{k+1} does not.*

Proof. Throughout the proof we write $n = 2^u t + 1$ where t is odd.

Suppose that n is a strong pseudoprime to base a. Either $a^t \equiv 1 \bmod n$, so that $a^t \equiv 1 \bmod p$ for each prime factor p of n, and thus each $\ell_a(p)$ is odd

---

[8] that is, the order of a in the group $(\mathbb{Z}/n\mathbb{Z})^*$

(giving $k = 0$ above). Or there must exist some integer k in the range $1 \leq k \leq u$ for which $a^{2^{k-1}t} \equiv -1 \bmod n$. But then $a^{2^{k-1}t} \equiv -1 \bmod p$ for each prime p dividing n, and so 2^k is the exact power of 2 dividing each $\ell_a(p)$.

Suppose conversely that $a^{n-1} \equiv 1 \bmod n$ and that 2^k is the exact power of 2 dividing $\ell_a(p)$ for each prime factor p of n. It is well known that for any prime power p^b, the order of a modulo p^b equals some power of p times $\ell_a(p)$. Since n is odd this means that 2^k is the exact power of 2 dividing $\ell_a(p^b)$ for each prime power p^b dividing n. However, since we already know that $a^{2^u t} \equiv a^{n-1} \equiv 1 \bmod p^b$ we thus deduce that $a^{2^k t} \equiv 1 \bmod p^b$, whereas $a^{2^{k-1}t} \equiv -1 \bmod p^b$ if $k \geq 1$. By the Chinese Remainder Theorem, this implies that $a^{2^k t} \equiv 1 \bmod n$, whereas $a^{2^{k-1}t} \equiv -1 \bmod n$ if $k \geq 1$, and so n is a strong pseudoprime to base a.

We shall apply Proposition 1.1 to special types of Carmichael numbers in the following way.

Corollary 1.2. *Suppose that n is a Carmichael number, and that every prime factor of n is $\equiv 3 \bmod 4$. Then a is not a witness for n if and only if the quadratic residue symbol $\left(\dfrac{a}{p}\right)$ takes the same value for each prime divisor p of n. In particular, the least witness for n is prime.*

Proof. Note that n divides a if and only if p divides a for every prime divisor p of n (since any Carmichael number n is squarefree by Korselt's criterion); and this is true if and only if $\left(\dfrac{a}{p}\right) = 0$ for each prime divisor p of n.

Otherwise we may assume n does not divide a, and so a is not a witness for n if and only if n is a strong pseudoprime to base a. Let p be any prime divisor of n, which must be $\equiv 3 \bmod 4$ by hypothesis. Since $\ell_a(p)$ divides $p - 1$ (which is divisible by 2 but not by 4), we see that the exact power of 2 dividing $\ell_a(p)$ can be either 2^0 or 2^1, but no higher power. However, if 2^0 is the exact power of 2 dividing $\ell_a(p)$, then $\left(\dfrac{a}{p}\right) \equiv a^{(p-1)/2} \equiv 1 \bmod p$ and so $\left(\dfrac{a}{p}\right) = 1$. Alternatively, if 2^1 is the exact power of 2 dividing $\ell_a(p)$, then $a^{(p-1)/2} \not\equiv 1 \bmod p$ and so $\left(\dfrac{a}{p}\right) = -1$. The result follows now directly from Proposition 1.1.

Let $\lambda(n)$ denote the largest order of any element of the group $(\mathbb{Z}/n\mathbb{Z})^*$; note that $a^{\lambda(n)} \equiv 1 \bmod n$ for any integer a which is coprime to n, and that $\lambda(n)$ is the least such integer. Thus, $\lambda(n)$ is the lcm of the numbers $\lambda(p^a)$, where p^a runs over the prime power divisors of n, and $\lambda(p^a) = p^{a-1}(p-1)$ if $p > 2$ or $p^a = 2$ or 4, and $\lambda(2^a) = 2^{a-2}$ if $a \geq 3$. Known as Carmichael's function[9], $\lambda(n)$ is intimately connected with Carmichael numbers: a composite number n is Carmichael if and only if $\lambda(n)$ divides $n - 1$.

---

[9] Gauss discovered Carmichael's function over a hundred years before Carmichael: see article 92 of '*Disquisitiones Arithmeticae*' where Gauss discussed the function whilst classifying those moduli for which there is a primitive root.

Proposition 1.3. *Suppose n and k are coprime integers with $n > 2$ and S is a set of primes not dividing n which are all of the form $dk+1$, where d is a divisor of n. If $\#S > \lambda(n)\log n$ then there is a nonempty subset of S whose product is a Carmichael number.*

Proof. Since n and k are coprime, $(\mathbb{Z}/n\mathbb{Z})^*$ is isomorphic to the subgroup of $(\mathbb{Z}/nk\mathbb{Z})^*$ of residues that are 1 mod k. Note that S is naturally embedded in this subgroup. Since $n > 2$, we have $\lceil\lambda(n)\log n\rceil > \lceil\lambda(n)(1 + \log(\varphi(n)/\lambda(n)))\rceil$, where $\varphi(n) = \#(\mathbb{Z}/n\mathbb{Z})^*$. From a result of van Emde Boas and Kruyswijk (see [AGP, Theorem 1.1]), there is a subset of $S\backslash\{nk+1\}$ whose product is 1 mod nk. But then this product is a Carmichael number by Korselt's criterion.

Corollary 1.4. *Suppose n and k are coprime integers with $n > 2$ and S is a set of primes not dividing n which are all 3 mod 4 and all of the form $dk+1$, where d is a divisor of n. If ℓ, t and Λ are integers for which $3^{-\ell}\#S > t > \Lambda > \lambda(n)\log n$, then for any set W of ℓ integers, there are at least $\binom{3^{-\ell}\#S}{t} / \binom{3^{-\ell}\#S}{\Lambda}$ distinct subsets of S, each containing $\leq t$ elements, such that the product of the elements in each such subset is a Carmichael number with no witness in W.*

Proof. Suppose that $W = \{w_1, \ldots, w_\ell\}$, and consider the function $\chi_W : S \to \{1, 0, -1\}^\ell$, where

$$\chi_W(p) = \left(\left(\frac{w_1}{p}\right), \left(\frac{w_2}{p}\right), \ldots, \left(\frac{w_\ell}{p}\right)\right) \quad \text{for every } p \in S,$$

and where $\left(\frac{w}{p}\right)$ is the Legendre sybmbol. Since there are only 3^ℓ possible values that $\chi_W(p)$ can take, there must be a subset S_0 of S, of order $\geq 3^{-\ell}\#S$, on which χ_W remains constant. But by Corollary 1.2, any Carmichael number formed from the primes in S_0 has no witness in W. The assertion about the number of such Carmichael numbers with $\leq t$ prime factors follows directly from Proposition 1.3 and [AGP, Proposition 1.2].

Our principal results follow from Corollary 1.4, but to make use of it, we must show how numbers n and k may be constructed satisfying the hypotheses. The next result, which is derived from [AGP, Theorem 3.1], provides a way.

Proposition 1.5. *There exists a constant $c_0 > 0$ such that for any given arithmetic progression $l \bmod m$ with $(l, m) = 1$, if x is sufficiently large (depending on the choice of m) and if n is a squarefree integer which is coprime to m, then there exists an integer $k \leq x^{3/5}$ such that*

$$\#\{p \text{ prime} : p \leq x, \ p = dk + 1 \text{ for some } d|n, \ p \equiv l \bmod m\}$$

$$> \frac{c_0}{\varphi(m)\log x}\#\{d|n : \ d \leq x^{2/5}\}.$$

Further, if n has $\leq x^{1/4}$ prime factors, and the sum of the reciprocals of the primes dividing n is $\leq 1/60$, then we may take k to be coprime to n.

Proof. We shall modify the proof of Theorem 3.1 in [AGP], taking $B = 2/5$ there[10] to simplify matters. Note that by definition every element of $\mathcal{D}_B(x)$ is $> \log x$, so if we take $x \geq e^m$ then no member of the set $\mathcal{D}_B(x)$ of exceptional moduli can divide m. Analogously to the proof of Theorem 3.1 in [AGP] we begin by forming a new number n', obtained by removing from n some prime factor of (d, n) for each $d \in \mathcal{D}_B(x)$, so that no member of $\mathcal{D}_B(x)$ divides mn' (since $(m, n) = 1$ and n is squarefree). Note that there are $\leq D_B$ prime factors of n/n'.

For every integer d coprime to m, let a_d be the congruence class mod dm which is $\equiv 1 \bmod d$ and $\equiv l \bmod m$. For $d|n'$, $d \leq x^{2/5}$, we are interested in counting the number of primes $p \leq dx^{3/5}$ with $p \equiv a_d \bmod dm$ and $((p - 1)/d, n) = 1$. We proceed as in the proof of Theorem 3.1 in [AGP], though replacing the various estimates for the number of primes $\equiv 1 \bmod D$ by the analogous estimates for the number of primes $\equiv a_D \bmod Dm$ (here $D = d$ or dq of [AGP])[11]. One difference is that there we assumed that n had no prime factor $q > x^{3/10}$; whereas here we shall bound the 'contribution' of all of the primes $q > x^{2/7}$ dividing n by using the trivial fact that the number of primes $\leq dx^{3/5}$ which are $\equiv a_{dq} \bmod dqm$, is less than the number of integers in this arithmetic progression, which is $\leq 1 + x^{3/5}/qm$. However, by (the extended) hypothesis we know that there are $\leq x^{1/4}$ such primes q, so their total contribution is $\leq x^{1/4}(1 + x^{3/5}/x^{2/7}m) \leq x^{3/5}/(9m \log x)$ if x is sufficiently large. Therefore there are at least

$$\frac{x^{3/5}}{3\varphi(m) \log x} \#\{d|n' : d \leq x^{2/5}\}$$

pairs (p, d), where d divides n' and $d \leq x^{2/5}$, and p is a prime $\equiv a_d \bmod dm$ with $p \leq dx^{3/5}$ and $((p - 1)/d, n) = 1$. Each such pair corresponds to an integer $k = (p - 1)/d$ which is coprime to n and $\leq x^{3/5}$. Thus there is some such k which corresponds to at least $\#\{d|n' : d \leq x^{2/5}\}/(3\varphi(m) \log x)$ such pairs (p, d). The result with k coprime to n now follows from (3.1) of [AGP], where $c_0 = 1/(3 \cdot 2^{D_B})$.

The arithmetic progressions mod dqm occurred in the proof solely to ensure that the integer k produced is coprime to n. If we remove this assertion from the theorem then it is easy to remove the restrictions placed on the number of prime factors of n, and the sum of their reciprocals, leading to our result in the case when k is not guaranteed to be coprime to n.

Let $\tau(n)$ denote the number of positive integers which divide n. Take $l = m = 1$ and $x = n^{5/2}$ in Proposition 1.5. Since $\tau(n) = \#\{d|n : d \leq n\}$ we have the following result with $c_1 = 2c_0/5$.

Corollary 1.6. *For any sufficiently large squarefree integer n, there is some positive integer $k \leq n^{3/2}$ for which*

$$\#\{d|n : p = dk + 1 \text{ is a prime }\} > \frac{c_1 \tau(n)}{\log n}.$$

---

[10] which is in the set $\mathcal{B}$ of [AGP] since $2/5 < 5/12$.

[11] and we still look at such primes $\leq dx^{3/5}$.

We make one further observation that will be implicitly used in the next section:

Lemma 1.7. *For any sufficiently large finite set of primes $\mathcal{P}$, if $\mathcal{P}'$ consists of the largest $[(\#\mathcal{P})/2]$ primes in $\mathcal{P}$, then the sum of the reciprocals of the primes in $\mathcal{P}'$ is $\leq 1/60$.*

Proof. Say the least prime in $\mathcal{P}'$ is p and say $\#\mathcal{P}' = k$. Then the sum of the reciprocals of the members of $\mathcal{P}'$ is majorized by k/p. But there are at least k primes below p (in $\mathcal{P}$), so p is larger than the k-th prime. By the prime number theorem, the k-th prime is $\sim k \log k$ as $k \to \infty$, so the sum of the reciprocals of the members of $\mathcal{P}'$ is $\leq (1 + o(1))k/k \log k = o(1)$ as $k \to \infty$, and the result follows.

2 Two constructions – three proofs

In this section we prove Theorems 1, 3 and 4. The emphasis in Theorems 1 and 3 is more on producing numbers with an extreme property and less on producing many such numbers, while the emphasis in Theorem 4 is the reverse. To accomplish these different goals we shall use two different, but related constructions.

The first construction. Let α and ε be arbitrary, but fixed numbers with $0 < 5\varepsilon < \alpha < 1$. We assume that y is sufficiently large depending on the choice of α and ε. Let N be the product of the primes up to y, and let $X = N^{5/4}$. Let $L = M = 1$, and let $K \leq X^{3/5} = N^{3/4}$ be the number produced by Proposition 1.5 (taken with capital letters N, X, L, M, K, D). We now let n be the product of the larger half of the primes of the form $DK + 1 \leq X = N^{5/4}$, for which D divides N. As usual, let $\omega(n)$ denote the number of prime factors of n, $\tau(y)$ denote the number of primes up to y and $\theta(y)$ the sum of their logarithms. We have

$$N = e^{\theta(y)}, \quad \tau(N) = 2^{\pi(y)}, \quad (1/2)\tau(N) \geq \omega(n) > (2c_0/5)2^{\pi(y)}/\theta(y),$$

$$\lambda(n) \leq \acute{K}N \leq e^{(7/4)\theta(y)}, \quad \log n \leq \omega(n)\log(N^{5/4}) \leq \tau(N)\log N \leq e^{\varepsilon^2 y}.$$

Note that $\theta(y) \sim y$ as $y \to \infty$, so we may assume that $\theta(y) < 1.01y$.

Given a positive integer s, consider those divisors d of n with exactly s prime factors. The number of such divisors is $\binom{\omega(n)}{s}$, and each of these divisors is $\leq N^{(5/4)s} < e^{1.3ys}$. We shall let $s = [2^{\alpha\pi(y)}]$, and we take $x = N^{(25/8)s} < e^{3.2ys}$, so that $x^{2/5}$ is larger than every divisor d of n with exactly s prime factors. By Proposition 1.5 applied to n and x there exists an integer $k \leq x^{3/5}$, coprime to n, such that the set $\mathcal{S}$ of primes $p \leq x$ not dividing n for which $p \equiv 3 \bmod 4$ and $p = dk + 1$ for some divisor d of n is of order

$$\#\mathcal{S} \geq \frac{c_0}{2\log x}\#\{d|n : d \leq x^{2/5}\} - \omega(n) \geq \frac{c_0}{2(3.2ys)}\binom{\omega(n)}{s} - \omega(n)$$

$$\geq \frac{c_0}{6.4ys}\left(\frac{\omega(n)}{s}\right)^s - \omega(n) \geq 2^{(1-\alpha-\varepsilon)\pi(y)2^{\alpha\pi(y)}}.$$

We now will apply Corollary 1.4. Let $\Lambda = [e^{1.78y}]$ and $t = [e^{1.8y}]$, so that $t > \Lambda > \lambda(n)\log n$. We choose $\ell = s \ (= [2^{\alpha\pi(y)}])$. From the above calculation,

$$3^{-\ell}\#\mathcal{S} \geq 2^{(1-\alpha-2\varepsilon)\pi(y)2^{\alpha\pi(y)}}.$$

Thus, from Corollary 1.4, the number of Carmichael numbers which are the product of at most t primes from $\mathcal{S}$ and which do not have a witness in any given set $\mathcal{W}$ of ℓ integers is at least

$$\binom{3^{-\ell}\#\mathcal{S}}{t} \Big/ \binom{3^{-\ell}\#\mathcal{S}}{\Lambda} \geq \left(\frac{3^{-\ell}\#\mathcal{S}}{t}\right)^t (3^{-\ell}\#\mathcal{S})^{-\Lambda} = (3^{-\ell}\#\mathcal{S})^{t-\Lambda}t^{-t}$$

$$> 2^{(1-\alpha-3\varepsilon)\pi(y)2^{\alpha\pi(y)}e^{1.8y}}.$$

Further, each Carmichael number produced is bounded by

$$Y := x^t < e^{3.2yst} \leq e^{3.2y2^{\alpha\pi(y)}e^{1.8y}}.$$

We now rewrite ℓ and the number of Carmichael numbers produced in terms of Y. Since $\log\log\log Y \sim \log y$ as $y \to \infty$, and $\pi(y) \sim y/\log y$, the number of Carmichael numbers produced exceeds $Y^{(1-\alpha-4\varepsilon)(\log 2)/3.2\log\log\log Y}$. And since $\log\log Y \sim 1.8y$ as $y \to \infty$, we have that $\ell > 2^{(\alpha-\varepsilon)\log\log Y/1.8\log\log\log Y}$. We now choose $\alpha = .866$, $\varepsilon = .0001$, so that $\ell > (\log Y)^{1/3\log\log\log Y}$; and the number of Carmichael numbers produced below Y with no witness in any given set of ℓ integers exceeds $Y^{1/35\log\log\log Y}$. This concludes the proofs of both Theorems 1 and 3.

Remarks. By Proposition 1.3, there is some Carmichael number ν which is the product of a subset of the t largest primes in $\mathcal{S}$. Then $\nu \leq Y$ and ν has no prime factor below the t-th largest member of $\mathcal{S}$, which is $> \#\mathcal{S}/2 > e^s = e^\ell > \exp\left((\log Y)^{1/3\log\log\log Y}\right)$. That is, for each fixed k there are infinitely many Carmichael numbers ν with no prime factor below $\log^k \nu$. For such Carmichael numbers ν, the probability that $a^{\nu-1} \not\equiv 1 \bmod \nu$ for a random choice of a in $[1, \nu-1]$ is $< \log^{-k+1} \nu$. Thus, these numbers cannot be shown composite in polynomial time via the simple Fermat congruence, with either the deterministic test of choosing $a = 2, 3, \ldots$, or the probabilistic test of choosing a at random. Thanks are due to Neal Koblitz for steering us to these observations.

The constant "1/3" in Theorems 1 and 3 and in the remark above is not optimal. If one is willing to replace "1/35" with a smaller positive number and one replaces "2/5" in Proposition 1.5 with a number less than and arbitrarily close to 5/12, we may improve "1/3" to any number smaller than $(10/17)\log 2 = .4077\ldots > 2/5$.

The second construction is similar to the first, but with X and x chosen differently in the two applications of Proposition 1.5. We shall let α, β, ε be arbitrary, but fixed numbers with $0 < \alpha, \beta < 1$ and $0 < \varepsilon < 1/10$. We shall assume that y is sufficiently large, depending on the choice of α, β and ε. Let $u = [y^\beta/\log y]$, and, as before, let N be the product of the primes $\leq y$. The number of divisors D of N with $\omega(D) = u$ is at least $\binom{\pi(y)}{u} \geq y\mathrm{e}^{(1-\beta)y^\beta}$, since $\pi(y) \geq y/(\log y - 1)$. Note that each such D is $\leq y^u$. We apply Proposition 1.5 with N as above, $X = y^{(5/2)u} \leq \mathrm{e}^{(5/2)y^\beta}$, and $L = M = 1$. Let $K \leq X^{3/5} = \mathrm{e}^{o(y)}$ be the number produced by Proposition 1.5 and let n be the product of the larger half of the primes $DK + 1 \leq \mathrm{e}^{(5/2)y^\beta}$ for which D divides N. We have

$$\omega(n) \geq \frac{c_0}{2\log X} \#\{D|N : \omega(D) = u\} > \mathrm{e}^{(1-\beta)y^\beta},$$

$$\lambda(n) \leq KN \leq \mathrm{e}^{(1+\varepsilon)y}, \quad \log n \leq \tau(N)\log(KN + 1) \leq \mathrm{e}^{\varepsilon y}.$$

We now take $\ell = [\mathrm{e}^{\alpha(1-\beta)y^\beta}]$ and consider divisors d of n with exactly ℓ prime factors. Each such divisor d is at most $\mathrm{e}^{(5/2)y^\beta\ell}$. We apply Proposition 1.5 to n, with $x = \mathrm{e}^{(25/4)y^\beta\ell}$ and $l \bmod m = 3 \bmod 4$. Let $k \leq x^{3/5}$, $(k, n) = 1$, be the number produced by Proposition 1.5 and let S be the set of primes $p \leq x$ not dividing n with $p \equiv 3 \bmod 4$ and $p = dk + 1$ for some divisor d of n. Then

$$3^{-\ell}\#S \geq \frac{3^{-\ell}c_0}{2\log x} \#\{d|n : d \leq x^{2/5}\} - 3^{-\ell}\omega(n) > \frac{3^{-\ell}c_0}{13y^\beta\ell}\binom{\omega(n)}{\ell} - \omega(n)$$

$$\geq \frac{c_0}{13y^\beta\ell}\left(\frac{\omega(n)}{3\ell}\right)^\ell - \omega(n) \geq \exp\left((1-\varepsilon)(1-\alpha)(1-\beta)y^\beta\mathrm{e}^{\alpha(1-\beta)y^\beta}\right).$$

We choose $\Lambda = [\mathrm{e}^{(1+3\varepsilon)y}]$ and $t = [\mathrm{e}^{(1+4\varepsilon)y}]$. From Corollary 1.4, the number of Carmichael numbers which are a product of at most t primes from S and which do not have a witness in any given set W of ℓ integers is at least

$$\left(3^{-\ell}\#S\right)^{t-\Lambda} t^{-t} > \exp\left((1-2\varepsilon)(1-\alpha)(1-\beta)y^\beta\mathrm{e}^{\alpha(1-\beta)y^\beta}\mathrm{e}^{(1+4\varepsilon)y}\right).$$

Since each prime in S is $\leq x$, these Carmichael numbers are all $\leq x^t \leq Y :=$ $\exp\left((25/4)y^\beta\mathrm{e}^{\alpha(1-\beta)y^\beta}\mathrm{e}^{(1+4\varepsilon)y}\right)$. In terms of the new variable Y we have $\ell >$ $\exp\left((1-4\varepsilon)\alpha(1-\beta)(\log\log Y)^\beta\right)$; and the number of Carmichael numbers produced below Y, with no witness in any given set of ℓ integers, is at least Y^c, where $c = (4/25)(1-2\varepsilon)(1-\alpha)(1-\beta)$. Taking ε and α so that $(1-2\varepsilon)(1-\alpha) = 3/4$ and letting $\beta = 1 - \delta$ completes the proof of Theorem 4 (with $c_6 \gg \delta$).

Remarks. By taking ε, α and β small, the above argument implies that up to Y there are at least Y^c Carmichael numbers, for any $c < 4/25$, for Y exceeding some number $Y_0(c)$. Since the number "2/5" in Proposition 1.5 may be replaced by any number smaller than $5/12$, we have the above for any $c < 25/144$, which is Theorem 5 in [AGP], though with a somewhat different proof. This result is effective, in that a value for $Y_0(c)$ may be computed in principle. Using the main theorem from [F], we were able to prove in [AGP] the non-effective result that there are more than $Y^{2/7}$ Carmichael numbers up to Y, for Y sufficiently large. That construction is not particularly amenable to producing composite numbers with no witness in a large predetermined set, but does at least give that for any number ρ with $0 < \rho < 2$ and for any predetermined set of size $(\log\log Y)^{1+\rho}$, there are more than $Y^{(2-\rho)/7}$ Carmichael numbers up to Y with no witness in the set, provided Y is sufficiently large, depending on the choice of ρ.

3 Upper bounds and heuristics

Proposition 3.1. *For any integer $x \geq 1$ there is a set $\mathcal{W}$ of at most $(6/5)\log x$ integers $\leq x$ such that every odd, composite integer $\leq x$ has a witness in $\mathcal{W}$.*

Proof. A slightly stronger version of the result of Monier [M] and Rabin [R] mentioned in the introduction is that if $n > 9$ is an odd composite, then at least $3/4$ of the integers in $[1, n]$ are witnesses for n. Since $a + kn$ is a witness for n whenever a is, we see that the number of witnesses up to x for any odd composite n with $9 < n \leq x$ is at least the maximum of $(3/4)n\lfloor x/n\rfloor$ and $x - (1/4)n\lceil x/n\rceil$. A simple calculation shows that this is at least $(3/5)x$; and that for $n = 9$ as well, there are at least $(3/5)x$ witnesses up to x.

Suppose $\mathcal{S}$ is a set of odd composites in $[1, x]$. There are two ways to count the number of pairs w, s, where $w \leq x$, $s \in \mathcal{S}$ and w is a witness for s. The first way is to count the number of w's for each $s \in \mathcal{S}$ and the second way is to count the number of s's for each $w \leq x$. From the first way we learn that there are at least $(3/5)x \cdot \#\mathcal{S}$ pairs w, s. Thus from the second way of counting we learn that there is some $w \leq x$ which is a witness for at least $3/5$ of the members of $\mathcal{S}$.

We apply this principle iteratively starting with $\mathcal{S}$ the set of all odd composites in $[1, x]$, to find some $w_1 \leq x$ which is a witness for at least $3/5$ of the members of $\mathcal{S}$, then to find some $w_2 \leq x$ which is a witness for at least $3/5$ of the remaining members of $\mathcal{S}$ for which w_1 is not a witness, and so on. Thus if $(2/5)^k x/2 \leq 1$, there is some choice of $w_1, \ldots, w_k \leq x$ such that every odd composite has a witness in $\{w_1, \ldots, w_k\}$, and the Proposition follows from a simple calculation.

Remarks: The proof shows that we may replace the number $6/5$ in the Proposition with any constant c larger than $1/\log(5/2) = 1.09135\ldots$, provided x is sufficiently large depending on the choice of c. An argument based on showing that the witnesses for an odd composite are roughly uniformly distributed allows one to do this for any $c > 1/\log 4 = .72134\ldots$. It is likely that one could do a little better using Theorem 4 in [DLP]. We suggest below that there is a limit to these improvements and that $o(\log x)$ is not attainable in the Proposition.

Lemma 3.2. *Suppose $m \equiv 150 \bmod 180$ and $m+1$, $5m+1$, $9m+1$ are all prime. Then the Carmichael number $n = (m+1)(5m+1)(9m+1)$ has no witness $\leq Q$ if and only if $\left(\frac{m+1}{q}\right) = \left(\frac{5m+1}{q}\right) = \left(\frac{9m+1}{q}\right)$ for each prime q with $7 \leq q \leq Q$.*

Proof. From the hypothesis it is easy to see that n is a Carmichael number, each of its prime factors is 3 mod 4 and its prime factors are congruent mod 8. Thus $\left(\frac{2}{m+1}\right) = \left(\frac{2}{5m+1}\right) = \left(\frac{2}{9m+1}\right)$, so Corollary 1.2 implies 2 is not a witness for n. Suppose q is an odd prime. Since $m+1$, $5m+1$, $9m+1$ are all 3 mod 4, the law of quadratic reciprocity and Corollary 1.2 imply that q is not a witness for n if and only if $\left(\frac{m+1}{q}\right) = \left(\frac{5m+1}{q}\right) = \left(\frac{9m+1}{q}\right)$. It remains to note that $m+1$, $5m+1$, $9m+1$ are each quadratic residues mod q for $q = 3$ and 5, since 15 divides m.

The proof of Theorem 2: We shall consider numbers $n = (m+1)(5m+1)(9m+1)$ as described in Lemma 3.2. For every prime $q \leq Q$ we will only allow $m \equiv 0 \bmod q$, so that $\left(\frac{m+1}{q}\right) = \left(\frac{5m+1}{q}\right) = \left(\frac{9m+1}{q}\right) = 1$. Therefore the least witness for n is $> Q$ by Lemma 3.2. The above congruence conditions, when combined, fix m in a congruence class mod $N = 6\prod_{q \leq Q} q$. We thus apply the 'Uniform prime triplets conjecture' with $a_1 = N, a_2 = 5N, a_3 = 9N$, and deduce that there is such a prime triplet with $m \leq \gamma_3 N (45N^3)^{A_3}$. The resulting Carmichael number n is $\leq c_2 N^{9A_3+3}$, and this is $\leq x$ for $Q = \log x/(9A_3 + 4)$ by the prime number theorem.

In the above argument we could actually have chosen any $m \bmod q$, for which $\left(\frac{m+1}{q}\right) = \left(\frac{5m+1}{q}\right) = \left(\frac{9m+1}{q}\right)$, for each prime $7 \leq q \leq Q$. One can use Weil's theorem[12] to show that this holds for $q/4 + O(\sqrt{q})$ of the congruence classes $m \bmod q$. We thus can construct $N/4^{(1+o(1))\pi(Q)}$ such prime triplets, and apply the 'Uniform prime triplets conjecture' to each such triplet. This gives the second part of the theorem

One could state a plausible variant of the prime triplets conjecture with the criterion $\left(\frac{m+1}{q}\right) = \left(\frac{5m+1}{q}\right) = \left(\frac{9m+1}{q}\right)$ for each prime $7 \leq q \leq Q$, in the hypothesis. Instead we wish to be a little more general[13]: Consider all triplets of the form $m+1, 5m+1, 9m+1$ where $m \equiv 150 \bmod 180$. As above, for each integer $w \leq x$ we have $\left(\frac{w}{m+1}\right) = \left(\frac{w}{5m+1}\right) = \left(\frac{w}{9m+1}\right)$ for roughly a quarter or more of the congruence classes mod w. Thus, for any given set $\mathcal{W}$ of ℓ integers $\leq x$, we expect that $\left(\frac{w}{m+1}\right) = \left(\frac{w}{5m+1}\right) = \left(\frac{w}{9m+1}\right)$ for all $w \in \mathcal{W}$, for at least $c_3 x^{1/3}/4^\ell$ values of $m \leq x^{1/3}/3$. Now, the usual heuristic is that a proportion $\gg 1/(\log x)^3$ of these triples will be simultaneously prime. We thus expect that

There are at least $x^{1/5}$ Carmichael numbers up to x without a witness from any given set of $\ell = \left[\frac{1}{11}\log x\right]$ integers $\leq x$.

Combining this with Proposition 3.1 it seems likely that

The smallest set of reliable witnesses in $[1, x]$ for the odd composites up to x has size $\sim c \log x$ for some positive constant c.

If we take our set $\mathcal{W}$ to be the set of primes q up to $Q = \frac{1}{12}\log x \log\log x$ then, by Lemma 3.2 and arguing as above we expect that

There are at least $x^{1/5}$ Carmichael numbers $n \leq x$, each of whose least witness is $\geq \frac{1}{12}\log n \log\log n$.

On the other hand (cf. [BH]), we conjecture that

Every odd composite number n has a witness $\leq (1/\log 4 + o(1))\log n \log\log n$.

Since the non-zero residue classes mod n which are not witnesses for n lie in a subgroup of $(\mathbb{Z}/n\mathbb{Z})^*$ of index at least 4, the "probability" that each of the first k primes lies in this subgroup is 4^{-k}; so that if $k \geq (1/\log 4 + \epsilon)\log n$, then

---

[12] That is, the 'Riemann Hypothesis for curves'
[13] and hopefully still plausible

we expect there are at most finitely many such n. This heuristic implies the conjecture, though we suspect that the constant "$1/\log 4$" is not best possible.

Carmichael numbers which are the product of three primes $\equiv 3 \bmod 4$ are among the integers which have the fewest witnesses; which is why we used them to (heuristically) argue that there are some integers with an extraordinarily large least witness. However, in some cases it is possible to find a small witness for such numbers: for example, if the three primes are not all congruent mod 8, then 2 is a witness. Suppose now that n is a Carmichael number with exactly three prime factors. Then n is the product of a prime triplet $am+1$, $bm+1$, $cm+1$, where a, b, c are pairwise coprime, and m is in that residue class mod abc such that abc divides $m(bc+ca+ab)+a+b+c$. If we also want the three primes to be 3 mod 4 and congruent mod 8, then we need to add the conditions that a, b, c are congruent mod 4 and that $m \equiv 2 \bmod 4$. For fixed a, b, c, the prime triplets conjecture implies that there are infinitely many m in the prescribed residue class mod $4abc$ with $am+1$, $bm+1$, $cm+1$ all prime. We now investigate whether some fixed number w can be a witness for every number n in this family.

Proposition 3.3. *Assume the prime triplets conjecture. Suppose that a, b and c are pairwise coprime integers which are all congruent mod 4. There is no reliable witness for all the Carmichael numbers given as the product of 3 primes $(am+1)(bm+1)(cm+1)$ with $m \equiv 2 \bmod 4$ if and only if $(*)$ $\left(\frac{-bc}{p}\right) = 1$ for each prime p dividing a, $\left(\frac{-ca}{q}\right) = 1$ for each prime q dividing b, and $\left(\frac{-ab}{r}\right) = 1$ for each prime r dividing c.*

Proof. Suppose $\left(\frac{-bc}{p}\right) = -1$ for some prime divisor p of a. Then $\left(\frac{am+1}{p}\right) = 1$. But since $bm+1 \equiv -b/c \bmod p$, $\left(\frac{bm+1}{p}\right) = -1$, so by Corollary 1.2 and quadratic reciprocity, p is a witness for all such Carmichael numbers. A similar argument works for prime divisors of b and c with appropriate Legendre symbol $= -1$.

On the other hand suppose w is a reliable witness for the family, yet $(*)$ holds. Let w_0 be the largest odd divisor of w. Thus, $\left(\frac{w}{um+1}\right) \big/ \left(\frac{w_0}{um+1}\right)$ has the same value for $u = a, b, c$. By quadratic reciprocity, $\left(\frac{w_0}{um+1}\right) \big/ \left(\frac{um+1}{w_0}\right)$ has the same value for $u = a, b, c$. Let w_1 be the largest divisor of w_0 coprime to abc. Then $\left(\frac{um+1}{w_0/w_1}\right) = 1$ for $u = a, b, c$, by $(*)$. Consider those Carmichael numbers with $m \equiv 0 \bmod w_1$ (of which there are infinitely many by the prime triplets conjecture). We have $\left(\frac{um+1}{w_1}\right) = 1$ for $u = a, b, c$. Thus, by Corollary 1.2, w is a witness for none of these Carmichael numbers, so $(*)$ must fail.

Remark: These criteria appear in a seemingly unrelated theorem of Legendre: Suppose that a, b and c are given pairwise coprime integers, not all having the same sign. Then there exist non-zero integer solutions x, y, z to the equation $ax^2 + by^2 + cz^2 = 0$ if and only if there is a solution to $ax^2 + by^2 + cz^2 \equiv 0 \bmod 8$ and $(*)$ holds for the odd prime factors of abc. Surely this is a coincidence?

4 Further remarks

Theorem 4.1. *For any fixed non-zero integer a, there exist infinitely many squarefree, composite integers n for which $p - a$ divides $n - 1$ for every prime p dividing n.*

In [AGP] we claimed we could prove this result. We provide a short argument below. Note how the condition in the theorem generalizes Korselt's criterion in a natural way. The case $a = -1$ is of particular interest in the Lucas probable prime test, and related tests.

We will need the following Lemma which is proved exactly as Theorem 3.1 in [AGP][14].

Lemma 4.2. *If x is sufficiently large and if m is any squarefree integer, coprime to a, which is not divisible by any prime bigger than $x^{3/10}$, such that the sum of the reciprocals of the primes dividing m is $\leq 1/60$, then there exists a positive integer $k \leq x^{3/5}$, coprime to m, such that*

$$\#\{d|m : dk + a \text{ is a prime } \leq x\} \geq \frac{c_0}{\log x}\#\{d|m : 1 \leq d \leq x^{2/5}\}.$$

Proof of Theorem 4.1. We again modify the proof in [AGP]. By the main theorem of [F], there are $\gg y^3/\log y$ primes q which do not divide a, in the range $y^{5/2} \leq q \leq y^3$, for which the largest prime factor of $q - 1$ is $\leq y$. Let m be the product of $[y^2/\log y]$ of these primes, so that $m < e^{3y^2}$, and $\lambda(m) \leq e^{(3+o(1))y}$.

We apply Lemma 4.2 with $x = m^{5/2}$. There exists an integer $k \leq m^{3/2}$, coprime to m, such that the number of primes of the form $dk + a$ where $d|m$ is

$$\geq 2c_0\tau(m)/5\log m \geq 2^{y^2/2\log y} \geq \lambda(m)\log m.$$

By Theorem 1.1 of [AGP] (modified analogously to our Proposition 1.3) there is some non-trivial subset of these primes with product $n \equiv 1 \bmod mk$. For each prime factor p of n, we have $p - a = dk$ for some divisor d of m, so that $p - a$ divides $n - 1$.

There are two related questions that highlight the depth of our ignorance on this topic, and provide interesting problems for further research:
1. *Are there infinitely many squarefree integers n for which $p^2 - 1$ divides $n - 1$ for every prime p dividing n ?*
The least such number n has the prime factorization $17 \cdot 31 \cdot 41 \cdot 43 \cdot 89 \cdot 97 \cdot 167 \cdot 331$ and was found by Richard Pinch. For this question we have no idea how to prove the necessary analogue of Proposition 1.5 (or Theorem 3.1 in [AGP]).
2. *Are there infinitely many squarefree composite integers n for which $p + 1$ divides $n + 1$ for every prime p dividing n ?*

---

[14] except that now we need to use the same bounds for $\pi(dx^{3/5}, D, a)$ (with $D = d$ or dq); and we shall again pick $B = 2/5$.

The least such number n is 399. For this question we have no idea how to prove the necessary analogue of Proposition 1.3 (or Theorem 1.1 in [AGP]).

Acknowledgements: The second and third authors wish to acknowledge support from an NSF grant. The second author is an Alfred P. Sloan Research Fellow. Thanks are due to Eric Bach, Ronnie Burthe, Paul Erdős, Neal Koblitz and Sergei Konyagin for valuable comments concerning this paper.

References

[Ad]   L. M. Adleman, *Two theorems on random polynomial time*, Proc. IEEE Symp. Found. Comp. Sci., **19** (1978), 75–83.

[AGP]   W. R. Alford, A. Granville and C. Pomerance, *There are infinitely many Carmichael numbers*, Annals Math., to appear.

[Ar]   F. Arnault, *Rabin-Miller primality test: composite numbers which pass it*, Math. Comp.; to appear.

[B]   E. Bach, *Analytic methods in the analysis and design of number-theoretic algorithms*, MIT Press, Cambridge, Mass., 1985.

[BH]   E. Bach and L. Huelsbergen, *Statistical evidence for small generating sets*, Math. Comp. **61** (1993), 69–82.

[BBCGP] P. Beauchemin, G. Brassard, C. Crépeau, C. Goutier and C. Pomerance, *The generation of random integers that are probably prime*, J. Cryptology **1** (1988), 53–64.

[C]   M. D. Coleman, *On the equation $b_1 p - b_2 P_2 = b_3$*, J. reine angew. Math. **403** (1990), 1–66.

[DLP]   I. Dåmgard, P. Landrock and C. Pomerance, *Average case error estimates for the strong probable prime test*, Math. Comp. **61** (1993), 177–194.

[D]   J. D. Dixon, *Factorization and primality tests*, Amer. Math. Monthly **91** (1984), 333–352.

[E]   P. Erdős, *On pseudoprimes and Carmichael numbers*, Publ. Math. Debrecen **4** (1956), 201–206.

[F]   J. B. Friedlander, *Shifted primes without large prime factors*, in *Number Theory and Applications* (ed. R. A. Mollin), (Kluwer, NATO ASI, 1989), 393–401.

[HB]   D. R. Heath-Brown, *Zero-free regions for Dirichlet L-functions, and the least prime in an arithmetic progression*, Proc. London Math. Soc. (3) **64** (1992), 265–338.

[HL]   G. H. Hardy and J. E. Littlewood, *Some problems on partitio numerorum III. On the expression of a number as a sum of primes*, Acta Math. **44** (1923), 1–70.

[J]   G. Jaeschke, *On strong pseudoprimes to several bases*, Math. Comp. **61** (1993), 915–926.

[Leh]   D. H. Lehmer, *Strong Carmichael numbers*, J. Austral. Math. Soc. Ser. A **21** (1976), 508–510.

[Len]   H. W. Lenstra, Jr., private communication.

[M]   L. Monier, *Evaluation and comparison of two efficient probabilistic primality testing algorithms*, Theoret. Comput. Sci. **12** (1980), 97–108.

[P]   C. Pomerance, *On the distribution of pseudoprimes*, Math. Comp. **37** (1981), 587–593.

[PSW]   C. Pomerance, J. L. Selfridge and S. S. Wagstaff, Jr., *The pseudoprimes to* $25 \cdot 10^9$, Math. Comp. **35** (1980), 1003–1026.

[R]   M. O. Rabin, *Probabilistic algorithm for primality testing*, J. Number Theory **12** (1980), 128–138.

[SS]   R. Solovay and V. Strassen, *A fast Monte-Carlo test for primality*, SIAM J. Comput. **6** (1977), 84–85; *erratum*, ibid. **7** (1978), 118.

Density computations for
real quadratic 2-class groups

Wieb Bosma[1] and Peter Stevenhagen[2]

[1] School of Mathematics and Statistics, University of Sydney,
Sydney NSW 2006, Australia
[2] Faculteit Wiskunde en Informatica, Universiteit van Amsterdam,
Plantage Muidergracht 24, 1018 TV Amsterdam, The Netherlands

We present an algorithm to compute the 2-part of the class group of quadratic orders of large, factored discriminants and its application to density questions concerning the norm of the fundamental unit in real quadratic orders.

If $D \equiv 0, 1 \bmod 4$ is a large integer that is not a square, then it is computationally not feasible to calculate the strict class group $\mathcal{C}$ of the quadratic order $\mathcal{O}$ of discriminant D. However, if the factorization of D is known, it is possible to compute the 2-part of the finite abelian group $\mathcal{C}$ in random polynomial time [1]. This is done by an algorithm that essentially goes back to Gauss. It is based on the fact that the factorization of D provides us with a set of generators for the 2-torsion subgroup $\mathcal{C}[2]$ of $\mathcal{C}$ and an explicit description of the quadratic characters on $\mathcal{C}$. Using an algorithm to extract square roots in $\mathcal{C}$ that uses the reduction of ternary quadratic forms, one can now find the complete 2-Sylow subgroup of $\mathcal{C}$.

As our algorithm finds a non-trivial relation between the generators of $\mathcal{C}[2]$, it computes the norm of the fundamental unit in $\mathcal{O}$ without computing the unit itself, which cannot even be written down in polynomial time.

We have assembled numerical data for a large number of discriminants with an implementation of the algorithm using the computer algebra system MAGMA. The results enable us to check numerically the conjectural densities in [2], which predict that a fraction $1 - \prod_{j \text{ odd}} (1 - 2^{-j}) \approx .58$ of all real quadratic fields without discriminantal prime divisors congruent to 3 mod 4 will have a fundamental unit of norm -1. More precisely, we use a theorem of Rédei to generate a large family of discriminants for which $\mathcal{C}$ has a fixed 4-rank e, for several values of e, and show that, exactly as predicted in [2], approximately 1 out of $2^{e+1} - 1$ fields in such a family has a unit of norm -1.

References

1. Lagarias, J.C., *On the computational complexity of determining the solvability or unsolvability of the equation* $X^2 - DY^2 = -1$, Trans. Amer. Math. Soc. **260** (1980) 485–508.
2. Stevenhagen, P., *The number of real quadratic fields having units of negative norm*, Exp. Math. **2** (1993) 121–136.

Lattice sieving and trial division

Roger A. Golliver[1], Arjen K. Lenstra[2], Kevin S. McCurley[3]

[1] Intel SSD, MS CO1-05, 15201 NW Greenbrier Parkway,
Beaverton, OR 97006, U. S. A
E-mail: roger@ssd.intel.com
[2] MRE-2Q334, Bellcore, 445 South Street,
Morristown, NJ 07960, U. S. A
E-mail: lenstra@bellcore.com
[3] Organization 1423, Sandia National Laboratories,
Albuquerque, NM 87185-1110, U. S. A
E-mail: mccurley@cs.sandia.gov

Abstract. This is a report on work in progress on our new implementation of the relation collection stage of the general number field sieve integer factoring algorithm. Our experiments indicate that we have achieved a substantial speed-up compared to other implementations that are reported in the literature. The main improvements are a new lattice sieving technique and a trial division method that is based on lattice sieving in a hash table. This also allows us to collect triple and quadruple large prime relations in an efficient manner. Furthermore we show how the computation can efficiently be shared among multiple processors in a high-band-width environment.

1 Introduction

Throughout this paper we assume that the reader is familiar with the number field sieve (NFS) integer factoring algorithm [8]. We restrict ourselves to the relation collection stage of NFS.

Let $n > 1$ be an odd integer which is not a prime power, and let m be an integer such that $f(m) \equiv 0 \bmod n$ for an irreducible polynomial $f \in \mathbf{Z}[X]$ of degree $d > 1$. We do not discuss how m and f might be found. In the relation collection stage of NFS we attempt to find sufficiently many *relations*, i.e., pairs (a, b), with $b > 0$, satisfying the following conditions (cf. [9: 2.8, 7.3]):

(1.1) $\gcd(a, b) = 1$,

(1.2) on the *rational side* the number $a - bm$ is B_1-smooth, except for at most two additional prime factors which should be $< B_3$, for some *rational smoothness bound* B_1 and *rational large prime bound* $B_3 < B_1^2$,

(1.3) on the *algebraic side* $N(a, b) = b^d f(a/b)$ is B_2-smooth, except for at most two additional prime factors which should be $< B_4$, for some *algebraic smoothness bound* B_2 and *algebraic large prime bound* $B_4 < B_2^2$,

where, as usual, an integer is called x-smooth is its prime factors are $< x$.

Notice that we allow two large primes on either side, unlike [9] or [2] where at most one large prime per side was used, and unlike [3] where at most one large prime was used on the rational side. One of the advantages of our new implementation is that it finds these *double large prime relations* without the loss of efficiency that was reported in [9: 7.3] or that would follow from a straightforward extension of the method from [2]. Another advantage is that it is substantially faster than implementations previously reported in the literature (cf. [2, 3, 9]), even though it considers far more reports (see below) and produces far more relations; of course it is even faster without the double large primes, but then it produces just the same (and much smaller) amount of relations as were found in [9] and [2].

In [9] the traditional method was used in the relation collection stage: for $b = 1, 2, \ldots$ in succession a large interval of a-values is *sieved* to find pairs (a, b) for which both $a - bm$ and $N(a, b)$ are likely to be smooth, and these candidates are tested for actual smoothness using *trial division*, until sufficiently many relations have been found. In [12] Pollard suggested a faster sieving method: for a large enough collection of medium sized primes $q < B_1$ sieve in the sublattice L_q of the (a, b)-plane consisting of the pairs (a, b) for which q divides $a - bm$, where the sieve (and the trial division) on the rational side can be restricted to the primes $< q$. Pollard describes two ways to do the sieving in L_q: *sieving by rows*, which requires only a small amount of memory but which is inefficient for large primes, and *sieving by vectors*, which is much faster if a large amount of memory is available. In [2] the first method is used, because of the small processors used there. In [3] a more efficient variation of the first method was used, also because not enough memory was available to make an efficient implementation of the second method possible (cf. [14]).

Our implementation makes use of the second method, for the sieving and for the trial division, though our q's are medium sized primes on the algebraic rather than the rational side. So, our L_q is the sublattice of the (a, b)-plane consisting of the pairs (a, b) for which q divides $N(a, b)$. The speed-up that we achieve depends for the sieving on the amount of memory that is available, but our trial division is an order of magnitude faster than previous methods, almost irrespective of the available memory. Our sieving implementation is explained in Section 2, the trial division in Section 3.

Our relation collection program can be executed on any multiple-instruction multiple data machine that has enough memory per node, by assigning different q's to different nodes. Our program can also be executed on two smaller nodes simultaneously if a high-bandwidth communication between the two nodes exists. Various parallelization issues are discussed in Section 4.

Relations with one or two large primes can be combined in the usual way to turn them into useful relations (cf. [9: 2.10]). The algorithm to find these combinations can be generalized to find most combinations among relations with at most three large primes, as shown in [4]. This approach gets stuck, however, at four large primes. Satisfactory methods to find all combinations among relations with any (small) number of large primes are described in [5]. As shown in [5] the relations with four large primes give rise to an unusually large number of combinations, when combined with relations with fewer large primes. This could lead to considerable savings in the relation collection stage compared to approaches that use fewer than four large primes. It could also lead to substantially more work in the final stages of NFS, the matrix reduction and the square root computation. It remains to investigate how these effects can best be balanced. We refer to [5] for a further discussion of this point.

In Section 5 we give the run times and yields of our relation collection program for various q's for a 129-digit n, and we derive some estimates how long the complete relation collection might take.

2   Sieving

Let the notation be as above, and let P be the set of all pairs (p, r) with p a prime $< B_1$ and $r \equiv m \bmod p$ with $0 \le r < p$. Similarly, let Q be the set of all pairs (p, r) with p a prime $< B_2$ and $f(r) \equiv 0 \bmod p$ with $0 \le r < p$. Notice that a particular prime can occur up to d times in Q because f may have d distinct roots modulo p. Assume that the elements of Q are ordered in some well-defined way such that $(p_1, r_1) < (p_2, r_2)$ if $p_1 < p_2$. In practice $\#Q$ will be close to $\pi(B_2)$, the number of primes $\le B_2$. The sets P and Q are called the *rational* and the *algebraic factor base*, respectively.

For any $u = (p, r) \in P \cup Q$ let L_u be the lattice generated by the vectors $(p, 0)$, $(r, 1)$ in the (a, b)-plane. Notice that L_u corresponds to the pairs (a, b) for which p divides $a - bm$ for $u \in P$, and for which p divides $N(a, b)$ and $r = a/b \bmod p$ for $u \in Q$.

2.1. *Preparing q.* Throughout this section and the next we fix a pair $(q, s) \in Q$ with $q > B_0$, for some $B_0 < B_2$. Abbreviate $L_{(q,s)}$ to L_q. We apply a straightforward lattice reduction method to the basis $(q, 0)$, $(s, 1)$ of L_q in the (a, b)-plane, and find two short vectors $V_1 = (a_1, b_1)$, $V_2 = (a_2, b_2)$ generating L_q, such that the Euclidean length $\|V_1\|$ of V_1 is $\ge \|V_2\|$. Each point of L_q can be written as $c \cdot V_1 + e \cdot V_2 \in \mathbf{Z}^2$, for $c, e \in \mathbf{Z}$, and each point in the (c, e)-plane corresponds to a pair (a, b) for which q divides $N(a, b)$ and $s = a/b \bmod q$:

$$(2.2) \qquad (a, b) = (c \cdot a_1 + e \cdot a_2, c \cdot b_1 + e \cdot b_2).$$

Sieving will take place in the (c, e)-plane. Let $Q_q = \{u : u \in Q, u < (q, s)\}$.

2.3. *Preparing the primes.* Let D be a crossover value between 'small' and 'large' primes. Its value depends on n, q and the implementation (cf. Section 5). Define for all $u = (p, r) \in P \cup Q_q$ the sublattice L_{qu} of L_q as $L_q \cap L_u$. A basis for L_{qu} in the (c, e)-plane is given by

$$\bar{U}_{u1} = (p, 0), \qquad \bar{U}_{u2} = \left(\frac{a_2 - r \cdot b_2}{r \cdot b_1 - a_1} \bmod p, 1 \right).$$

For all u's for which this basis is well-defined, $\bar{U}_{u2} \not\equiv (0, 1) \bmod p$, and $p \ge D$, we compute a reduced basis U_{u1}, U_{u2} for L_{qu} in the (c, e)-plane, with $\|U_{u1}\| \ge \|U_{u2}\|$; for all other u we call u *exceptional* and set both U_{u1} and U_{u2} equal to $\bar{U}_{u2}$ (but see Remarks 2.4 and 2.5).

2.4. *Remark.* In our implementation we have a total of 65 bits available to store U_{u1} and U_{u2}: one parity bit, two bits for signs, and 2×17 and 2×14 bits for the absolute values of the entries of U_{u1} and U_{u2}, respectively (cf. Remark 2.12). If the basis U_{u1}, U_{u2} for a certain u does not fit in this format we declare u to be exceptional and treat it as such. This happens only occasionally for the B_1 and B_2 that we have been using; see Section 5 for examples. For exceptional u's the vectors U_{u1} and U_{u2} can easily be stored in 65 bits.

2.5. *Remark.* In the above initialization of U_{u1} and U_{u2} for the exceptional u's we assume that the smallest e-value to be sieved equals 1. If the first e-value to be sieved equals e_1, then U_{u1} should be initialized as $(e_1 \cdot \bar{U}_{u2}) \bmod p$.

2.6. *Partitioning the sieve.* Suppose that we want to sieve the (c, e)-plane for $|c| \leq C$ and $0 < e \leq E$, where C and E depend on n; see Section 5 for examples. Because in general the required $(2C + 1) \cdot E$ sieve-locations do not fit in memory, we partition the sieve into t pieces $S_1, S_2, \ldots, S_t$, with $S_i = \{(c, e) : |c| \leq C, E_{i-1} < e \leq E_i\}$ for appropriately chosen $0 = E_0 < E_1 < \ldots < E_t = E$, and sieve over S_i for $i = 1, 2, \ldots, t$ in succession as described in 2.7, 2.8, and 2.9 (cf. Remark 2.13). See Section 5 for examples of $\#S_i$ used in practice.

2.7. *The algebraic sieve.* To find the pairs $(c, e) \in S_i$ for which the corresponding $N(a, b)/q$ is likely to be q-smooth we first set all $S[c, e]$ to zero, where S is an array with indices $(c, e) \in S_i$ (but see Remark 2.10 and 2.13). Next we do the following for all $u = (p, r) \in Q_q$.

For the exceptional u's we sieve 'by rows': for $e = E_{i-1} + 1, E_{i-1} + 2, \ldots, E_i$ in succession first replace $S[c, e]$ by $S[c, e] + [0.5 + \log p]$ for all c with $|c| \leq C$ that are modulo p equal to the first coordinate of U_{u1}, and next replace U_{u1} by $(U_{u1} + U_{u2}) \bmod p$. Notice that the u with $p \leq 2C + 1$ should be treated differently from the u with larger p. For exceptional u's with very small p sieving by rows is quite slow. In practice the small p are therefore replaced by appropriately chosen powers (and the roots are changed accordingly).

For the non-exceptional u's we sieve 'by vectors': for all $\mathbf{Z}$-linear combinations (c, e) of U_{u1} and U_{u2} with $(c, e) \in S_i$ replace $S[c, e]$ by $S[c, e] + [0.5 + \log p]$. The appropriate linear combinations can be found efficiently by considering the relevant inequalities involving U_{u1}, U_{u2}, E_{i-1}, E_i, and C; we do not elaborate.

The vectors U_{u1} and U_{u2} remain unchanged in this step, unlike the vectors for the exceptional u's which get updated from one e-row to the next. So, for the non-exceptional u's the S_i can be processed in any order, but for the exceptional u's they have to be processed in order, unless U_{u1} is given the correct initial value for each S_i, using Remark 2.5. Obviously this step takes at least a constant amount of time per u; it follows that $\#S$ should not be too small compared to p to make 'sieving by vectors' efficient.

2.8. *Checking sieve locations and rational sieve.* After sieving with all $u \in Q_q$, we inspect all values stored in S (cf. Remark 2.10), remember the coprime pairs (c, e) and $S[c, e]$'s for which the latter is larger than some fixed *algebraic report bound* and replace those $S[c, e]$'s by zero. Next we sieve with all $u \in P$, as described in 2.7 (but with the obvious changes), to find the $(c, e) \in S_i$ for which the corresponding $a - bm$'s are likely to be B_1-smooth.

2.9. *Collecting reports.* Finally we re-inspect the same pairs and S-values that were remembered after the algebraic sieve, and keep the *reports*: the coprime (a, b) for which the corresponding $S[c, e]$ (cf. (2.2)) is larger than some fixed *rational report bound*, for which $S[c, e]$ is sufficiently close to a floating point approximation of $\log(a - bm)$, and for which the value of the first sieve is sufficiently close to a floating point approximation of $\log(N(a, b)/q)$. Notice that these tests get successively more discriminating and more expensive. The resulting pairs (a, b) have a good chance to satisfy (1.2) and (1.3) if the report bounds are set correctly. See Section 5 for examples of report bounds, of numbers of pairs to be remembered, and of numbers of reports.

After $S_1, S_2, \ldots, S_t$ have been sieved all reports (a, b) have to be inspected, i.e., the corresponding $a - bm$ and $N(a, b)/q$ have to be trial divided (but see Remark 3.4). Our implementation of the trial division is described in Section 3.

2.10. *Remark.* As usual we use single bytes to represent the $S[c, e]$. Instead of initializing the $S[c, e]$ as zero for the first sieve and testing $S[c, e] \geq x$ for some bound x, we initialize the $S[c, e]$ as $-x$ and test $S[c, e]$ for non-negativity. On most architectures this can be done several bytes at a time, which is often much faster. The initialization of S can be speeded up similarly.

2.11. *Remark.* Because for general n the number $N(a, b)/q$ can be expected to be substantially larger than $a - bm$ we sieve over the algebraic side before we sieve over the rational side. This minimizes the number of pairs and values to be remembered after the first sieve. For special n (where $a - bm$ is often larger than $N(a, b)/q$) we change the order of sieving.

2.12. *Remark.* Per $u = (p, r) \in P \cup Q_q$ we use 16 consecutive bytes, i.e., 128 bits, of storage: one byte for the difference (divided by 2) with the previous p, three bytes for r, a floating point approximation of $1/p$ in 31 bits, and 65 bits for the vectors U_{u1} and U_{u2} (cf. Remark 2.4). The approximation of $1/p$ is used to speed up the computations modulo p in the sieving and in the trial division. Evidently, this limits the general applicability of our implementation, but for n well beyond our current range of interest this format suffices.

This dense data structure was chosen for two reasons. First, by having the factor base's C data structure size equal to a small power of two, we can be sure that the compiler will efficiently map the structure in memory, avoiding miss-aligned data access penalties and wasted memory. Secondly, since the factor base is accessed sequentially, having a single large array for the factor base, requires only one active TLB entry for accessing the structure, thus freeing more TLB entries to map the large sieve array.

2.13. *Remark.* Because only the $u \in P \cup Q_q$ have to be stored during the sieving and trial division for a certain q, the value of t increases with q, without affecting the total memory used, i.e., the smaller q, the more memory can be devoted to S. During the sieving the (c, e) for which both c and e are even can be avoided at no extra cost. Because these points are not needed (cf. (1.1)) we do not include them in S. This slightly affects the description given above.

3 Trial division

The straightforward approach to process the reports (a, b) is to apply trial division successively to each $a - bm$ with the elements from the rational factor base and, if necessary, to each $N(a,b)/q$ with the elements $\leq (q, s)$ from the algebraic factor base. Because up to two additional primes are allowed per side, these trial divisions might be followed, per report, by at most two applications of for instance, Shanks's 'squfof' to factor the remaining composite cofactors.

In [9] and [2] at most one additional prime factor per side is allowed in the relations. Consequently there are far fewer reports, and there are no remaining composites to be factored after the trial divisions. At most two trial divisions per report works thus reasonably well, although the trial division time becomes a substantial fraction of the time spent on the relation collection stage (varying from a quarter to a third, according to [2: Table 1]).

There are several ways to make the trial divisions go faster. Obviously, it is always better to check that $a \equiv br \bmod p$ for $(p, r) \in P \cup Q$, before the actual trial division on $a - bm$ or $N(a, b)/q$ is carried out. We can also keep track of the sum of the logarithms of the divisors found, compare this to the sieve values, and try to use this (cheap) information to jump ahead in the factor bases; this saves some time. We could use *early aborts*: give up on reports for which not enough divisors have been found after processing certain fixed fractions of the factor bases. This also saves some time, but some relations get lost which is something we try to avoid in NFS (unlike Quadratic Sieve).

Because we have substantially more reports than [9] and [2], but also than [3], none of these approaches works satisfactorily for us. Therefore we decided to do the trial division simultaneously for all reports per (q, s) by sieving in a hash table. The pairs (c, e) corresponding to the reports (a, b) are the hash keys, and for each key the hash table H has storage for the key, a counter $C_{(c,e)}$, and an array $A_{(c,e)}$ of, say, 50 integers (cf. Remark 3.4). We also use flags F_c and F_e for $|c| \leq C$ and $0 < e \leq E$ (cf. 2.6). We use the space allocated for S to store H, the F_c's, and the F_e's.

3.1. *Rational trial sieving.* Initially we set all flags to 0. For all reports (a, b) we store the corresponding (c, e) in H, we set $C_{(c,e)}$ to -1 and set F_c and F_e to 1. For all non-exceptional $u = (p, r) \in P$ we do the following: for all Z-linear combinations (c, e) of U_{u1} and U_{u2} for which F_c and F_e are both equal to 1 and for which (c, e) is in H, we increase $C_{(c,e)}$ by 1, and we store p in $A_{(c,e)}[C_{(c,e)}]$. This is similar to what we did in 2.7, except that we now remember the primes instead of simply accumulating the logarithms of the primes, and that we process all S_i for $1 \leq i \leq t$ simultaneously (but see Remark 3.4).

3.2. *Rational factorization.* For all reports (a, b) we remove all factors p from $a - bm$, first for the primes p stored in $A_{(c,e)}[0]$ through $A_{(c,e)}[C_{(c,e)}]$, and next for all exceptional $u = (p, r) \in P$. If the remaining cofactor of $a - bm$ is either $\geq B_3^2$, or both $\geq B_3$ and $< B_1^2$, or not easily proved to be composite (cf. [7]), then (a, b) is removed from the list of reports. We keep the reports (a, b) for which we have found the complete factorization of $a - bm$, except for at most one factor which is either prime and $< B_3$, or composite and $< B_3^2$.

3.3. *Algebraic trial sieving and factorization.* For the remaining reports we repeat 3.1 with the $u \in Q_q$ instead of the $u \in P$. Next we remove all factors p from $N(a,b)/q$, by retrieving the p's stored in $A_{(c,e)}$ followed by trial division with the p's for which $u = (p,r) \in Q_q$ is exceptional. If the remaining cofactor is 1, prime and $< B_4$, or composite and $< B_4^2$, then we use Shanks's 'squfof' (followed by the elliptic curve method (cf. [7]) if 'squfof' did not work) to factor both this cofactor and the cofactor of $a - bm$, if needed, and report the resulting relation if it satisfies (1.2) and (1.3). Notice that $N(a,b)$ might have one or two prime factors p with $q < p < B_2$ (cf. [2: Section 12]); these (a,b) are reported as well, but since they might also be found for a larger q (i.e., one of those p's) we have to remove duplicate relations before processing them.

3.4. *Remark.* Because efficient look-up in H is essential, we assume that the number of reports is small compared to the number of locations #H in H. After processing a piece of the (c,e)-plane (i.e., after each completion of 2.9), we therefore check if the total number of reports found so far exceeds #H$/x$, for some small x (2 or 3, say), and if that is the case we perform 3.1 to process them. Of course, we also perform 3.1 after the sieving for a particular (q,s) is completed if new reports have been found after the last execution of 3.1.

3.5. *Remark.* Because sieving with the exceptional u's is relatively slow, and because there are typically only a few thousand of them, we do not resieve with them, but find the corresponding primes using root checking and trial division.

4 Parallelization issues

The code was originally developed to run as a single UNIX®* process. That version can be executed in an "embarrassingly parallel" fashion by assigning disjoint sets of q's to different processes and collecting the relations. We wanted to run this code on the Intel Paragon [6] at Sandia National Laboratories. This machine currently has 528 nodes with 32MB of memory and 1312 nodes with 16MB of memory, for a total of 1840 computational nodes and 37GB of memory. Other nodes in the machine are responsible for providing interactive user and I/O services. The nodes are all connected by a high speed/low latency communication network as a 16×120 mesh. Each node has two i860XP processors, although the present software dedicates one of the processors to handling communication (in the future we plan to utilize the second processor for computation). We chose to use the SUNMOS operating system** [11] on the compute nodes rather than the standard OSF operating system. The reasons for this are:

- SUNMOS consumes only 240K of physical memory on a node, whereas OSF consumes approximately 5 megabytes,

---

* UNIX is a registered trademark of UNIX Software Laboratories, Inc.
** SUNMOS stands for Sandia UNM Operating System, named after the organizations that collaborated on its development.

- SUNMOS offers message passing bandwidth of 170 megabytes per second, whereas OSF currently peaks at about 35 megabytes per second,
- SUNMOS user 4 megabyte physical memory pages for it's memory addressing, while OSF uses 4 kilobyte pages. The larger memory page size allows yields approximately 35% faster speed in the sieving code due to more efficient use of TLB entries.

Experiments are currently underway with SUNMOS to allow the second node processor to be used for computation.

4.1. *Large nodes.* Our first step at parallelization for the Paragon followed the "embarrassingly parallel" UNIX model. Each 32MB Paragon node ran one process which was given a distinct set of q's. To optimize the performance the only change required was to reduce the startup I/O by having a single root node read the factor base data from disk and broadcast the data to the other nodes over the communication network. The problem of load balancing was resolved by assigning each node a number of q's per job. Each node appended the relations it found to a file which is named to help the management of sieving over a large number of q's and possibly a number of (c, e) ranges.

4.2. *Small nodes.* We wanted to be able to use the entire Sandia Paragon, but the majority of nodes in the machine have only 16MB of memory, and this is too small to hold the factor base and the sieve array in physical memory. Paging would extract an extremely heavy performance penalty, so we overcame this problem with a second level of parallelization. The 16MB nodes were treated as even and odd node pairs. The factor base was distributed between the node pairs. The even node got all the even indexed factor base elements, and the odd processor got all the odd ones, with the obvious implications for the 'differences' from (2.12). This distribution of the factor bases was chosen over others, because it had better load balancing properties for the division of work between the node pairs. All other data structures including the sieve array were duplicated between the node pair.

The lattice sieving and trial division process is done as described in Sections 2 and 3. The sieving is done in parallel with the partitioned factor base, and the node pairs use the high speed communication network to combine and exchange the sieving results. Both nodes in the pair duplicate the initialization steps. The algebraic sieving is done in parallel by each node, using their own portion of the factor base, after which the node pair synchronizes and combine their sieve results. This communication step is the most intensive of the parallelization, since the entire sieve array is exchanged between the pair of nodes.

Each node examines the sieve array for the interesting locations as in 2.8. These are re-initialized and each node sieves over their own half of the rational factor base. The node pair then synchronizes again and exchanges the values of the interesting sieve locations remembered after the algebraic sieve. The nodes in the pair each examine disjoint parts of the possible reports and save those which pass the screening described in 2.9. The nodes then synchronize and exchange the reports which will be saved for the trial division stage.

The trial division sieving parallelized as follows. First each node sieves with their half of the rational factor base as in 3.1. For each report the node pairs exchange and remove the prime factors thus found, after which they both remove their exceptional primes from the remaining cofactors and exchange these factors too. The even node then checks to see if the report might satisfy condition (1.2) and should be checked for condition (1.3) (i.e., no attempt is made yet to factor a remaining composite cofactor $\geq B_1^2$ and $< B_3^2$). Algebraic trial sieving is parallelized in a similar way to process the remaining reports.

The node pair parallelization required the modification of about 100 lines of the 4000 line sieve program, and the inclusion of about 800 lines for the data partitioning, synchronization, and communications operations. Our experiments using the same size sieve array, show that the time to sieve over an equivalent $q \times (c, e)$ range on a 32MB node vs. a 16MB node pair is about 1.8 to 1. Thus this second level of parallelization allows us to use the 16MB nodes with only a small loss of efficiency and greatly increasing the number of nodes which can be used. It should be noted that this level of parallelization to use small memory nodes heavily depends on the existence of a high-bandwidth, reliable network between the nodes.

5 Run times and parameter choices

At the time of writing the relation collection stage of a factoring attempt of n=RSA-129 (cf. [13]) has just been completed on the Internet, using the quadratic sieve method (QS) and the familiar electronic mail approach [10]. This computation took approximately 5000 to 6000 MIPS years [1]. We list some timings and parameter choices for our program applied to the same n and running on a Sparc-10 workstation, estimated as 33 MIPS. Our numbers suggest that relation collection for NFS for this n could be completed in a fraction of the time needed by QS.

We use $m = 8000000099160814875591$ from which f follows by writing n in the symmetric base m; these m and f were found by James B. Shearer. Based on results from [2], we chose $B_1 = 3 \cdot 10^6$, $\#P = 216815$, $B_3 = 11932088$, $\#Q = 783185 = 10^6 - \#P$, $B_2 = B_4 = 2^{30}$. We used $D = 50000$ (cf. 2.3). Logarithms on the algebraic side have base 2, with report bound 83; on the rational side these numbers are 1.6 and 42.

We tried low range $(5 \cdot 10^6)$, medium range $(7 \cdot 10^6)$, and high range $(11 \cdot 10^6)$ q's, all of them with $C = 2^{13}$ and $E = 30000$ (cf. 2.6), which are realistic choices. The number of exceptional primes $\geq D$ on the rational side usually varied between 10 and 20; on the algebraic side the number grows with q, but never became more than 250. The total number of exceptional primes on either side was therefore never substantially more than $\pi(D) + 250 \approx 5300$.

With a bound of 23.5MB on the total memory use by our program, we could devote 14MB to S for the low range q's, 12MB for mid range q's, and 8MB for the high range. The 23.5 was chosen to make it easier to compare performance with both the 32MB and the pairs of 16MB Paragon nodes. If more than 23.5MB can efficiently be used it is in general a good idea to do that. Our choice led to average sieving times of 1810, 1900, and 2300 seconds per q, and average trial division times of 360, 420, and 500 seconds per q, for the low, medium, and high range, respectively. Sieving and trial division for 500000 q's and our C and E can therefore be done in 1400 MIPS years.

After the algebraic sieve we typically have to remember $<= 120000$ pairs per S_i. The yield after the rational sieve decreases with i, usually from ≈ 800 to ≈ 80. On average slightly less than $1/4$ of the accumulated reports survives 3.2, and a much smaller fraction of the survivors leads to a relation. The yields per q are roughly 109, 120, and 143 relations for the three ranges. The q's combined should therefore lead to more than $5 \cdot 10^7$ relations. According to data collected for another number in [5] it is very likely that this will lead to far more than 10^6 combinations. We refer to [5] for details, also concerning the distribution of the various types of relations.

References

1. D. Atkins, M. Graff, A. K. Lenstra, P. C. Leyland, Title to be announced, in preparation

2. D. J. Bernstein, A. K. Lenstra, A general number field sieve implementation, 103–126 in: [8]

3. J. Buchmann, J. Loho, J. Zayer, An implementation of the general number field sieve, Advances in Cryptology, Proceedings Crypto'93, Lecture Notes in Comput. Sci. **773** (1994) 159–165

4. J. Buchmann, J. Loho, J. Zayer, Triple-large-prime variation, manuscript, 1993

5. B. Dodson, A. K. Lenstra, NFS with four large primes: an explosive experiment, in preparation

6. Intel Corporation, Paragon(tm) XP/S Product Overview, 1991

7. A. K. Lenstra, H. W. Lenstra, Jr., Algorithms in number theory, Chapter 12 in: J. van Leeuwen (ed.), Handbook of theoretical computer science, Volume A, Algorithms and complexity, Elsevier, Amsterdam, 1990

8. A. K. Lenstra, H. W. Lenstra, Jr. (eds), The development of the number field sieve, Lecture Notes in Math. **1554**, Springer-Verlag, Berlin, 1993

9. A. K. Lenstra, H. W. Lenstra, Jr., M. S. Manasse, J. M. Pollard, The number field sieve, 11–42 in: [8]

10. A. K. Lenstra, M. S. Manasse, Factoring by electronic mail, Advances in Cryptology, Eurocrypt '89, Lecture Notes in Comput. Sci. **434** (1990) 355–371

11. B. Maccabe, K. S. McCurley, R. Riesen, SUNMOS for the Intel Paragon: A Brief User's Guide, Sandia National Laboratories Technical Report # SAND 93-1024

12. J. M. Pollard, The lattice sieve, 43–49 in: [8]

13. RSA Data Security Corporation Inc., sci.crypt, May 18, 1991; public information available by sending electronic mail to `challenge-rsa-list@rsa.com`

14. J. Zayer, personal communication, September 1993

A Subexponential Algorithm for Discrete Logarithms over the Rational Subgroup of the Jacobians of Large Genus Hyperelliptic Curves over Finite Fields

Leonard M. Adleman, Jonathan DeMarrais, and Ming-Deh Huang

Department of Computer Science, University of Southern California, Los Angeles CA 90089

Abstract. There are well known subexponential algorithms for finding discrete logarithms over finite fields. However, the methods which underlie these algorithms do not appear to be easily adaptable for finding discrete logarithms in the groups associated with elliptic curves and the Jacobians of hyperelliptic curves. This has led to the development of cryptographic systems based on the discrete logarithm problem for such groups [12, 7, 8]. In this paper a subexponential algorithm is presented for finding discrete logarithms in the group of rational points on the Jacobians of large genus hyperelliptic curves over finite fields. We give a heuristic argument that under certain assumptions, there exists a $c \in \Re_{>0}$ such that for all sufficiently large $g \in Z_{>0}$, for all odd primes p with $\log p \leq (2g + 1)^{.98}$, the algorithm computes discrete logarithms in the group of rational points on the Jacobian of a genus g hyperelliptic curve over GF(p) within expected time:

$$L_{p^{2g+1}}[1/2, c]$$

where $c \leq 2.181$.

1 Introduction

Let K be a field, let $f \in K[x]$ be a polynomial of odd degree $n = 2g + 1$ without multiple roots, let H be the smooth projective curve associated with $y^2 - f$, then H is a hyperelliptic curve of genus g. Let $K(H)$ be the function field of H over K, let $J = J_K(H)$ be the divisors of degree 0 modulo the principal divisors of $K(H)$, then J is isomorphic as a group to the group of K-rational points on the Jacobian of H. If $g = 1$ then H is an elliptic curve.

For the rest of this paper, we will assume that K is a finite field. In this case, it is known that $J = J_K(H)$ is a finite group. We will analyze the running time of the algorithm only in the case that K is a finite prime field of characteristic unequal to 2. However, there appear to be no significant barriers to generalizing this analysis to arbitrary finite fields.

Given two elements α, β in J the discrete logarithm problem is to calculate an $r \in Z_{\geq 0}$ (if such exists) such that $r\alpha = \beta$.

The problem of solving the discrete logarithm problem for elliptic curves has proven difficult. None of the subexponential techniques useful for solving discrete logarithms over finite fields have been successfully applied to elliptic curves. This has led to the invention of cryptosystems whose security is based on the difficulty of discrete logarithms over the groups associated with elliptic curves[12, 7], and more recently the groups associated with the Jacobians of hyperelliptic curves[8].

Somewhat surprisingly, the high genus hyperelliptic curve case may be easiest solved. We present an algorithm that is subexponential for curves of high genus. For the elliptic curve case our algorithm will be slower than the naive algorithm. We give a heuristic argument that under certain assumptions, there exists a $c \in \Re_{>0}$ such that for all sufficiently large $g \in Z_{>0}$, for all odd primes p with $\log p \leq (2g + 1)^{.98}$, the algorithm computes discrete logarithms in the group of rational points on the Jacobian of a genus g hyperelliptic curve over $\mathrm{GF}(p)$ within expected time:

$$L_{p^{2g+1}}[1/2, c]$$

where $c \leq 2.181$.

The algorithm which follows proceeds in two steps. The structure of the group is determined, and the discrete logarithm is calculated. To determine the group structure of $J_K(H)$, we proceed in a manner analogous to that used to calculate the structure of a class group of a number field. First random elements of the form $a(x)y + b(x) \in K[x, y]/(H)$ are generated and tested for "smoothness". Such an element is smooth if and only if its norm, $a(x)^2 f(x) - b(x)^2$, is the product of irreducible polynomials $g(x) \in K[x]$ of small degree. Each such small degree polynomial is associated with a divisor and hence each such smooth $a(x)y + b(x)$ yields a relationship on the group $J_K(H)$. When enough such relationships are obtained, the structure of $J_K(H)$ can be determined by linear algebra. In particular, a representation of $J_K(H)$ is obtained as the direct sum of $Zx_1 \oplus Zx_2 \oplus \ldots \oplus Zx_n$ for elements $x_1, \ldots x_n \in J_K(H)$. For particular divisors $\alpha, \beta \in J_K(H)$, one now represents $\alpha = \sum t_i x_i$ and $\beta = \sum s_i x_i$. Given the group structure of the Jacobian and representations for α and β, one can solve for r.

2 Preliminaries

In this section some basic facts are presented.

Solving for the Structure of a Group[5].

Let G be a finite Abelian group generated by $g_1, ..., g_n$. Let A be the free Z-module with basis $e_1, ..., e_n$ and let $\theta : A \to G$ be the homomorphism such that $\theta(\sum_{i=1}^n a_i e_i) = \sum_{i=1}^n a_i g_i$. Let K be the kernel of θ, then $G \equiv A/K$.

Given a relation in G, $\gamma = \sum_{j=1}^n m_j g_j = 0$, then $\delta = \sum_{j=1}^n m_j e_j \in K$. Given a set of m such relations $\gamma_i = \sum_{j=1}^n m_{ij} g_j = 0$, $i = 1, 2, ..., m$, let M be the matrix (m_{ij}). The crucial step of finding the structure of G is to find matrices $P = (p_{ij}) \in GL(n, Z)$, $Q = (q_{ij}) \in GL(m, Z)$, and a diagonal $m x n$ matrix $D = (d_i)$ with $d_i \mid d_{i+1}, 1 \leq i \leq n - 1$ such that $QMP^{-1} = D$.

From the theory of modules over PIDs [5], P, D, Q exist and can be calculated from M using elementary row and column operations. Assuming that $\{\delta_i\}$ generates K, then each $d_i \neq 0$ since G is finite. For $i = 1, 2, ..., n$, let $x_i = \sum_{j=1}^n p_{ij} g_j$, then x_i has order $|d_i|$, $\#G = \prod_{i=1}^n |d_i|$ and $G = Zx_1 \oplus Zx_2 \oplus ... \oplus Zx_n$ (direct sum) (whenever d_i is a unit, x_i is the identity and the corresponding summand is trivial).

Given an element $\alpha \in G$ where $\alpha = \sum_{i=1}^n \alpha_i g_i$, then $\alpha = \sum_{i=1}^n \alpha'_i x_i$, where $(\alpha_1, \alpha_2, ..., \alpha_n)P^{-1} = (\alpha'_1, \alpha'_2, ..., \alpha'_n)$.

Given elements $\alpha, \beta \in G$ where $\alpha = \sum_{i=1}^n \alpha_i x_i$, and $\beta = \sum_{i=1}^n \beta_i x_i$, to solve the discrete logarithm problem, one needs to find an r such that $r\alpha = \beta$. This can be done by solving the congruences $r\alpha_i \equiv \beta_i \bmod d_i$, $i = 1, 2, ..., n$.

Some remarks on the computational aspects of the preceding methods are in order. When $\{\delta_i\}$ does not generate K, there are several possible situations. In the first case the matrix M may have rank $m < n$. When this occurs, the $d_i = 0$, for $m < i \leq n$. In this case it can be easily determined that the set $\{\delta_i\}$ does not generate K and that more relations are needed. A second possibility is that the matrix M has rank n, but the set $\{\delta_i\}$ generates a subgroup K' of K. As above we may calculate $P, Q, D, x_1, .., x_n$. In this case $\prod_{i=1}^n d_i$ will be a multiple of $\#G$. We may express $\alpha = \sum_{i=1}^n \alpha_i x_i$, and $\beta = \sum_{i=1}^n \beta_i x_i$ and determine if there exists an r such that $r\alpha_i \equiv \beta_i \bmod d_i$, $i = 1, 2, ..., n$. If such an r exists then $r\alpha = \beta$ as desired (though the r found may not be the least positive one). However, these congruences may fail to have a solution despite the fact that there exists an r such that $r\alpha = \beta$. If one knows *a priori* that such an r exists then one can collect further relations on G until it is found. Unfortunately, if no such r exists then no amount of relations will reveal this. For this reason our algorithm will find discrete logarithms when they exist but will run forever when they do not. If $\#G$ is known or can be efficiently approximated to within a factor of 2, then our algorithm can be modified to recognize when no solution to the discrete logarithm problem exists.

Representing and Adding Elements of the Jacobian.

Let C be a smooth projective curve defined over a field K where K is either a finite field or the algebraic closure of a finite field. We will denote by $K(C)$ the function field of C over K. A *prime* of $K(C)$ is the prime ideal of a discrete valuation ring of $K(C)/K$. If K is algebraically closed, it is well-known that there is a one-to-one correspondence between the set of primes of $K(C)$ and the set of points of C [3]. A *K-rational divisor* (or simply a *K-divisor*) D of C is a formal sum $D = \sum_P m_P P$ where P ranges over all primes of $K(C)$, $m_P \in \mathbf{Z}$, and $m_P = 0$ for all but finitely many primes P. The degree of D, denoted $\deg(D)$, is defined by $\deg(D) = \sum_P m_P[k(P) : K]$ where $k(P)$ denote the residue class field of P. The *support* of D, denoted $\mathrm{supp}(D)$, is the set of primes P in $K(C)$ with $m_P \neq 0$. When K is algebraically closed, by identifying each prime P with its corresponding point on C, we can view D as a formal sum of points on C. Let $\mathrm{Div}_K(C)$ denote the set of K-divisors of C. For $D_1 = \sum_P m_P P$ and $D_2 = \sum_P n_P P$ in $\mathrm{Div}_K(C)$, define the sum of D_1 and D_2, denoted $D_1 + D_2$, by $D_1 + D_2 = \sum_P (m_P + n_P)P$. Then $\mathrm{Div}_K(C)$ is a group under the formal-sum operation with the 0-divisor, $\sum_P 0P$, as the identity. Let $\mathrm{Div}_K^0(C)$ denote the subgroup of $\mathrm{Div}_K(C)$ of divisors of degree 0. For $f \in K(C)$, the *divisor of f*, denoted $\mathrm{div}(f)$, is defined by $\mathrm{div}(f) = \sum_P \nu_P(f)P$, where P ranges over all primes of $K(C)$ and ν_P denotes the discrete valuation for P such that $\nu_P(t) = 1$ if t is a principal generator of P. Let $\mathrm{Div}_K^l(C) = \{\mathrm{div}(f) : f \in K(C)\}$. Then $\mathrm{Div}_K^l(C)$ is a subgroup of $\mathrm{Div}_K^0(C)$. Two divisors $D, D' \in \mathrm{Div}_K(C)$ are *linearly equivalent* iff $D - D' \in \mathrm{Div}_K^l(C)$. The *Jacobian group* of C over K, denoted $J_K(C)$, is the quotient group $\mathrm{Div}_K^0(C)/\mathrm{Div}_K^l(C)$. For all $D \in \mathrm{Div}_K^0(C)$, let $[D]$ denote the class of D in $J_K(C)$.

Suppose K is a finite field. Let $\bar{K}$ denote the algebraic closure of K. For $D = \sum_P m_P P \in \mathrm{Div}_K(C)$, let $\hat{D} = \sum_P m_P \hat{P}$ where $\hat{P}$ denotes the formal sum of all distinct primes of $\bar{K}(C)$ containing P.

Let H be a hyperelliptic curve of genus g with an affine model $y^2 - f$ over a finite field K, where $f \in K[x]$ is of degree $2g + 1$ without multiple roots. Let $C = \{(x, y) \mid x, y \in \bar{K}, \text{ and } y^2 = f(x)\}$. Then the set of points on H over $\bar{K}$ can be identified with C together with an additional K-rational point which we shall call *the point at infinity*, denoted by ∞ [11]. We shall also denote by ∞ the prime in $K(C)$ corresponding to the rational point at infinity. The function field $K(H)$ can be identified with the quadratic extension $K(x, y)$ over the purely transcendental extension $K(x)$ with $y^2 = f(x)$. Let σ denote the automorphism of $K(H)$ over $K(x)$ sending y to $-y$. For $z \in K(H)$, let $N(z)$ denote the norm of z relative to $K(x)$.

For all $P = (x, y) \in C$, define its opposite P' to be the point $(x, -y)$. Let $D \in \mathrm{Div}_{\bar{K}}^0(H)$. Then $D = \sum_P m_P P - (\sum_P m_P)\infty$ where P ranges over all points in C, $m_P \in \mathbf{Z}$, and $m_P = 0$ for all but finitely many $P \in C$. D is *semi-reduced* if and only if for all $P \in C$, $m_P \geq 0$, and when $P' \neq P$ either $m_P = 0$

or $m_{P'} = 0$ and when $P = P'$, $m_P = 0$ or $m_P = 1$. D is *reduced* if and only if D is semi-reduced and $\sum_{P \in C} m_P \leq g$. For all $D \in \text{Div}_K^0(H)$, D is semi-reduced (reduced) if and only if $\hat{D}$ is semi-reduced (reduced).

All semi-reduced $D \in \text{Div}_K^0(H)$ can be uniquely represented as the g.c.d. of two principal divisors of functions of the form a and $b - y$, with $a, b \in K[x]$ such that if $\hat{D} = \sum m_i(x_i, y_i) - (\sum m_i)\infty$, then $a = \prod(x - x_i)^{m_i}$ and b is the unique polynomial of degree less than the degree of a such that $b(x_i) = y_i$ for each i, and $a \mid b^2 - f$. We will use the notation $D = \text{div}(a, b)$.

Given two reduced degree 0 K-rational divisors, $D_1 = \text{div}(a_1, b_1)$ and $D_2 = \text{div}(a_2, b_2)$, then we can find a reduced divisor $D_3 = \text{div}(a_3, b_3)$ for $[D_1 + D_2]$ in polynomial time using Cantor's algorithm [2].

3 Technical Preparations for the Algorithm

Prime Divisors

Let u be an irreducible polynomial in $K[x]$. Then u is a principal generator of a prime of $K(x)$. We will also denote by u the prime associated with it. We say that a prime P in $K(H)$ *lies over* u if $u \in P$. We call u a *splitting* prime if u does not divide f and $y^2 - f$ splits mod u. In this case, u splits into two primes in $K(H)$ each with ramification index and residue class degree 1 over u, with $y - v$ and $y + v$ respectively as a principal generator for all $v \in K[x]$ which are roots of $y^2 - f$ mod u. We call u *inert* if u does not divide f and $y^2 - f$ does not split mod u. In this case, there is one prime P in $K(H)$ containing u; u is a principal generator for P; and P is unramified with residue class degree 2 over u. We call u *ramified* if u divides f. In this case, there is one prime P in $K(H)$ containing u. P has ramification index 2 and residue class degree 1 over u and $y \in P$. We shall call a prime in $K(H)$ splitting, inert, or ramified respectively if it lies over a prime in $K(x)$ that is splitting, inert, or ramified respectively.

For all primes P in $K(H)$, let $D_P = P - [k(P) : K]\infty \in \text{Div}_K^0(H)$. Suppose u is the prime in $K(x)$ contained in P. If P is splitting, then $D_P = \text{div}(u, v)$ where $v \in K[x]$ is of degree less than u such that $y - v \in P$. If P is ramified, then $D_P = \text{div}(u, 0)$. If P is inert, then $D_P = \text{div}(u)$.

Let $D \in \text{Div}_K^0(H)$. Then $D = \sum_P m_P D_P$ where P ranges over all primes in $K(H)$ not equal to ∞. Suppose D is semi-reduced. Then for all $P \in \text{supp}(D)$, D_P must be semi-reduced.

We observe that for inert primes P, D_P is not semi-reduced. Consequently the support of a semi-reduced divisor does not contain an inert prime. In fact, let u be the prime of $K(x)$ contained in P. As u does not divide f and $y^2 - f$ does not split mod u, P splits into $\deg(u)$ pairs of primes in $\bar{K}(H)$, each corresponding to a point $(x_i, y_i) \in C$, with x_i a root to u, and its opposite $(x_i, -y_i)$. It follows that $\hat{D}_P$ is not semi-reduced.

Decomposing a Linear Divisor Into Prime Divisors

Let $D = \operatorname{div}(Ay + B)$ where $A, B \in K[x]$ and are relatively prime. Let u be a prime in $K(x)$. Then $\operatorname{supp}(D)$ contains a prime over u iff $u \mid N(Ay + B) = B^2 - A^2 f$. Suppose $u \mid N(Ay + B) = B^2 - A^2 f$.

First observe that u cannot be inert. Otherwise, let P be the prime over u in $K(H)$. Since $P^\sigma = P$, $-Ay + B \in P$ and $Ay + B \in P$, it follows that $B \in P$, so $u \mid B$. Also $2Ay \in P$ and since $y \notin P$, $A \in P$, hence $u \mid A$, contradicting the assumption that A, B are relatively prime.

Suppose u is ramified and P is the prime over u in $K(H)$. Then $u \mid f$ implies $u \mid B$. Since $y \in P$ it follows that $\nu_P(Ay + B) \geq 1$ and since $P = P^\sigma$ that $\nu_P(Ay - B) \geq 1$. Since A and B are relatively prime, u does not divide A and it follows that $u \parallel B^2 - A^2 f = N(Ay + B)$. Since u is ramified $\nu_P(u) = 2$ and hence $\nu_P(Ay - B * Ay + B) = 2$. Therefore the coefficient of D_P in D is 1.

Suppose u is splitting. Let P be a prime over u in $K(H)$ with $y - v$ as a principal generator where $v \in K[x]$. Then $Ay + B \in P$ iff $Av + B \in P$ iff $u \mid Av + B$. Suppose this is the case then $Ay + B \notin P^\sigma$. Otherwise $-Ay + B \in P$, hence $2Ay, B \in P$. Since $y \notin P$, it follows that $A, B \in P$ hence u divides both A and B, a contradiction. Let ν denote the normalized valuation associated with P. Then the coefficient of P in D is $\nu(Ay + B) = \nu((Ay + B)(-Ay + B)) = \nu(B^2 - A^2 f)$. As P is unramified, $\nu(B^2 - A^2 f)$ is the number m such that $u^m \parallel B^2 - A^2 f$.

From the preceding discussion we see that the decomposition of D can be done in expected time polynomial in the degree of A, B and $\log(\#K)$. It also follows from the discussion that D is semi-reduced.

Decomposing a Semi-reduced Divisor Into Prime Divisors

Let $D = \operatorname{div}(a, b)$ be a semi-reduced divisor.

Let u be a prime in $K(x)$. Then $\operatorname{supp}(D)$ contains a prime over u iff $u \mid a$, and in that case u is not inert, as noted above. Suppose $u \mid a$. If u is splitting, letting $v = b \pmod{u}$, then the prime over u which is in $\operatorname{supp}(D)$ is the prime P that contains $y - v$; and the coefficient for P in D is m such that $u^m \parallel a$. If u is ramified, letting P be the prime over u in $K(H)$, then $P \in \operatorname{supp}(D)$ with coefficient 1.

In fact $\sum_P D_P$ where P ranges over all ramified primes in $\operatorname{supp}(D)$ is $\operatorname{div}(d, 0)$

where $d = gcd(a, f)$. Hence we can decompose D as $\text{div}(d, 0) + D_1$, where $D_1 = \sum_P m_P D_P$ with P ranging over all splitting primes in $\text{supp}(D)$. From the preceding discussion we see that such decomposition can be done in expected time polynomial in the degree of a and $\log(\#K)$.

Removing Ramified Primes From a Semi-reduced Divisor

Let D, a, b, d, D_1 be as in the preceding subsection.

It follows from the above, that $\text{div}(y - d) = \text{div}(d, 0) + D'$ for some D' with only splitting primes in the support. Hence $D - \text{div}(y - d) = D_1 - D'$ has only splitting primes in the support. From the decomposition of D and $\text{div}(y - d)$ into prime divisors, and using the fact that for splitting primes P, $D_P + D_{P^\sigma} \in \text{Div}_K^l(H)$, we can easily transform $D_1 - D'$ into a linearly equivalent semi-reduced divisor $\tilde{D}$ whose support contains only splitting primes. Further, this can be done in expected time polynomial in the degree of a and $\log(\#K)$.

To summarize, we have shown:

- Given a semi-reduced $D = \text{div}(a, b)$, we can find, in expected time polynomial in the degree of a, a linearly equivalent semi-reduced divisor $\tilde{D} = \sum_P m_P D_P$ where P ranges over the splitting primes.
- Consequently, $J_K(H)$ is generated by $\{[D_P] : P$ is a splitting prime in $K(H)\}$.

In what follows we will assume that $\{[D_P] : P$ is a splitting prime in $K(H)$ lying over a prime $u \in K(x)$ of degree less than or equal to $\log_p L_{p^n}[1/2, 1/2]\}$ generates $J_K(H)$ whenever $\log p \leq n^{.98}$. This is stronger than what would follow immediately from the Čebotarev density theorem for function fields, but nonetheless seems plausible.

Smoothing a Prime Divisor

Let G be a set of splitting primes in $K(x)$. A polynomial $k \in K[x]$ is G-smooth iff all irreducible factors of k are in G. A divisor $D \in \text{Div}_K^0(H)$ is $G - smooth$ iff the non-infinity primes in $\text{supp}(D)$ all lie over primes in G.

Let P be a splitting prime in $K(H)$ lying over a prime u in $K(x)$ with $y - v$ as a principal generator for some $v \in K[x]$. Then $D_P = \text{div}(u, v \bmod u)$.

To find a linearly equivalent divisor of D_P which is G-smooth, it is sufficient to find relatively prime $A, B \in K[x]$ such that $u \mid B^2 - A^2 f$ and $(B^2 - A^2 f)/u$ factors into primes in G. From earlier discussions on the decomposition of linear divisor, we see that these conditions imply $\text{div}(Ay + B)$ and similarly $\text{div}(-Ay + B)$ are semi-reduced, with all primes in their supports except P or P^σ lying over primes in G. Moreover, if $u \mid Av + B$ then $P \parallel Ay + B$ otherwise $P \parallel -Ay + B$. Hence if $u \mid Av + B$ then $\text{div}(Ay + B) - D_P$ is G-smooth; otherwise $\text{div}(-Ay + B) - D_P$ is G-smooth. The decomposition of the resulting divisor into prime divisors can be computed using the procedure described earlier.

4 Algorithm

Let K be a finite field. Let H be a hyperelliptic curve over K with an affine model $y^2 - f$ where $f \in K[x]$ is irreducible of odd degree and without multiple roots.

Let D_α and D_β be two reduced divisors in $\text{Div}_K^0(H)$ such that there exists some $r \in \mathbf{Z}_{>0}$ with $r[D_\alpha] = [D_\beta]$. Our goal is to find such an r.

Let $n = \deg f$. We will choose a bound S (to be determined later), and let G be the set of splitting primes in $K(x)$ of degree bounded by S.

We first construct for D_α a linearly equivalent semi-reduced divisor $\tilde{D}_\alpha = \sum_P m_P D_P$ where P ranges over splitting primes (not necessarily lying over primes in G). This can be done in random polynomial time as discussed earlier.

For each $P = \text{div}(u, v) \in \text{supp}(\tilde{D}_\alpha)$, we construct for D_P a linearly equivalent semi-reduced divisor S_P such that S_P is G-smooth. As discussed earlier it is sufficient to find relatively prime $A, B \in K[x]$ such that $u \mid B^2 - A^2 f$ and $(B^2 - A^2 f)/u$ factors into primes in G. This is done by randomly generating pairs $A, B \in K[x]$ such that $B \cong -Av \bmod u$.

In this manner we construct for D_α a linearly equivalent $E_\alpha = \sum_i e_i D_{P_i}$ where P_i lies over a prime in G. We construct in like manner for D_β a linearly equivalent E_β which is G-smooth.

We then randomly choose many relatively prime pairs (A, B) with $A, B \in K[x]$ of degree bounded by S' and test $Ay + B$ for G-smoothness. Each pair which passes the test gives a relation on $J_K(H)$. As described in the preliminary section, when sufficiently many such relations have been obtained we can determine the group structure, a set of generators for the group decomposition, and express $[E_\alpha]$ and $[E_\beta]$ with respect to these generators. We then solve congruence relations to find r. If no such r is found another pair is chosen and the step is repeated. That this procedure works within a reasonable number of iterations is an unproven assumption and is discussed in greater detail in the analysis below.

For the case where K is a finite prime field of characteristic unequal to 2, the algorithm is described in greater detail below.

Algorithm

1. Input $p, f, a_\alpha, b_\alpha, a_\beta, b_\beta$, where $p \neq 2$ is prime, $f \in K[x]$, (where $K = Z/pZ$), of odd degree n and having no multiple roots, $a_\alpha, b_\alpha, a_\beta, b_\beta \in K[x]$, such that there exist reduced degree 0 K-rational divisors $D_\alpha, D_\beta \in \text{Div}_K^0(H)$ of the curve H associated with $y^2 - f$, with $D_\alpha = \text{div}(a_\alpha, y - b_\alpha)$, and $D_\beta = \text{div}(a_\beta, y - b_\beta)$

2. Choose the bounds S and S' (to be determined later). Find $G = \{g | g \in K[x]$ monic irreducible, deg $g \leq S$, $g \nmid f$ and f a square mod $g\}$. Let $w = \#(G)$. Let $g_1, ..., g_w$ be an ordering of G

3. For $i = 1, ..., w$ there are two roots of $y^2 - f$ mod g_i. Select one (choice will not matter) and denote it y_i. Then $\text{div}(g_i, y_i) = D_{P_i}$ where P_i is a splitting prime in $K(H)$ lying over g_i.

4. As described earlier, construct for D_α a linearly equivalent semi-reduced divisor $\tilde{D}_\alpha = \sum_{i=1}^m q_i D_{Q_i}$ where the Q_I are splitting primes (not necessarily lying over primes in G). Similarly construct $\tilde{D}_\beta$ for D_β, where $\tilde{D}_\beta = \sum_{i=1}^{m'} r_i D_{R_i}$.

5. For $i = 1, ..., m$, let $D_{Q_i} = \text{div}(u_i, v_i)$ and repeat the following.

   (a) Choose random polynomials $A, a \in K[x]$ of degree less than or equal to S', let $B' = -Av_i \mod u_i$, and let $B = B' + au_i$. Repeat until A and B are relatively prime.

   (b) If $(B^2 - A^2 f)/u_i$ is not G-smooth go to Stage 5a. Otherwise $(B^2 - A^2 f)/u_i = \prod_{j=1}^w g_j^{e_j}$. For $j = 1, ..., w$, if $Ay_j + B \equiv 0 \mod g_j$ then $\rho_{i,j} = -e_j$, else $\rho_{i,j} = e_j$. Let ρ_i be the vector $(\rho_{i,1}, \rho_{i,2}, ..., \rho_{i,j})$

6. For $i = 1, ..., m'$, let $D_{R_i} = \text{div}(u_i', v_i')$ and repeat the following.

   (a) Choose random polynomials $A, a \in K[x]$ of degree less than or equal to S', let $B' = -Av_i' \mod u_i'$, and let $B = B' + au_i'$. Repeat until A and B are relatively prime.

   (b) If $(B^2 - A^2 f)/u_i'$ is not G-smooth go to Stage 6a. Otherwise $(B^2 - A^2 f)/u_i' = \prod_{j=1}^w g_j^{f_j}$. For $j = 1, ..., w$, if $Ay_j + B \equiv 0 \mod g_j$ then $\rho_{i,j}' = -f_j$, else $\rho_{i,j}' = f_j$. Let ρ_i' be the vector $(\rho_{i,1}', \rho_{i,2}', ..., \rho_{i,j}')$

7. Let ρ_α be the vector $\sum_{i=1}^m q_i \rho_i$, and ρ_β be the vector $\sum_{i=1}^{m'} r_i \rho_i'$

8. Repeat until the discrete logarithm is found

   (a) Repeat until an additional γ_j's is produced

      i. Randomly choose polynomials $A, B \in K[x]$ of degree less than or equal to S'. Repeat until A and B are relatively prime. Let $g = A^2 f - B^2$.

      ii. If $g = \prod_{i=1}^w g_i^{e_i}$ (i.e. if G is G-smooth): for $i = 1, ..., w$ if $Ay_i + B \equiv 0 \mod g_i$ then $e_i' = e_i$, else $e_i' = -e_i$. Let γ_j be the vector $(e_1', e_2', ..., e_w')$

   (b) Form the matrix $M = (m_{jk})$ with m_{jk} equal to the kth component of γ_j. If the rank of M is less than w then go to Stage 8a.

(c) As described in the preliminary section, find the matrices $P, D = (d_i), Q$, and calculate $x_1, \ldots, x_n$ such that $J_K(H) = Zx_1 \oplus Zx_2 \oplus \ldots \oplus Zx_n$.

(d) Let $(t_1, \ldots, t_w) = \rho_\alpha P^{-1}$, and $(s_1, \ldots, s_w) = \rho_\beta P^{-1}$.

(e) If there exists an $r \in Z_{\geq 0}$ such that $rt_j = s_j \bmod d_j$ for all $j = 1, 2, \ldots, w$, then output r. Otherwise go to Stage 8a.

5   Analysis of Algorithm

Let $q = p^n$, $S = \log_p L_q[1/2, a]$, and $S' = \log_p L_q[1/2, b]$ (where $a, b \in \Re_{>0}$ will be determined later).

Our analysis will make use of the results of Lovorn [10] on the number of smooth polynomials of bounded degree. Let $N_p(n, m)$ denote the number of polynomials over $GF(p)$ of degree at most n, which have no irreducible factors of degree greater than m. Then it follows from Lovorn that when $\log p \leq n^{.98}$:

$$N_p(n + 2S', S) = p^{n+2S'} L_q[1/2, -1/(2a)]$$

We will henceforth assume that $\log p \leq n^{.98}$.

In the algorithm several matrix operations are performed. These include the calculation of the determinant of M, the calculation of P, Q, D, P^{-1} and the calculation of P^{-1} times a vector. By [6] these can be performed in polynomial time. Let $k \in \Re_{\geq 1}$ be the least such that they can be performed within time w^k.

Stage 1 Stage 1 will not add significantly to the running time.

Stage 2 For Stage 2, the number of polynomials of degree less than S, is $p^S = L_q[1/2, a]$. We determine whether f is a square modulo each of these polynomials. This can be done in random polynomial time using Berlekamp's algorithm[1]. Thus the expected running time for Stage 2 is at most $L_q[1/2, a]$.

Stage 3 In Stage 3, a square root modulo g_i is calculated. This can also be done with Berlekamp's algorithm in random polynomial time. Since w is bounded by $p^S = L_q[1/2, a]$, the expected running time for Stage 3 is at most $L_q[1/2, a]$.

Stage 4 It follows from the fact that D_α and D_β are reduced that the degree of a_α and the degree of a_β are less than n. Hence as described the technical preparation section, Stage 4 can be done in expected time polynomial in $\#K$ and n. Hence the expected running time for Stage 4 is negligible.

Stage 5 Since the degree of a_α is less than n, it follows from the technical preparation section than there are at most n splitting primes in $\operatorname{supp}(D_\alpha)$ and that each such prime P is such that $D_P = \operatorname{div}(u, v)$ with degree of u less than n. Letting $d = gcd(a_\alpha, f)$, it follows from the technical preparations section that than there are at most n splitting primes in $\operatorname{supp}(\operatorname{div}(y - d))$ and that each such prime P is such that $D_P = \operatorname{div}(u, v)$ with degree of u less than n. From this we can conclude that $m \leq 2n$ and the degree of $u_i \leq n$, $i = 1, 2, ..., m$.

In Stage 5a-b, there are $L_q[1/2, 2b]$ pairs A, B to try. We will assume that with high probability such pairs are relatively prime and hence that the number of relatively prime pairs A, B is also $L_q[1/2, 2b]$.

It follows from the above that in Stage 5a-b, the degree of $B^2 - A^2 f/u_i$ is at most $n + 2S'$. Using the usual heuristic that the chance of $B^2 - A^2 f/u_i$ being smooth is the same as the chance of a random polynomial of the same degree being smooth, we get from Lovorn's results above that the chance is $L_q[1/2, -1/(2a)]$. It follows that the expected number of A, B pairs that must be tried is $L_q[1/2, 1/(2a)]$. Hence we have the constraint that $2b \geq 1/(2a)$

Since factoring polynomials can be done in random polynomial time[1], it follows that the expected running time for Stage 5 is at most $L_q[1/2, 1/(2a)]$.

Stage 6 As in Stage 5, the expected running time for Stage 6 is at most $L_q[1/2, 1/(2a)]$.

Stage 7 Stage 7 will not add significantly to the running time.

Stage 8a As in the analysis of Stage 5, the degree of $B^2 - A^2 f$ is at most $n + 2S'$. Again using the heuristic that the chance of $B^2 - A^2 f$ being smooth is the same as the chance of a random polynomial of the same degree being smooth, we get from Lovorn's results above that the chance is $L_q[1/2, -1/(2a)]$. It follows that the expected number of A, B pairs that must be tried until a new γ is obtained is $L_q[1/2, 1/(2a)]$.

Since factoring polynomials can be done in random polynomial time[1], it follows that the expected running time for a single pass through Stage 8a is at most $L_q[1/2, 1/(2a)]$.

Stage 8b The rank of M can be calculated within $w^k = L_q[1/2, ka]$.

Stage 8c Solving for P, D, Q, P^{-1} can be done in time $w^k = L_q[1/2, ka]$. Thus the expected running time for a single pass through Stage 8c is at most $L_q[1/2, ka]$.

Stage 8d As indicated above, the expected running time of this stage is at most $L_q[1/2, ka]$.

Stage 8e Since $d_i \mid d_{i+1}$ for $i = 1, 2, ..., w-1$, it is enough to see if an $r \in Z_{\geq 0}$ exists for which $rt_j \equiv s_j \bmod d_w$ for $j = 1, 2, ..., w$. The expected running time for Stage 8e is thus at most $L_q[1/2, a]$.

Stage 8 The number of repeats of Stage 8 required is quite unclear. Essentially, it is the question of when enough relations on the class group of the function field have been found to define the class group entirely. This is a problem which also arises in considering algorithms that calculate the class group of algebraic number fields ([14, 9]). Intuition suggests that the number of γ's needed should be slightly more than the number of elements in the "factor base", G or roughly $L_q[1/2, a]$. We will henceforth assume that this is the number of repeats of Stage 8 which are required.

Since as argued in Stage 8a, the expected number of A, B pairs that must be tried until a new γ is obtained is $L_q[1/2, 1/(2a)]$. And $L_q[1/2, a]$, total γ's are needed, it follows that we have the constraint that $2b \geq a + (1/(2a))$.

The expected time for one pass of Stage 8 is $\max(L_q[1/2, 1/(2a)], L_q[1/2, ka])$. To minimize take $a = 1/\sqrt{2k}$. Thus the total running time is for Stage 8 is $L_q[1/2, (k+1)/\sqrt{2k}]$. Using the constraints in Stages 5 and Stages 8a we can take $b = (k+1)/(2\sqrt{2k})$.

The entire algorithm is dominated by the running time of Stage 8, so the total expected running time is at most $L_q[1/2, (k+1)/\sqrt{2k}]$.

Acknowledgments

Research supported by NSF CCR-9214671 for Leonard M. Adleman and Jonathan DeMarrais, and NSF CCR-8957317 for Ming-Deh Huang.

6 Discussion

It is worth noting that under certain assumptions our algorithm above can be used to calculate the cardinality and structure of $J_K(H)$.

There are several areas in which this paper could be strengthened. We are unable to argue rigorously that the set of primes lying above those in the set G generate $J_K(H)$. This seems likely to be true but appears to be difficult to prove. We give no formal justification that the structure $J_K(H)$ can be determined by the relationships found by the algorithm within a reasonable amount of time. A strong heuristic argument or experimental evidence would be an improvement.

Further, the algorithm cannot determine if the discrete logarithm problem has no solution. This is not a serious problem for cryptographic applications since in that setting one is usually guaranteed that a solution does exist. One way to solve this problem would be to find a subexponential algorithm for computing a bound on $\#J_K(H)$ to within a factor of 2.

The more efficiently the matrix operations in the algorithm can be performed the better the overall running time. From the results of Iliopoulos ([4]), it appears that $k = 7.376$ is provable. This would give an expected running time of $L_q[1/2, 2.181]$. However this method does not take into account the sparseness of the matrices which occur in our algorithm. $k = 3$ is reasonable if the coefficients do not get too large, in which case, the algorithm runs in expected time

$L_q[1/2, 4/\sqrt{6}]$. If sparse matrix methods can be brought to bear, then $k = 2$ might be possible, which would result in an expected time of $L_q[1/2, 3/2]$. Further, if one assumes that the required number of γ's needed is $w + c$ for some c polynomial in the size of the input, then the analysis may be modified giving an expected running time of $L_q[1/2, 3/\sqrt{6}]$ or $L_q[1/2, 1]$ when k is assumed to be 3 or 2 respectively.

Though our analysis was only for discrete logarithms over $J_K(H)$ where the finite field K is a prime field of characteristic different from 2, it seems likely that our analysis could be generalized in a straightforward manner for the case that K is an arbitrary finite field.

Probably the most interesting open question is whether the techniques in this paper can be successfully applied to give a subexponential algorithm for elliptic curves.

References

1. Berlekamp, E.R., Factoring Polynomials over Large Finite Fields, *Math. Comp.* 24 (1970),713-715.
2. Cantor, D., Computing in the Jacobian of a Hyperelliptic Curve, *Math. Comp.*, Num. 177, 1987, pp. 95-101
3. Hartshorne, R., *Algebraic Geometry*, Springer-Verlag, New York, 1977.
4. Iliopoulos, C., Worst case complexity bounds on algorithms for computing the canonical structure of finite abelian groups and the Hermite and Smith Normal forms of an integer matrix, *SIAM J. Comput.*, 18, 1989, pp. 658- 669
5. Jacobson, N., *Basic Algebra I*, W. H. Freeman and company, 1974
6. Kannan, R. and Bachem A., Polynomial algorithms for computing the Smith and Hermite normal forms of an integer matrix, *SIAM J. Comput.*, 8, 1979, pp. 499-507
7. Koblitz, N., Elliptic curve cryptosystems, *Math. Comp.*, 1987,
8. Koblitz, N., Hyperelliptic Cryptosystems, *J. Cryptography* 1, 1989, pp. 139-150.
9. Lenstra H. W., Algorithms in Algebraic Number Theory, Preliminary version, May 20 1991.
10. Lovorn R., Rigorous, subexponenial algorithms for discrete logarithms over finite fields, PhD Thesis, University of Georgia, May 1992
11. Mumford, D., *Tata Lectures on Theta II*, Birkhäuser, Boston, 1984.
12. Miller, V., The use of elliptic curves in cryptography. *Advances in Cryptography*, Ed. H.C. Williams, Springer-Verlag, 1986, 417-426.
13. Rabin, M.O.,Probabilistic Algorithms in Finite Fields, *Siam J. Comput.* 9 (1980) 273-280
14. Zantema, H., Class Numbers and Units, *Computational Methods in Number Theory*, Part II, 1987, 213-234

Computing Rates of Growth of Division Fields on CM Abelian Varieties

Bruce A. Dodson, Matthew J. Haines

Lehigh University, Department of Mathematics, Bethlehem, PA 18015

Abstract. We report on algorithmic aspects of the problem of explicitly computing the rate of growth of the field of N^k-th division points on an n-dimensional simple Abelian variety with Complex Multiplication. Two new examples are discussed.

For a given Abelian variety with complex multiplication, let $d_k = d_{k,N}$ be the degree of the field extension generated by the N^k-th division points. The rate of growth of d_k, as a function of k, is polynomial and independent of N. Determining the degree of this growth rate depends upon computing the dimension of a certain rational vector space. Starting from n, the description of all possible rates depends upon the collection of permutation groups of degree n, data that specifies certain associated permutation groups of degree $2n$, and then a search through 2^n isogeny types of Abelian varieties, as described in B. Dodson, *The Structure of Galois Groups of CM-fields*, Trans. AMS, **283** (1984), p. 1-32 (cf. J. Algebra **111** (1987), p. 57-58).

The possible rates strongly depend upon the prime factorization of n; and computation depends upon explicit information on the minimal transitive permutation groups. As our first example, we use the recent classification of permutation groups of degree 12 to give analyses of $n = 6, 2n = 12$ (a complete analysis) and $n = 12, 2n = 24$ (a partial analysis, pending information on groups of degree 24). We use G. Royle's Cayley library.

We also consider the cyclic group of order $2n$, where n is the product of three distinct odd primes. The second author applies a method due to H. W. Lenstra, Jr. to give a complete analysis. There are initially 128 possible rates; and the result depends upon whether 3 is one of the prime factors. For example, for simple Abelian varieties with complex multiplication (over $\mathbb{C}$) by the field of 211-th roots of unity, isogeny classes are constructed for 49 different rates; while the remaining 79 rates are ruled out. When 3 is not a factor, 50 rates are constructed, with the remaining 78 ruled out. The slowest possible rate when $n = pqr$ with $3 < r < q < p$ is polynomial of degree $n + 1 - t$ for "corank" $t = \phi(pqr) + \phi(pq) + \phi(p) + \phi(q)$, where $\phi(j)$ denotes the Euler ϕ-function. When 3 is a divisor, the terms $\phi(pq) + \phi(q)$ are replaced by $\phi(3p) + \phi(3)$. These rates improve upon the original construction of Ribet and Lenstra, for which corank $\phi(pqr)$ was achieved.

Algorithms for CM-Fields

Sachar Paulus[*]

Institute of Experimental Mathematics, University of Essen, Ellernstrasse 29,
45326 Essen, Germany

Motivated by the search for a systematic way to produce cryptographically useful hyperelliptic curves, we were led to the need for algorithms solving the following problems in a CM-Field K (that is, $K = K_0(\sqrt{D})$, K_0 a totally real number field, $D \in K_0$ totally negative):

1. solve a relative norm equation in K/K_0 and
2. compute representatives of the ideal classes of K in terms of the ring of integers O_{K_0} of K_0.

Since a CM-field is in some sense a generalization of an imaginary quadratic field, there was hope that one could generalize the algorithms working there, at least for K_0 having some special properties. We proved:

Theorem 1. *If O_{K_0} is euclidean and the euclidean function ϕ_{K_0} is known, then a relative norm equation in K/K_0 can be solved by the Gaussian Reduction Algorithm applied to ϕ_{K_0} and $\|(x,y)\| := N_{K/K_0}(x + \sqrt{D}y)$.*

Let U_{K_0} be the group of units of O_{K_0}, $U_{K_0}^+ := \{\epsilon \in U_{K_0} : \epsilon \gg 0\}$, $U_{K_0}^N := \{\epsilon \in U_{K_0} : \epsilon = N_{K/K_0}(\xi), \ \xi \in K_0\}$, $U_{K_0}^N := \{\epsilon^2 : \epsilon \in U_{K_0}\}$; furthermore, let $\mathcal{F}$ be the set of binary quadratic forms over O_{K_0} and $\mathcal{F}^+ := \{f \in \mathcal{F} : f \text{ totally positive definite}\}$.

Theorem 2. *If O_{K_0} is a principal ideal ring, then*

$$\mathcal{F}\big/PSL_2(O_{K_0}) \cong \mathcal{CL}(K) \times \left(U_{K_0}\big/U_{K_0}^N \right)$$

$$\mathcal{F}^+\big/PSL_2(O_{K_0}) \cong \mathcal{CL}(K) \times \left(U_{K_0}^+\big/U_{K_0}^N \right) \ .$$

Corollary 3. *If the number of equivalence classes of ideals in the narrow sense is 1, then*

$$\mathcal{F}^+\big/PSL_2(O_{K_0}) \cong \mathcal{CL}(K) \ .$$

If O_{K_0} is euclidean and ϕ_{K_0} is known, we can develop a reduction algorithm for binary quadratic forms over O_{K_0} which can be carried over to a reduction algorithm for ideals of K. This allows us to perform computations in the class group of K in terms of O_{K_0}.

This work is in process; the Ph. D. thesis of the author will be extended to contain explicit algorithms, complexity statements and a report on implementation of these algorithms using C++ and PARI.

---

[*] sachar@exp-math.uni-essen.de

Schoof's algorithm and isogeny cycles

Jean-Marc Couveignes[1] and François Morain[2]

[1] U. M. R. d'Algorithmique Arithmétique
Université de Bordeaux, 351 Cours de la Libération, F-33400 Talence, France
& GRECC, D.M.I., E.N.S., 45 rue d'Ulm, F-75230 Paris Cedex 05, France
[2] Laboratoire d'Informatique, École Polytechnique
F-91128 Palaiseau Cedex, France

Abstract. The heart of Schoof's algorithm for computing the cardinality m of an elliptic curve over a finite field is the computation of m modulo small primes ℓ. Elkies and Atkin have designed practical improvements to the basic algorithm, that make use of "good" primes ℓ. We show how to use powers of good primes in an efficient way. This is done by computing isogenies between curves over the ground field. A new structure appears, called "isogeny cycle". We investigate some properties of this structure.

1 Introduction

Let E be an elliptic curve over a primitive finite field $\mathbb{F}_p$ where p is a large prime integer. (We are not dealing with the case of small characteristic here.) The curve is given by some equation $\mathcal{E}(X, Y) = 0$ in Weierstrass form

$$\mathcal{E}(X, Y) = Y^2 - X^3 - AX - B$$

so that a generic point on the curve is given by $(X, Y) \bmod \mathcal{E}$. Let m be the number of points of E. It is well known that $m = p + 1 - t$, with t an integer satisfying $|t| < 2\sqrt{p}$. If p is small the problem of computing the cardinality of E is easy: one can simply enumerate all the points on E. When p is moderately large, say $p \approx 10^{30}$ (see [5]), one can use Shanks's baby-steps giant-steps method. When p is larger, say p up to 10^{200}, one must use Schoof's algorithm, or more precisely the improvements of Atkin and Elkies to the basic scheme.

As a matter of fact, Schoof's algorithm computes $t \bmod \ell$ for sufficiently many small primes ℓ, performing arithmetic modulo polynomials of degree $(\ell^2 - 1)/2$. The basic algorithm can be extended to the case of prime powers ℓ^n as well. In Elkies's improvements, a prime ℓ can be either good or bad. When ℓ is good, one can compute $t \bmod \ell$ more rapidly than in Schoof's basic approach, performing arithmetic modulo polynomials of degree $(\ell - 1)/2$. Moreover, one can in this case compute $t \bmod \ell^n$ pretty much as in Schoof's case. However, one can do better in this case. The purpose of this paper is to explain how one can compute $t \bmod \ell^n$ within the same time complexity as the original $t \bmod \ell$, in the case of good primes. For this, we need to review first Schoof's algorithm, then we give a rough explanation of the improvements of Elkies and Atkin. After that, we explain the role of isogenies and deduce from that an algorithm that enables one to compute $t \bmod \ell^n$. We note that our method has some common points with that of [10], but in a different context.

From a historical point of view, we note that Atkin gave some improvements to Schoof's algorithm as early as 1986 [1], coming up with the match and sort approach in 1988 [2]. In 1989, Elkies [8], described the use of good primes, some details of which were given in [6]. Then, in 1992, Atkin [3] gave the major improvements to Elkies's scheme and made it very practical, his record (March 1994) being computing the cardinality of $E_I : Y^2 = X^3 + 105X + 78153$ modulo $10^{275} + 693$. Morain has also recently implemented the algorithm and obtained similar results using a distributed implementation [12]; his record (December 1993) is the computation of the cardinality of $E_X : Y^2 = X^3 + 4589X + 91128$ modulo $10^{249} + 1291$. We give a table explaining this record at the end of the paper. Recently, Schoof has written an account of the relevant theory in [14]. Some algorithmic details are given in [11].

2 A rough description of the Schoof-Atkin-Elkies ideas

2.1 The basic scheme

We refer to [13]. Let ℓ be some small prime number. For theoretical reasons we need that $\ell < p$. (In practice, p is around 10^{200} while ℓ is always smaller than 500.)

We recall that if π denotes the Frobenius action on the curve, π induces an automorphism of the ℓ-torsion space $E[\ell]$ which extends to Tate's module $T_\ell(E)$. The ring of endomorphisms of the curve contains $\mathbb{Z}[\pi]$ and π satisfies the following degree 2 equation

$$\pi^2 - t\pi + p = 0,$$

where t is related to the cardinality of the curve by

$$\#E = p + 1 - t.$$

Of course, the same equality holds if we consider π as an element of $GL(E[\ell])$ or $GL(T_\ell(E))$. This remark leads to Schoof's idea: compute t modulo ℓ by looking at the action of π on the ℓ-torsion.

To achieve this goal, one first needs to compute the ℓ-torsion polynomial of E, $f_\ell^E(X)$, using the recurrence formulae. Then, a non zero ℓ-torsion point on E is given by

$$(X, Y) \bmod (\mathcal{E}(X, Y), f_\ell^E(X)),$$

so that, for any $\lambda \bmod \ell$ a residue modulo ℓ, one can test whether the trace of π is λ by checking the following identity, written in homogeneous coordinates:

$$(X^{p^2}, Y^{p^2}, 1) \ominus [\lambda](X^p, Y^p, 1) \oplus [p](X, Y, 1) = (0, 1, 0) \bmod (\mathcal{E}, f_\ell^E).$$

For some λ the above equality will hold thus giving $t \bmod \ell$. If one does the same computation for enough primes ℓ_i (i.e., such that $\prod_i \ell_i > 4\sqrt{p}$), then one knows the cardinality of E.

This leads to a polynomial time algorithm. From a practical point of view, the problem is the size of the torsion polynomials. Indeed, $f_\ell^E(X)$ is of degree $(\ell^2 - 1)/2$. In practice one cannot hope to compute $t \bmod \ell$ for $\ell > 31$.

2.2 Elkies's ideas

The whole theoretical background for this section can be found in [9], particularly chapters 12 and 13.

The center of Elkies's ideas [8] is that if $\mathrm{disc}(\pi) = t^2 - 4p$ is a non-zero square modulo ℓ (the zero case works as well but in a slightly different way) then π has two rational distinct eigenvalues τ_1 and τ_2 in $\mathbb{F}_p$ and even in $\mathbb{Z}_p$. Then, Tate's module decomposes as a sum of the two corresponding rational eigensubspaces

$$T_\ell(E) = T_1^E \oplus T_2^E$$

and the ℓ-torsion as well. Such a prime ℓ is called good, and bad in the other case.

We know that there exist $\ell + 1$ isogenies of degree ℓ

$$E \xrightarrow{I_u} E_u \ , \ 1 \le u \le \ell + 1$$

and we are looking for some explicit knowledge about these isogenies, such as their field of definition or their kernel for example. The kernel of those isogenies are the one dimensional subspaces of the ℓ-torsion. Furthermore, their definition field is the definition field of their kernel. Indeed, E_u is just defined to be the quotient of E by the corresponding linear subspace. So, the existence of two rational eigenvalues for the Frobenius implies the existence of two isogenies defined over the base field. Namely, the ℓ-torsion polynomial will have two (non necessarily irreducible) factors h_1 and h_2 of degree $(\ell - 1)/2$, each corresponding to a eigenvalue. We have two isogeneous curves E_i, for $i = 1, 2$, given by some equations $\mathcal{E}_i(X, Y) = 0$ where

$$\mathcal{E}_i(X, Y) = Y^2 - X^3 - A_i X - B_i$$

together with two isogenies $I_1 : E \to E_1$ and $I_2 : E \to E_2$, with kernel $T_1^E \cap E[\ell]$ and $T_2^E \cap E[\ell]$. And for $P = (X, Y) \bmod \mathcal{E}$ a point on E,

$$I_i(P) = \left(\frac{k_i(X)}{h_i^2(X)}, \frac{g_i(X)}{h_i^3(X)} \right) \bmod \mathcal{E}_i$$

for $i = 1, 2$.

All along the paper, we represent the ℓ-torsion on some elliptic curve as a parallelogram with sides the "rational directions". The picture for $\ell = 5$ is given in Figure 1.

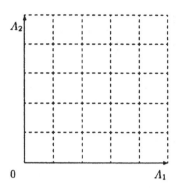

Fig. 1. The 5-torsion structure

A non zero point in $T_1^E \cap E[\ell]$ is given by $(X,Y) \bmod (\mathcal{E}(X,Y), h_1(X))$, which is much nicer than the above, because of the degree of h_1. In view of those considerations, one would like to replace, in Schoof's algorithm, the torsion polynomial by some rational factor h_i when it exists. Or, more conceptually, the $[\ell]$-isogeny by some isogeny of degree ℓ.

We now need to compute the I_i's, and firstly the h_i's. Brute force factorization of f_ℓ^E would be even more difficult than the whole Schoof's method since we would need to compute

$$X^{(\ell^d - 1)/2} \bmod f_\ell^E$$

for some integer d. Nevertheless, the coefficients of h_1 and h_2 are modular functions over $\Gamma_0(\ell)$ and thus can be computed from analytic evaluation in $\mathbb{C}$. Indeed, one considers their Fourier expansion at infinity to find out some modular equation of degree $\ell + 1$. The coefficients of those equations being integers can be reduced modulo p. The existence of some rational eigenvalues to the Frobenius implies the existence of some roots in $\mathbb{F}_p$ to the modular equations. In fact, we have even better, since the modulo p decomposition type of such modular equations gives the permutation type of π seen as a permutation of $\mathbb{P}_1(\mathbb{F}_p)$ thus providing some knowledge about the (non necessarily rational) eigenvalues: the multiplicative order of their quotient. This is the original remark of Atkin. One gets conditions over the residues modulo ℓ_i of the cardinality and then tries to glue up all this knowledge thanks to a sieving process. Note that this is heavier but it works all the time, even if all the small primes we choose are bad.

Note that the whole method splits in two steps:

- Look for some rational root modulo p of the degree $\ell + 1$ modular equations, and build h_1 from it if there is some. Otherwise factor the modular equation completely and deduce the (several) possible values of $t \bmod \ell$ (bad case).
- If you have found some h_1, compute $(X^p, Y^p) \bmod (\mathcal{E}, h_1)$ and then, look for some $\tau \bmod \ell$ such that $(X^p, Y^p) = [\tau](X,Y) \bmod (\mathcal{E}, h_1(X))$ which gives the actual value of $t \bmod \ell$ (good case).

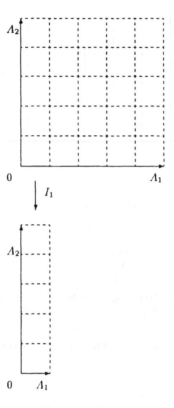

Fig. 2. Isogenies for $\ell = 5$

Note that in both steps we are dealing with polynomials of degree $\ell + 1$ and $(\ell - 1)/2$ which is much smaller than $(\ell^2 - 1)/2$.

Remark: We are not very explicit here about which equation to use. One may think about using the classical modular equation (or rather its quotient by Atkin-Lehner's involution). In this case, the solutions to those equations stand for the isogeneous curves and *not* for the isogenies themselves. It may be that there are two isogenies with distinct kernel and of the same degree, going to the same isogeneous curve. In this case, the endomorphism ring of E must have a non-integer element of norm ℓ^2, which is rather unlikely. Anyway, we can then use another modular equation, such as those given in [6].

2.3 Computing $h_1(X)$

Over $\mathbb{C}$. Let ℓ be a fixed odd prime. Suppose first that we are dealing with a complex curve $E = \mathbb{C}/(\mathbb{Z} + \tau\mathbb{Z})$, of invariant $j(\tau)$ with $\Im(\tau) > 0$. The equation of E is $Y^2 = X^3 + AX + B$. Let $\wp(z)$ denote the Weierstrass function of E:

$$\wp(z) = \frac{1}{z^2} + \sum_{k=1}^{\infty} c_k z^{2k}$$

where the c_k are in $\mathbb{Q}(A,B)$: $c_1 = -A/5, c_2 = -B/7$, and for $k \geq 3$:

$$c_k = \frac{3}{(k-2)(2k+3)} \sum_{h=1}^{k-2} c_h c_{k-1-h}.$$

The ℓ-th division polynomial is then simply

$$f_\ell^E(X) = \ell \prod_{\substack{1 \leq r \leq (\ell-1)/2 \\ 0 \leq s < \ell}} (X - \wp((r+s\tau)/\ell))$$

and is in fact in $\mathbb{Q}(A,B)[X]$. This polynomial has a factor

$$h_1(X) = \prod_{r=1}^{(\ell-1)/2} (X - \wp(r/\ell))$$

which has coefficients in an extension of degree $\ell + 1$ of $\mathbb{Q}(A,B)$. We let

$$p_k = \sum_{r=1}^{(\ell-1)/2} \wp(r/\ell)^k.$$

Elkies shows how to compute all p_k's using only p_1, p_2 and p_3. He also shows that p_1 can be obtained as a root of a degree $\ell+1$ equation, whereas p_2 and p_3 can be obtained from the coefficients A_1 and B_1 of the curve $E_1 = \mathbb{C}/(\frac{1}{\ell}\mathbb{Z} + \tau\mathbb{Z})$ which is isogenous to E. We make the important remark that the periods of E_1 are the image of that of E by the Atkin-Lehner involution, $W_\ell(F(\tau)) = F(-1/\ell\tau)$ for any function F, and in particular $W_\ell(j(\tau)) = j(-1/\ell\tau) = j(\ell\tau)$.

In Atkin's approach, one first determines a modular equation for $X_0(\ell)$, that is to say an algebraic relation

$$\Phi_\ell(X,Y) = 0$$

which relates a function $F(q)$ on $\Gamma_0(\ell)$ and the modular invariant $j(q)$ (with $q = \exp(2i\pi\tau)$). One knows that

$$\Phi_\ell(X,Y) = \sum_{r=1}^{\ell+1} C_r(Y) X^r$$

where the C_r's have integer coefficients and $C_{\ell+1}(Y) = 1$. Starting from

$$\Phi_\ell(X,j(\tau)) = 0$$

one can compute $F(\tau)$ and then all quantities p_1, A_1 and B_1 can be deduced from this in an algebraic way.

Remark. Atkin distinguishes between two types of modular equations: the "canonical" one and the "star" one. In the first case, one uses the function $\mathcal{F}_\ell(\tau) = \ell^s(\eta(\ell\tau)/\eta(\tau))^{2s}$ where $s = 12/\gcd(12, \ell-1)$. As Atkin shows, with this function, it is easy to compute $j_1 = j(\ell\tau)$ using $F_1 = \mathcal{F}_\ell(\tau)$ without finding the roots of $\Phi_\ell(W_\ell(F_1), Y) = \Phi_\ell(\ell^s/F_1, Y)$, but on the other hand the valence of $\mathcal{F}_\ell$ grows linearly as a function of ℓ. In the star case, one uses a function with smallest possible valence on $X_0^*(\ell) = X_0(\ell)/W_\ell$. This has the advantage of having a very small valence, but we have then to compute the roots of $\Phi_\ell(W_\ell(F_1), Y) = \Phi_\ell(F_1, Y)$.

Over $\mathbb{F}_p$. Now, modulo p, if ℓ is a good prime, then $\Phi_\ell(X, j(E)) \equiv 0 \bmod p$ has (in general) two distinct roots and we can use the previous algebraic relations modulo p and deduce from this an isogeneous curve E_1 and the polynomial $h_1(X)$ which is the desired factor of $f_\ell^E(X)$ modulo p.

3 Walking along the rational cycles of isogeneous curves

3.1 Theory

We now suppose that $\pi \in GL(T_\ell)$ has two distinct rational eigenvalues τ_1 and τ_2. We notice that, since the two isogenies I_1 and I_2 are rational, they commute with π. This implies that on the isogeneous curves, the eigenvalues of the Frobenius are the same. Since the eigenspaces T_1^E and T_2^E are independent, I_1 induces a bijection between T_2^E and the corresponding eigensubspace on E_1 and reciprocally I_2 induces a bijection between T_2^E and the corresponding eigensubspace on E_2.

The existence of two distincts rational eigenvalues has another interesting consequence. It is that E_1 again has two rational isogenies of degree ℓ, one associated to each of the two eigenvalues. We call I_{11} and I_{12} those isogenies and E_{11} and E_{12} the image curves. On the other hand, we know that, since I_1 is rational, the dual isogeny I_1^* must be rational as well (by uniqueness of it). Therefore I_1^* either equals I_{11} either equals I_{12}. By consideration of the restriction to T_1^E we see that

$$I_1^* = I_{12}.$$

We could express that by saying that the two rational directions are not only independent but dual.

We show all that on Figure 1.

Now, if E is a curve over $\mathbb{F}_p$ such that $t^2 - 4p$ is a non zero square mod ℓ we can build two periodic sequences of isogeneous curves over $\mathbb{F}_p$. These sequences define two permutations $\mathcal{I}_1$ and $\mathcal{I}_2$ on the set of elliptic curves over $\mathbb{F}_p$, classified up to $\mathbb{F}_p$-isomorphisms. The permutation $\mathcal{I}_i$ is generated by the quotient of E by τ_i and the two permutations are inverse of each other.

$$E \xrightarrow{I_1} E_1 \xrightarrow{I_{11}} E_{11} \xrightarrow{I_{111}} \cdots$$

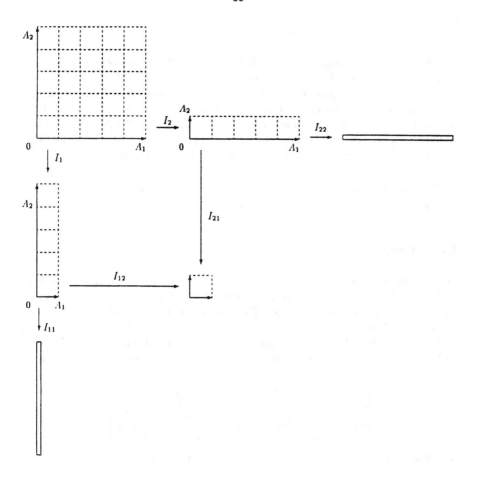

Fig. 3. Action of the isogenies

$$E \xrightarrow{I_2} E_2 \xrightarrow{I_{22}} E_{22} \xrightarrow{I_{222}} \cdots$$

These series are computed in the following way. We use some modular equation $\Phi_\ell(X, Y)$. Let's call j_0 the invariant of E and let's solve $\Phi_\ell(X, j_0) = 0$ over $\mathbb{F}_p$. If we are in the "good case" we have two rational distinct simple roots F_1 and F_2, from which we compute two curves E_1 and E_2 of respective invariants j_1 and j_2. Let's now solve the equation $\Phi_\ell(X, j_1) = 0$ over $\mathbb{F}_p$. We find two rational distinct simple roots, one of them being $W_\ell(F_1)$ and corresponding to the dual isogeny I_1^*. We choose the other one and call it F_{11}, yielding E_{11}. We go on, solving the equation $\Phi_\ell(X, j_{11}) = 0$, etc.

Since the field is finite, the two series of curves are periodic and they provide an explicit description of the two rational subspaces of Tate's module.

3.2 Example

Let $p = 101$ and consider all the (classes of) curves E for which have $(p + 1) - \#E = t = 3$. There are 8 of them and the following table gives their invariant and a representative for each class. These curves were obtained by brute force, but they could have been obtained by noting that $3^2 - 4 \times 101 = -395$, implying that all curves have complex multiplication by the ring of integers of $\mathbb{Q}(\sqrt{-395})$ and therefore their j-invariant are the roots of the 8-degree Weber polynomial as in [4].

j	E	j	E	j	E	j	E
2	$[68, 79]$	10	$[19, 59]$	15	$[56, 41]$	20	$[27, 18]$
34	$[13, 51]$	56	$[3, 2]$	82	$[53, 37]$	90	$[49, 100]$

Starting from $E_0 = [68, 79]$, $J_0 = 2$, using $\ell = 7$, one first finds

$$\Phi_7^c(X, 2) \equiv (F + 84)(F + 64)(F^6 + 82F^5 + 81F^4 + 49F^3 + 32F^2 + 34F + 68) \bmod 101.$$

We choose $F_1 = 17$ and find $\tau_1 = 6$. The permutation $\mathcal{I}_1$ is then given in

E	$j(E)$	$F(E)$
$[68, 79]$	2	17
$[27, 68]$	82	14
$[50, 89]$	56	33
$[31, 28]$	10	9
$[45, 15]$	34	20
$[47, 87]$	90	100
$[42, 63]$	20	43
$[97, 32]$	15	45
$[56, 31]$	2	

The other permutation starts using $F_1 = 37$ and corresponds to $\tau_2 = 4$.

3.3 Application to Schoof's algorithm

For example, the factor of $f_\ell^E(X)$ corresponding to $T_1^E \cap E[\ell]$ is h_1, the denominator of I_1. Now, if we want the factor of $f_{\ell^2}^E$ corresponding to $T_1^E \cap E[\ell^2]$, we proceed in the following way. We first compute the polynomial h_{11} which is the denominator of I_{11}, in the same way we computed h_1 except that we replace E by E_1 and pay attention not to confuse I_{11} with $I_1^* = I_{12}$. Indeed we consider the isogeny from E_1 associated with τ_1. We then note that $T_1^E \cap E[\ell^2] = I_1^{-1}(T_1^{E_1} \cap E_1[\ell])$ so that the factor we are looking for is obtained by plugging I_1 into h_{11}. And so on ...

In this way one can compute factors of degree $\ell^{k-1}(\ell-1)/2$ to the polynomial $f_{\ell^k}^E$ and then, using Schoof's idea compute the cardinality of E modulo ℓ^k rather than just ℓ. This allows us to take more advantage of the small good primes.

4 Implementation

4.1 Computing h_1 and I_1

The way Atkin's approach works, one first solves $\Phi_\ell(X, j_0) \equiv 0 \bmod p$ for a root F_1 and then one computes j_1 as a root of

$$\Phi_\ell(W_\ell(F_1), Y) \equiv 0 \bmod p.$$

Each solution y yields a putative factor $h_y(X)$ of $f_\ell^E(X)$. We check this factor by checking that $[\ell](X, Y) = 0 \bmod (\mathcal{E}(X, Y), h_y(X))$. Once we know a proper factor h_1 of f_ℓ^E, we proceed to find I_1. We know that

$$I_1(X, Y) = \left(\frac{k_1(X)}{h_1(X)^2}, \frac{g_1(X)}{h_1(X)^3} \right)$$

where $k_1(X)$ is a polynomial of degree ℓ with coefficients in $\mathbb{F}_p$. Let $\wp_1(z)$ denote the Weierstrass function of E_1. Then

$$\wp_1(z) = \frac{k_1(\wp(z))}{h_1(\wp(z))^2}.$$

Replacing $\wp$, $\wp_1$ and h_1 by their value, one deduces easily from this the coefficients of k_1.

4.2 Examples

Let us work out an example. Let's take $E : Y^2 = X^3 + 2X + 3 \bmod 97$. We use the so-called "canonical" equation of $X_0(5)$, namely the relation between $\mathcal{F}_5(x) = 5^3(\eta(5\tau)/\eta(\tau))^6 = 125(x + 6x^2 + \cdots)$ and $j(x)$, which is

$$\Phi_5(X, Y) = 125 - YX + X^6 + 30X^5 + 315X^4 + 1300X^3 + 1575X^2 + 750X.$$

One computes $j_0 = j(E) = 36$ and $\Phi_5(X, 36)$ factors as

$$(X + 25)(X + 10)(X^4 + 92X^3 + 46X^2 + 67X + 49).$$

We choose $F_1 = 87$ and find easily that j_1 is the root of

$$\Phi_5(5^3/F_1, Y) \bmod p$$

that is $j_1 = 48$, from which we deduce from that $E_1 : Y^2 = X^3 + 96X + 83$. We also find that

$$h_1(X) = X^2 + 16X + 30.$$

Now, one has

$$\wp(z) = z^{-2} + 19z^2 + 55z^4 + 88z^6 + 91z^8 + O(z^{10}),$$
$$\wp_1(z) = z^{-2} + 39z^2 + 2z^4 + 22z^6 + 83z^8 + O(z^{10})$$

so that

$$\wp_1(z)h_1(\wp(z))^2 = z^{-10} + 32z^{-8} + 43z^{-6} + 83z^{-4} + 93z^{-2} + 76 + O(z^2)$$

from which we recognize that

$$k_1(X) = X^5 + 32X^4 + 45X^3 + 92X^2 + 18X + 35.$$

A factor of $f_{25}^E(X)$ is then the numerator of $h_1(I_1(X))$ namely

$$X^{10} + 48X^9 + 77X^8 + 54X^7 + 38X^5 + 36X^4 + 40X^3 + 3X^2 + 90X + 5.$$

Now, we want to compute E_{11} and so we want to solve

$$\Phi_5(X, j_1) \equiv (X + 61)(X + 5)(X^4 + 61X^3 + 58X^2 + 13X + 2) \equiv 0 \bmod p.$$

We note that a solution to this is $W_\ell(F_1) = \ell^3/F_1 \equiv 36 \bmod p$. We must discard this one, since we would go back to E_0. So, we take $F_{11} = 92$ and find E_{11} : $Y^2 = X^3 + 95X + 66$, together with

$$I_{11}(X) = \frac{X^5 + 65X^4 + 75X^3 + 85X^2 + 6X + 71}{X^4 + 65X^3 + 36X^2 + 28X + 72}.$$

4.3 An improved strategy

It is easy to see that the algorithm works also if we replace h_1 by a factor of h_1. In that case, a factor of degree d of $f_\ell^{E_1}$ can be lifted to a factor of degree $d\ell$ of $f_{\ell^2}^E$. A good strategy for using the isogeny idea is given in the following algorithm. We suppose that E is given modulo p and ℓ is an odd prime.

1. find the roots of $\Phi_\ell(X, j(E))$;
2. if Φ_ℓ has two distinct rational roots then
   (a) compute a factor h_1 of f_ℓ^E;
   (b) find the eigenvalue τ_1 and deduce τ_2 from it;
   (c) find the order d_i of τ_i modulo ℓ; if d_i is even, divide by 2; [d_i is now the smallest field $\mathbb{F}_{p^d}$ containing abscissa of points of ℓ-division]
   (d) let $d = \min(d_1, d_2)$ and τ the associated eigenvalue; [d is now the degree of a factor of f_ℓ^E of minimal degree, see [3]]
   (e) if d is small enough, then factor the factor associated with d and compute $t \bmod \ell^n$ for small n using arithmetic modulo a polynomial $h_{\ell^n}(X)$ of degree $d\ell^{n-1}$, using the fact that the eigenvalue k is congruent to τ modulo ℓ.

4.4 Experimental results

The second author has implemented the Schoof-Elkies-Atkin algorithm in C, using the **BigNum** package. The details of this implementation will be described in a forthcoming article [12].

His latest record (March 1994) concerns the curve:

$$E_X : Y^2 = X^3 + 4589 * X + 91128$$

modulo $p = 10^249 + 1291$. Its cardinality is $m = p + 1 - t$ where t is

812863330901169485115745076523086320636188340265983567
8383607032008620595243247600658124603970833311581801435393008665561929.

It took 1027 CPU hours on several DecAlpha's to perform the job, 641 of which were needed for the computation of various $X^p \bmod f(X)$. We give in Table 1 the ℓ-primes used, together with a code, which says that ℓ was an **A**tkin prime, an **E**lkies prime of a **S**choof prime (a Schoof prime is a small Atkin prime for which the original algorithm could be used). If ℓ is an Elkies prime, the third column contains the values $(t \bmod \ell^n, k_1, o_1, k_2, o_2)$ where k_1 and k_2 are the eigenvalues and o_1 and o_2 their respective orders. If ℓ is an Atkin prime, then we put the ratio of the possible number of t modulo ℓ versus $\ell-1$. (See [3] for the importance of that quantity.) More details will be given in [12].

5 The case $\ell = 2$

5.1 The equation $X^2 - tX + p \equiv 0 \bmod 2^n$

Let us first consider the set of solutions $\mathcal{X}_n$ of the equation

$$(R_n) \quad X^2 - tX + p \equiv 0 \bmod 2^n \tag{1}$$

with p odd. The following tables give the solutions of this equation for small n.

Lemma 1. *Equation* (R_1) *has solutions modulo 2 if and only if* $t \equiv 0 \bmod 2$ *and in this case* $\mathcal{X}_1 = \{1\}$.

Lemma 2. *Equation* (R_2) *has solutions if and only if* $t \equiv p+1 \bmod 4$, *in which case* $\mathcal{X}_2 = \{1, 3\}$.

As is customary, one wants to compute the solutions of (R_{n+1}) starting from (R_n). Let x_n be a solution of (R_n) and put

$$t_n \equiv t \bmod 2^n, 0 \leq t_n < 2^n, \quad p_n \equiv p \bmod 2^n, 0 \leq p_n < 2^n.$$

We look for a solution $x_{n+1} = x_n + \xi 2^n$, $\xi \in \{0,1\}$. The following result is immediate.

ℓ^n	type		ℓ^n	type		ℓ^n	type	
2^5	E	(9)	89	A	0.27	211	A	0.50
3^3	S	(26)	97	A	0.44	223	A	0.03
5^2	S	(4)	101	E	(75, 66, 100, 9, 50)	227	E	(201, 160, 113, 41, 226)
7	S	(0)	103	A	0.02	233	A	0.31
11^3	E	(730, 1, 1, 3, 5)	107	A	0.17	239	E	(120, 46, 238, 74, 238)
13	S	(8)	109	E	(47, 9, 27, 38, 9)	257	E	(231, 196, 128, 35, 64)
17^3	E	(707, 11, 16, 16, 2)	113	E	(93, 27, 112, 66, 112)	269	E	(256, 76, 268, 180, 67)
19	S	(9)	127	E	(38, 119, 14, 46, 126)	281	E	(92, 42, 280, 50, 35)
23	S	(12)	131	A	0.15	283	E	(53, 123, 282, 213, 282)
29^2	E	(91, 24, 7, 9, 14)	137	E	(6, 110, 136, 33, 136)	293	E	(211, 81, 73, 130, 292)
31^2	E	(104, 29, 10, 13, 30)	139	E	(45, 7, 69, 38, 69)	311	E	(117, 51, 62, 66, 310)
37	A	0.50	149	A	0.27	317	E	(112, 308, 79, 121, 79)
41	A	0.30	151	A	0.48	331	E	(175, 19, 165, 156, 165)
43^2	E	(375, 11, 7, 20, 42)	157	E	(17, 71, 39, 103, 52)	347	E	(267, 56, 173, 211, 346)
47	E	(38, 8, 23, 30, 46)	163	E	(35, 155, 27, 43, 81)	353	E	(335, 303, 176, 32, 88)
53	E	(4, 18, 52, 39, 52)	167	E	(73, 30, 166, 43, 166)	373	E	(325, 266, 186, 59, 186)
59	E	(23, 3, 29, 20, 29)	173	E	(152, 56, 86, 96, 43)	379	E	(364, 333, 189, 31, 378)
61	A	0.50	179	E	(2, 64, 89, 117, 89)	383	E	(349, 308, 382, 41, 382)
67	E	(14, 40, 11, 41, 66)	181	A	0.40	431	E	(56, 171, 215, 316, 430)
71	E	(40, 4, 35, 36, 35)	191	A	0.17	439	E	(24, 200, 219, 263, 438)
73	A	0.50	193	A	0.50	443	E	(338, 72, 442, 266, 442)
79^2	E	(2660, 55, 3, 77, 78)	197	A	0.03	449	E	(333, 384, 448, 398, 112)
83	A	0.15	199	A	0.40			

Table 1. Data for E_X mod $10^{249} + 1291$

Proposition 3. *Let $n \geq 1$. Write*

$$p_{n+1} = p_n + \pi 2^n, \pi \in \{0, 1\}, \quad t_{n+1} = t_n + \tau 2^n, \tau \in \{0, 1\}$$

and $x_n^2 - t_n x_n + p_n = K 2^n$. Then x_{n+1} is a solution of (R_{n+1}) (for any choice of ξ) if and only if

$$K + \pi + \tau \equiv 0 \bmod 2.$$

For example, one obtains the following result for $n = 3$ starting from the solutions corresponding to $n = 2$.

Lemma 4. *For $n = 3$, one gets*

t	p	X_3	t	p	X_3
0	$\{1, 3, 5\}$	$\emptyset$	4	$\{1, 5, 7\}$	$\emptyset$
	7	$\{1, 3, 5, 7\}$		3	$\{1, 3, 5, 7\}$
2	$\{3, 7\}$	$\emptyset$	6	$\{3, 5\}$	$\emptyset$
	1	$\{1, 5\}$		1	$\{3, 7\}$
	5	$\{3, 7\}$		5	$\{1, 5\}$

It is clear from the result that if (R_n) has a solution, this does not imply that (R_{n+1}) does. In some cases, one can do better.

Proposition 5. *Assume $n \geq 3$. If $t_n \equiv 0 \bmod 4$ and x_n is a solution of (R_n), then there exists $\xi \in \{0, 1\}$ such that $x_n + 2^{n-1}\xi$ is a solution of (R_{n+1}).*

Proof. Writing $x_{n+1} = x_n + 2^{n-1}\xi$ and with the notations as above, one finds that $K + (1 - t_n/2)\xi + \rho + \tau$ should be 0 modulo 2, which always yields a solution in ξ if $t_n/2 \equiv 0 \bmod 2$. $\square$

5.2 Computing I_1 and h_1

We first note the important result.

Theorem 6. *If $X^2 - tX + p \equiv 0 \bmod 2^n$ has a solution, then $f_{2^n}^E(X)$ has a factor of degree 2^{n-2}.*

The methods described by Atkin enables one to compute the isogenous curve, but not the factor of the division polynomial. However, one can compute the Weierstrass function of the isogenous curve and deduce from this the isogeny I_1 as in [15] using continued fractions and thus h_1.

The results of the preceding section has important consequences for our purpose. As a matter of fact, using Atkin's approach, one has to find the roots of $\Phi_2(X, J_1)$ which is of degree 3, so has 1 or 3 roots. Since $X = 2^{12}/F_1$ is already a root, this leaves us with 0 or 2 roots for F_{11}. If there are two roots, we can proceed to find E_{11}, but we are not sure which one it is, and sometimes we have to backtrack. When there are no more roots for a certain depth, this means that $X^2 - tX + p \bmod 2^n$ has no roots for this n. This implies new restrictions on t. We will give examples next.

5.3 Example

Let $p = 101$, $E_0 = [77, 69]$. One finds that Φ_2 factors as

$$\Phi_2(X, 22) = (X^2 + 80X + 74)(X + 69) \bmod 101$$

and thus $F_1 = 32$. One finds $E_1 = [58, 34]$, $J_1 = 98$ and the isogeny is

$$I_1 = \frac{X^2 + 4X + 24}{X + 4}$$

and $X + 4$ is indeed a factor of $X^3 + 77X + 69$. We compute

$$\Phi_2(X, 98) = (X + 74)(X + 98)(X + 78) \bmod 101.$$

We discard $X = 27 = 2^{12}/F_1$ as usual and we have to choose between 3 and 23. It turns out that we must take $F_{11} = 23$, thus obtaining $E_{11} = [42, 43]$ and

$$I_{11} = \frac{X^2 + 50X + 84}{X + 50}.$$

Now, we compute the numerator of $I_1 + 50$ and find it is $(X + 27)^2$ and $X + 27$ is indeed a factor of $f_4^{E_0}$. After that, $F_{111} = 54$, $E_{111} = [85, 11]$ and a factor of $f_4^{E_1}$ is $X + 86$ so that a factor of $f_8^{E_0}$ is $X^2 + 90X + 65$.

In some other cases, we have to do more computations, as shown in the following. Take $E = [1, 3]$ modulo $p = 1009$. One finds that

$$\Phi_2^c(X, 269) \equiv (X - 484)(X - 994)(X - 492) \bmod p.$$

In what follows, we list the depth of the search, followed by the value of F. Here is the beginning of it

$$0(484), 1(198), 2(--), 1(446), 2(--), 0(492), 1(483), 2(--), 1(281), 2(--),$$

$$0(994), 1(225), 2(649), 3(289), 4(644), 5(--), 4(233), 5(--), \ldots$$

As a matter of fact, we cannot go deeper than 6 levels. This means we can compute $t \bmod 2^7$, but not $t \bmod 2^8$, which is coherent with the fact that $t = -50$ and that $X^2 + 50X + 1009 \bmod 2^n$ has no roots for $n \geq 8$.

The above example shows that the implementation of this part of the algorithm is rather tricky: we first find the longest path in our tree, then compute the isogenies and then the division polynomials. We also use the informations we have gathered to perform the final computations.

6 Conclusions

We have shown how to use small prime powers in Schoof's algorithms. This raises interesting questions concerning isogeny cycles. Our approach should also work for the new approach used by the first author for extending Atkin's ideas to small characteristic [7].

Also, we never considered the degenerate cases where the modular equation has only one root modulo p, or splits completely modulo p. These cases can also be treated, in some cases as in the $\ell = 2$ case. We will describe this somewhere else.

Acknowledgments. The first author is a member of the "Option Recherche du Corps des Ingénieurs de l'Armement; the second author is on leave from the French Department of Defense, Délégation Générale pour l'Armement.

References

1. A. O. L. Atkin. Schoof's algorithm. Preprint, 1986.
2. A. O. L. Atkin. The number of points on an elliptic curve modulo a prime. Preprint, 1988.
3. A. O. L. Atkin. The number of points on an elliptic curve modulo a prime (ii). Preprint, 1992.
4. A. O. L. Atkin and François Morain. Elliptic curves and primality proving. *Math. Comp.*, 61(203):29–67, July 1993.

5. Johannes Buchmann and Volker Müller. Computing the number of points of elliptic curves over finite fields. In S. M. Watt, editor, *ISSAC '91*, pages 179–182, 1991. Proceedings of the International Symposium on Symbolic and Algebraic Computation, July 15–17, Bonn, Germany.

6. Leonard S. Charlap, Raymond Coley, and David P. Robbins. Enumeration of rational points on elliptic curves over finite fields. Draft, 1991.

7. Jean-Marc Couveignes. Thesis. Manuscript, 1994.

8. Noam D. Elkies. Explicit isogenies. Manuscript, 1991.

9. D. Husemöller. *Elliptic curves*, volume 111 of *Graduate Texts in Mathematics*. Springer, 1987.

10. J.-F. Mestre. La méthode des graphes. Exemples et applications. In *Proc. of the International Conference on class numbers and fundamental units of algebraic number fields*, pages 217–242, Nagoya, 1986. Nagoya Univ. Katata (Japan).

11. François Morain. Calcul du nombre de points sur une courbe elliptique dans un corps fini : aspects algorithmiques. Submitted for publication of the Actes des Journées Arithmétiques 1993, March 1994.

12. François Morain. Implantation de l'algorithme de Schoof-Elkies-Atkin. Preprint, January, 1994.

13. R. Schoof. Elliptic curves over finite fields and the computation of square roots mod p. *Math. Comp.*, 44:483–494, 1985.

14. René Schoof. Counting points on elliptic curves over finite fields. Submitted for publication of the Actes des Journées Arithmétiques 1993, March 1994.

15. H. M. Stark. Class-numbers of complex quadratic fields. In W. Kuyk, editor, *Modular functions of one variable I*, volume 320 of *Lect. Notes in Math.*, pages 155–174. Springer Verlag, 1973. Proceedings International Summer School University of Antwerp, RUCA, July 17-Agust 3, 1972.

Integer Points on Rational Elliptic Curves

Nelson Stephens

Department of Computing Mathematics
University of Wales College of Cardiff
Cardiff, CF2 4YN, UK.

Abstract. Let E be a rational elliptic curve in global minimal Weierstrass form and let P be a point on E with integer co-ordinates. We consider the problem of determining the multiples of P which also have integer co-ordinates. This is a difficult problem for the case where P is on the infinite real component of E.

Let Z_m^2 denote the denominator of the x-co-ordinate of mP. For certain curves we show that the sequence $\{Z_m\}$ satisfies the elliptic divisibility recurrence relation studied by Morgan Ward:

$$Z_{m+n}Z_{m-n} = Z_{m+1}Z_{m-1}Z_n^2 - Z_{n+1}Z_{n-1}Z_m^2.$$

Thus the problem is equivalent to determining all indexes m for which $Z_m = \pm 1$. It is shown that this determination can sometimes be solved using elementary p-adic techniques on the sequence modulo a prime power.

For example, on the curve E of conductor 43, the rational points form an Abelian group of rank one and no torsion. If P denotes the basic generator of this group, the sequence $\{Z_m\}$ modulo 2^t is purely periodic with period $5 \times 2^{t-1}$. It is shown using 2-adic analysis that the only multiples of P having integer co-ordinates are 1, 2, 3, 4 and 7.

Counting the Number of Points on Elliptic Curves over Finite Fields of Characteristic Greater than Three

Frank Lehmann, Markus Maurer
Volker Müller*, Victor Shoup

FB Informatik
Universität des Saarlandes
Postfach 1150
D-66041 Saarbrücken
Germany

Abstract. We present a variant of an algorithm of Oliver Atkin for counting the number of points on an elliptic curve over a finite field. We describe an implementation of this algorithm for prime fields. We report on the use of this implementation to count the number of points on a curve over $\mathbb{F}_p$, where p is a 375-digit prime.

1 Introduction

In this paper we study the problem of counting the number of points on an elliptic curve over a finite field. This problem is not only very interesting for number theorists but has recently gained a lot of attention among cryptographers. The use of elliptic curves in public key cryptography was suggested by Koblitz [Ko86] and Miller [Mi86]. The security of their elliptic curve cryptosystems is based on the intractability of the problem of computing discrete logarithms in the elliptic curve group. The best algorithms known for solving this problem for arbitrary elliptic curves are the exponential square root attacks [Od86] which have running time proportional to the largest prime factor dividing the group order. Consequently, in order to guarantee the security of the system it is necessary to find this group order and its prime factorization. Although Schoof [Scho85] proved that the cardinality of an elliptic curve group over a finite field can be computed in polynomial time, his algorithm is extremely inefficient in practice.

We present a short description of an variant of an algorithm of Oliver Atkin for counting points on elliptic curves over finite prime fields [At91]. We also describe an implementation of this algorithm and report on its effectiveness. With this implementation we were able to compute the group order of a curve defined over $\mathbb{F}_p$, where p is a 375-digit prime. The total time used was approximately 1765 MIPS-days (this time does not include the time for precomputations that are independent of the input).

---

* email: vmueller@cs.uni-sb.de

Background on the theory of elliptic curves can be found in [Si86]; a detailed description of the algorithm (including proofs of the theorems used) is given in [Mü94]. The implementation makes use of a library of routines for efficient computations with large polynomials over finite fields described in [Sho94].

2   Elliptic Curves and the Frobenius Endomorphism

Let $\mathbb{F}_q$ be the finite field with q elements, and let p be its characteristic. We shall assume that $p > 3$. Let E be an elliptic curve defined over $\mathbb{F}_q$ by the equation

$$y^2 = x^3 + ax + b, \tag{1}$$

where a and b are in $\mathbb{F}_q$, and the discriminant $4a^3 + 27b^2$ is nonzero.

Let $\overline{\mathbb{F}_q}$ be the algebraic closure of $\mathbb{F}_q$. For a field K, $\mathbb{F}_q \subset K \subset \overline{\mathbb{F}_q}$, the set $E(K)$ of K-rational points consists of the affine solutions $(x, y) \in K^2$ of (1), together with the point $\mathcal{O}$ "at infinity" obtained by considering the projective closure of (1). The set $E(K)$ has a group structure given by the well-known "tangent and chord method", with $\mathcal{O}$ acting as the identity element. The sum of two given points can be computed by simple formulas (see, e.g., [BuMü91]).

We will describe an algorithm that takes as input elements a and b defining an elliptic curve E as in (1), and produces as output the order of the group $E(\mathbb{F}_q)$. An important tool for solving this problem is the Frobenius Endomorphism for E.

Definition 1. The **Frobenius Endomorphism** for E is the map

$$\begin{aligned}
\Phi_E: \quad E(\overline{\mathbb{F}_q}) &\longrightarrow E(\overline{\mathbb{F}_q}) \\
(x, y) &\longmapsto (x^q, y^q) \ .
\end{aligned}$$

The connection between this map and the problem of computing the order $\#E(\mathbb{F}_q)$ of $E(\mathbb{F}_q)$ follows from following well-known Theorem of Hasse.

Theorem 2 (Hasse). *Let $c = q + 1 - \#E(\mathbb{F}_q)$. Then $|c| \leq 2\sqrt{q}$, and moreover the Frobenius Endomorphism Φ_E of E satisfies the equation $f(\Phi_E) = 0$ in the endomorphism ring of E, where*

$$f(X) = X^2 - cX + q \in \mathbb{Z}[X] \ .$$

The value c in this theorem is the trace of Φ_E. The algorithm of Atkin consists of two steps: (1) for small primes l, compute a small set of possible values for c mod l; (2) use the Chinese Remainder Theorem to get a set of possible values for c, and then determine c from among these possible values.

3 Computing Possible Values for $c \bmod l$

In this section we describe an algorithm for computing a small set of possible values for $c \bmod l$. Schoof showed in [Scho85] how to compute $c \bmod l$ exactly with the help of $(l^2 - 1)/2$-degree polynomials. We will make use of a polynomial of degree $l + 1$, but in general we are not able to compute $c \bmod l$ exactly.

Let E be a fixed elliptic curve over $\mathbb{F}_q$ and l an odd prime, $l \neq p$. The l-torsion group $E[l]$ of E is the subgroup of points $P \in E(\overline{\mathbb{F}_q})$ such that $l \cdot P = \mathcal{O}$. It is well-known that

$$E[l] \cong \mathbb{Z}/l\mathbb{Z} \times \mathbb{Z}/l\mathbb{Z} .$$

It follows that there are exactly $l + 1$ (cyclic) subgroups of $E(\overline{\mathbb{F}_q})$ of order l, which we denote by $C_1, \ldots, C_{l+1}$.

Since $\Phi_E(E[l]) \subset E[l]$, we can consider the restriction of the Frobenius Endomorphism to $E[l]$, yielding an automorphism on $E[l]$ whose characteristic polynomial is $\overline{f}(X) \in \mathbb{F}_l[X]$, obtained by reducing the coefficients of the polynomial $f(X)$ in Theorem 2 mod l.

We can compute information about $c \bmod l$ by examining the behavior of the groups C_i under the action of the Frobenius Endomorphism and its powers. The following theorem relates this behavior to the splitting type of $\overline{f}(X)$.

Theorem 3. *For $1 \leq i \leq l + 1$, let d_i be the least positive integer such that $\Phi_E^{d_i}(C_i) = C_i$.*

1. *If $\overline{f}(X) = (X - \alpha)^2$ with $\alpha \in \mathbb{F}_l$, then either $d_i = 1$ for all $1 \leq i \leq l+1$ or there exists exactly one j with $d_j = 1$ and $d_i = l$ for all $1 \leq i \leq l+1, i \neq j$.*
2. *If $\overline{f}(X) = (X - \alpha) \cdot (X - \beta)$ with $\alpha, \beta \in \mathbb{F}_l$ and $\alpha \neq \beta$, then there exist $i_1 \neq i_2$ with $d_{i_1} = d_{i_2} = 1$ and $d_i = d = ord(\alpha/\beta)$ for $1 \leq i \leq l+1, i \neq i_1, i_2$.*
3. *If $\overline{f}(X)$ is irreducible over $\mathbb{F}_l$, then $d_i = d$ for all $1 \leq i \leq l + 1$, where $d = ord(\alpha^{l-1})$ and α is a root of $\overline{f}(X)$ in $\mathbb{F}_{l^2}$.*

Thus we know the splitting type of $f(X) \bmod l$ if we know the behavior of the groups C_i under powers of the Frobenius Endomorphism. Possible values for $c \bmod l$ can then easily be determined: we obtain two possible values for $c \bmod l$ in case 1 and $\varphi(d)$ possible values in the cases 2 and 3.

The next problem is the determination of the values d_i in Theorem 3. Let C be a finite subgroup of $E(\overline{\mathbb{F}_q})$. Then it is known that there exists an elliptic curve (unique up to isomorphism) which we denote by E/C and an isogeny $\psi : E(\overline{\mathbb{F}_q}) \to (E/C)(\overline{\mathbb{F}_q})$ such that the kernel of ψ is C. We write j/C to denote the j-invariant of E/C. The following theorem gives the connection between such j-invariants and the values d_i in Theorem 3.

Theorem 4. *Let E be a non supersingular elliptic curve over $\mathbb{F}_q$ such that there exists no $\mathbb{F}_q$-isogeny (i.e. an non constant isogeny defined over $\mathbb{F}_q$) to an elliptic curve of j-invariant 0 or 1728. Then for $1 \leq i \leq l + 1$ we have*

$$d_i = \min_{k \in \mathbb{N}} \left\{ j/C_i \in \mathbb{F}_{q^k} \right\} .$$

For supersingular elliptic curves the group order is given by the Theorem of Waterhouse; the group order of elliptic curves which are $\mathbb{F}_q$-isogenous to curves with j-invariant 0 or 1728 can be computed by finding elements of given norm in the maximal order of some imaginary quadratic field. More details can be found in [Mü94]. Thus we can assume that the assumptions of Theorem 4 are fulfilled.

So to compute d_i, we have to find minimal extensions of $\mathbb{F}_q$ which contain the j-invariants j/C_i. This problem is solved by finding the splitting type of a special polynomial, as the following theorem shows.

Theorem 5. *Let E be an elliptic curve and $C_i, (1 \leq i \leq l+1)$ be all subgroups of $E(\overline{\mathbb{F}_q})$ of exact order l. There exists a **modular polynomial** $\Phi_l(X, Y) \in \mathbb{F}_p[X, Y]$ such that all roots of $\Phi_l(X, j(E)) \in \mathbb{F}_q[X]$ are given by the j-invariants $j/C_i, 1 \leq i \leq l+1$.*

Note that by Theorem 5 the degree of the l-th modular polynomial is exactly $l+1$. Combining all theorems of this section we obtain the following algorithm for computing some information about the trace c of the Frobenius Endomorphism of an elliptic curve E over $\mathbb{F}_q$ modulo some odd prime $l \neq p$:

1. compute the l-th modular polynomial $\Phi_l(X, Y) \in \mathbb{F}_p[X, Y]$.
2. substitute the j-invariant of E and obtain $\Phi_l(X, j(E)) \in \mathbb{F}_q[X]$.
3. compute the splitting type of this polynomial and obtain $\{d_i\}_{i=1,\ldots,l+1}$.
4. compute information about $c \bmod l$ with Theorem 3.

4 Computing Modular Polynomials

In the last section we described how to compute partial information about the order of $E(\mathbb{F}_q)$ modulo some odd prime l if we know the l-th modular polynomial for fields of characteristic p. There is also a notion of modular polynomials for elliptic curves defined over $\mathbb{C}$, the field of complex numbers (see [Si86]). We recall this briefly, and mention how it is related to the notion of modular polynomials over finite fields.

An elliptic curve defined over $\mathbb{C}$ is isomorphic to $\mathbb{C}/L$, where $L = \mathbb{Z} + \mathbb{Z}\tau$ ($\text{Im}(\tau) > 0$) is a two-dimensional lattice in $\mathbb{C}$. The j-invariant of an elliptic curve over $\mathbb{C}$ can be interpreted as the j-invariant $j(\tau)$ of the corresponding lattice L. The classical modular polynomial for the complex numbers is a polynomial $\Phi_l(X, Y) \in \mathbb{Z}[X, Y]$, such that the roots of $\Phi_l(X, j(\tau))$ are given by

$$j\left(\frac{\tau + n}{l}\right) \quad \text{for} \quad 0 \leq n < l \quad \text{and} \quad j(l\tau) .$$

It is known that a modular polynomial for $\mathbb{F}_p$ is equal to the reduction modulo p of the corresponding modular polynomial for $\mathbb{C}$.

Actually, we have chosen to work with a different modular polynomial: one that has the same splitting type as the above modular polynomial, but is slightly easier to compute. We now describe this polynomial.

Let $q_\tau = \exp(2\pi i\tau)$ and $\eta(\tau) = q_\tau^{1/24} \cdot \prod_{n=1}^{\infty}(1 - q_\tau^n)$ be the Dedekind η-function. Let $s \in \mathbb{N}$ be minimal such that $v = (s(l-1))/12 \in \mathbb{N}$ and define

$$f(\tau) = \left(\frac{\eta(\tau)}{\eta(l\tau)}\right)^{2s}.$$

Using the function $f(\tau)$ we can prove the following theorem (see [Mü94]).

Theorem 6. *There exist coefficients $a_{r,k} \in \mathbb{Z}$ such that*

$$\sum_{r=0}^{l+1} \sum_{k=0}^{v} a_{r,k} \cdot j(l\tau)^k \cdot f(\tau)^r = 0.$$

For an elliptic curve E over $\mathbb{F}_q$ the polynomial

$$G_l(X) = \sum_{r=0}^{l+1} \sum_{k=0}^{v} a_{r,k} \cdot j(E)^k \cdot X^r \in \mathbb{F}_q[X]$$

has the same splitting type as the l-th modular polynomial $\Phi_l(X, j(E))$ for $\mathbb{F}_q$.

Next we describe the ideas for computing the coefficients $a_{r,k}$ used in the definition of $G_l(X)$. We can show that the roots of the polynomial

$$H_l(X) = \sum_{r=0}^{l+1} \underbrace{\left(\sum_{k=0}^{v} a_{r,k} \cdot j(l\tau)^k\right)}_{=:k_r(\tau)} \cdot X^r$$

are given by

$$f\left(\tau + \frac{n}{l}\right), \quad 0 \le n < l \quad \text{and} \quad \frac{l^s}{f(l\tau)}.$$

Since we can expand the functions $f(\tau)$ and $f(l\tau)$ as power series in q_τ with some precision, we can compute the power series expansions of the power sums of all zeroes of $H_l(X)$. Note that the r-th power sum of the "first" l zeroes can be computed by changing the power series expansion of $f(\tau)^r$ appropriately. Then we can use Newton's identities to compute power series expansions of the coefficients $k_r(\tau)$ of the polynomial $H_l(X)$. Knowing the coefficients $k_r(\tau)$ we can compute the values of $a_{r,k}$ by comparing minimal powers in the power series expansions of $k_r(\tau)$ and powers of $j(l\tau)$.

The advantage of $G_l(X)$ in comparison to the original l-th modular polynomial is the smaller precision which we need in the computation of power series expansions. For $l \equiv 1 \bmod 12$ the precision of power series expansions we need for $G_l(X)$ is about $\frac{1}{12}$ of the precision of power series expansions for computing $\Phi_l(X)$.

The method explained above works for all odd prime numbers l, but for some l we have used yet other functions for computing an analogue to $G_l(X)$, but which are sometimes easier to compute.

5 Computing c mod l Exactly

One problem with our method as we have described it so far is that we obtain only partial information about c mod l. As we consider more and more primes l, the set of possible values for c grows exponentially in the number of primes.

Fortunately, for primes l such that $f(X)$ splits mod l, (which we expect to happen about half the time), we can compute the value of c mod l exactly. We now describe the method.

Assume that the modular equation $G_l(X)$ has at least one root in $\mathbb{F}_q$. By the theorems of section 3 there exists at least one group C_i which is invariant under the Frobenius Endomorphism. Consider the polynomial

$$f_l(X) \quad := \quad \prod_{\pm P \in C_i - \{\mathcal{O}\}} \left(X - x(P) \right) .$$

This polynomial has degree $(l-1)/2$, and since C_i is invariant under Φ_E, its coefficients lie in $\mathbb{F}_q$. For $p > l$ we can efficiently compute $f_l(X)$ with the help of one root of the modular equation. This computation is somewhat complicated; a detailed description can be found in [Mü94].

Assume that we know $f_l(X)$. By the invariance of C_i under Φ_E there exists a number $1 \le \alpha < l$ such that the Frobenius Endomorphism satisfies for all points $P \in C_i$

$$\Phi_E(P) \quad = \quad \alpha \cdot P.$$

Similar to Schoof's algorithm, we transform this equation in the Endomorphism ring of C_i into polynomial equations modulo $f_l(X)$. For $1 \le j < l$ we have to check whether

$$X^q \cdot \psi_j^2(X) \quad \equiv \quad X \cdot \psi_j^2(X) - \psi_{j-1}(X) \cdot \psi_{j+1}(X) \mod f_l(X)$$

and

$$4(X^3 + aX + b)^{(q+1)/2} \psi_j^3(X) \equiv \psi_{j+2}(X) \cdot \psi_{j-1}^2(X) - \psi_{j-2}(X) \cdot \psi_{j+1}^2(X) \mod f_l(X)$$

holds, where $\psi_j(X)$ is the reduced j-th division polynomial (compare [BuMü91]). For $j = \alpha$ both equations are true and we have found α. From Theorem 3 we know that

$$X^2 - \bar{c}X + \bar{q} \quad = \quad (X - \alpha) \cdot (X - \beta)$$

holds in $\mathbb{F}_l[X]$ and since we know α and $\bar{q}$, we can compute β and $\bar{c}$ by comparing coefficients.

6 Combining Possible Values

We will describe how we actually compute the order of the group $E(\mathbb{F}_q)$ after knowing possible values for c mod l_i for primes $l_1, \ldots, l_r$ with $\prod_{i=1}^{r} l_i > 4\sqrt{q}$. Using Chinese Remaindering we can compute moduli $m_i, i = 1, 2, 3$, a value c_3, and sets L_1, L_2 such that

- $c \equiv c_3 \bmod m_3$
- $c \equiv c_1 \bmod m_1$, where $c_1 \in L_1 = \{c_{1,1}, \ldots, c_{1,k_1}\}$.
- $c \equiv c_2 \bmod m_2$, where $c_2 \in L_2 = \{c_{2,1}, \ldots, c_{2,k_2}\}$.

Note that we do not know the values of c_1 and c_2. We use a so-called "baby step/giant step" strategy to compute these values. Write

$$c \;=\; c_3 + m_3 \cdot (m_1\, r_2 + m_2\, r_1),$$

where r_1, r_2 are numbers which satisfy

$$r_2 \equiv c_2\,(m_1\, m_3)^{-1} - c_3 \bmod m_2 \qquad \text{and} \qquad r_1 \equiv c_1\,(m_2\, m_3)^{-1} - c_3 \bmod m_1 \;.$$

If we choose c_3 as the least positive residue mod m_3 and the number r_1 as the least absolute residue mod m_1, we can show that $|r_2| \leq m_2$.

By Lagrange's Theorem, a random point $Q \in E(\mathbb{F}_q)$ satisfies $(q+1-c)\cdot Q = \mathcal{O}$ and—substituting the equation for c—we get

$$(q + 1 - c_3) \cdot Q - m_1\, m_3\, r_2 \cdot Q \;=\; m_2\, m_3\, r_1 \cdot Q \;. \tag{2}$$

Since we do not know c_1 and c_2, we compute corresponding numbers $r_{1,i}, r_{2,j}$ for all elements in the sets L_1 and L_2. Then we check (2) for all these possibilities $r_{1,i}$ and $r_{2,j}$ with the following algorithm:

1. for all $1 \leq j \leq k_2$
   (a) compute $H_j = (q + 1 - c_3) \cdot Q - m_1\, m_3\, r_{2,j} \cdot Q$.
   (b) if $H_j = \mathcal{O}$ then return $q + 1 - c_3 - m_1\, m_3\, r_{2,j}$.
   (c) store $(H_j, r_{2,j})$ in a table T.
2. for all $1 \leq i \leq k_1$
   (a) compute $K_i = m_2\, m_3\, r_{1,i} \cdot Q$.
   (b) if (K_i, x) exists in T for some number x, return $q + 1 - c_3 - m_1\, m_3\, x - m_2\, m_3\, r_{1,i}$.

Obviously this algorithm doesn't guarantee that the result really is the correct order of the group $E(\mathbb{F}_q)$; it returns only a multiple of the order of the point Q which lies in the interval given by Hasse's theorem. In practice the result is equal to the desired group order, which can be probabilistically checked by using a few more random points. Moreover there is an algorithm which really proves the correctness of the group order (see [Mü94]).

7 Computational Results and Implementation Details

We have recently implemented this algorithm and used it to compute the group order of a curve defined over $\mathbb{F}_p$, where p is a 375-digit prime. We chose $p = 10^{374} + 169$, and the curve

$$E: \quad y^2 \;=\; x^3 + 9051969\, x + 11081969 \;.$$

The group order was computed as

9999999999999999999999999999999999999999999999999999999999\
9999999999999999999999999999999999999999999999999999999999\
9999999999999999999999999999999999999999999999999999999933\
484119377057740549595798019340277299296194413840780451243105834\
253911258272979527787642571590910992330571399128986816096949176\
53999177002350595375701482255870650298836577293810050217802.

We could prove the correctness of the group order by factoring the group order of the quadratic twist (having a 362-digit prime factor). The factorization was done with the XFACT-tool developed at the University of Saarbrücken.

The total computation (not including the computation of modular polynomials, which is a precomputation that is independent of the input), took approximately 1765 MIPS-days, of which 1080 were used in computing the splitting type and 685 to compute the trace c modulo Elkies primes exactly. We found approximately 15 000 possible values; computing the correct order with the baby step/giant step algorithm took approximately 17 MIPS minutes.

In the computation, we used modular polynomials for l up to 839. Of the 137 used primes up to 839, we found that for this curve, 70 were so-called Elkies primes, i.e., primes for which the polynomial $f(X)$ splits, and we can quickly compute $c \bmod l$ exactly. Using the ideas of Couveignes and Morain [CoMo94] we could compute the trace modulo the square of a prime for 7 primes below 257.

We have computed modular polynomials for l up to 839, and have stored these on disk. The storage requirement is approximately 150 MBytes. Individual modular equations are relatively expensive to compute. For example, in our implementation, we used 833 MIPS-days to compute the modular polynomial for $l = 571$.

It is fairly easy to distribute much of the computation over a network of workstations. We have done so using LiPS [Li93]. This proved quite effective.

We now give some details of the implementation.

7.1 Computing Modular Polynomials

To compute the l-th modular polynomial, we compute it modulo several small primes, and then use Chinese remaindering to get the coefficients over $\mathbb{Z}$. These small primes are chosen so that each prime r fits in one computer word, and so that $r-1$ is divisible by a high power of 2. This allows us to multiply polynomials of very large degree modulo r very quickly using the FFT. Arithmetic modulo r is performed using floating point operations (which is both portable and fast).

We distributed the computation modulo different primes r over a network of workstations.

7.2 Computing the Splitting Type

Assuming we have precomputed all of the necessary modular polynomials, the most time consuming step is the computation of the splitting type of $G_l(X)$ modulo p, and to compute a root should it admit one.

These problems are special cases of the polynomial factorization problem, and the techniques used are closely related.

To compute the splitting type of the modular polynomial $G_l(X)$, we first compute $X^p \bmod G_l(X)$. Then we compute $\gcd(X^p - X, G_l(X))$. If this gcd is non-trivial, then we can compute a root of $G_l(X)$ modulo p, and from this we compute the value of $c \bmod l$ exactly. If this gcd is trivial, then we know that $G_l(X)$ factors as a product of distinct irreducibles, all of the same (unknown) degree d. We can still obtain partial information about $c \bmod l$ by computing this value d.

To carry out these computations, we need to perform polynomial arithmetic modulo $G_l(X)$. Multiplication of polynomials is done using a combination of Chinese remaindering and the FFT. Small primes r are chosen so that $r - 1$ is divisible by a high power of two, and the product of these primes is a bit bigger than p^2. To multiply two polynomials over $\mathbb{F}_p$, the coefficients (represented as nonnegative integers less than p) are reduced modulo the small primes; then we compute the product polynomial modulo each small prime via the FFT; finally, we apply the Chinese remainder algorithm to each coefficient, and reduce modulo p.

In practice, this runs much faster than the classical "school" method for the size of polynomials we are considering (the cross-over point being less than degree 50), and is critical in obtaining reasonable running times.

Division by $G_l(X)$ with remainder is done using a standard reduction to polynomial multiplication; however, as $G_l(X)$ remains fixed for many divisions, it pays to perform some precomputation on $G_l(X)$. With this precomputation, one squaring modulo $G_l(X)$ costs about 1.5 times the cost of simply multiplying two degree l polynomials.

Also, a new algorithm for computing the value d was employed which is quite efficient in practice—the number of polynomial multiplications is a small multiple of $l^{1/2} \log_2 l$.

We also remark that very good performance for these algorithms was obtained, despite the fact that the algorithms were implemented in C and are very portable. Details on these algorithms and their implementation can be found in [Sho94].

The computation for different primes l was distributed over a network of workstations.

7.3 Baby Step/Giant Step Computation

Several tricks were used to speed up the baby step/giant step computation in section 6.

1. We store only one computer word of the x-coordinate of points computed in the baby step part. After having done all baby steps we sort the table according to the "x-coordinate" (using quicksort). Thus we are able to use binary search for checking a match in the following giant step part. If we

find a match, we recompute the original baby step point (using the second component of the table entry) and compare.

2. For doing fast steps we sort the numbers $r_{1,i}, r_{2,j}$ such that we can do one step by adding some small multiple of Q. These different multiples of Q are computed with the following trick: precompute the table

$$1 \cdot Q, \; 2 \cdot Q, \ldots, L \cdot Q, \; 2L \cdot Q, \; 4L \cdot Q, \ldots, L^2 \cdot Q, \; 2L^2 \cdot Q, \; 4L^2 \cdot Q, \ldots, \; L^3 \cdot Q$$

for some constant $L \approx 200$. Since with $x \cdot Q$ we directly know the point $-x \cdot Q$, we can compute the point $x \cdot Q$ for all $|x| < L^3$ with at most three additions by expanding x as

$$x \;=\; x_1 + x_2 \cdot (2L) + x_3 \cdot (2L^2), \qquad |x_i| \leq L \;.$$

3. The most time consuming part of point addition is the inversion of some field element. We are doing several steps in this part in parallel so that we can use a trick of Montgomery which reduces the number of inversions on the cost of a few multiplications (see [Mo87]).

Acknowledgements

We are very grateful to Oliver Atkin for many fruitful discussions related to algorithmic problems of the algorithm. The third author thanks Johannes Buchmann for making him possible to visit Oliver Atkin at the University of Illinois at Chicago.

References

[At91]   A.O.L. Atkin: *The number of points on an elliptic curve modulo a prime*, unpublished manuscript

[BuMü91] J. Buchmann, V. Müller: *Computing the number of points on an elliptic curve over finite fields*, Proceedings of ISSAC 1991, 179-182

[CoMo94] J. M. Couveignes, F. Morain: *Schoof's algorithm and isogeny cycles*, Proceedings of ANTS 1994

[Ko86]   N. Koblitz: *Elliptic curve cryptosystems*, Mathematics of Computation, **48** (1987), 203-209

[Mi86]   V. Miller: *Uses of elliptic curves in cryptography*, Advances in Cryptology: Proceedings of Crypto '85, Lecture Notes in Computer Science, **218** (1986), Springer-Verlag, 417-426

[Mü94]   V. Müller: *Die Berechnung der Punktanzahl elliptischer Kurven über endlichen Körpern der Charakteristik größer 3*, Thesis, to be published

[Mo87]   P. L. Montgomery: *Speeding the Pollard and Elliptic Curve Methods for Factorization*, Mathematics of Computation, **48** (1987), 243 - 264

[Od86]   A. Odlyzko: *Discrete logarithms and their cryptographic significance*, Advances in Cryptology: Proceedings of Eurocrypt '84, Lecture Notes in Computer Science, **209** (1985), Springer-Verlag, 224-314

[Li93]   R. Roth, Th. Setz: *LiPS: a system for distributed processing on workstations*, University of Saarland, 1993

[Scho85]   R. Schoof: *Elliptic curves over finite fields and the computation of square roots mod p*, Mathematics of Computation, **44** (1985), 483-494

[Sho94]   V. Shoup: *Practical computations with large polynomials over finite fields*, in preparation (1994).

[Si86]   J. Silverman: *The arithmetic of elliptic curves*, Springer-Verlag, 1986

Straight-Line Complexity and Integer Factorization

Richard J. Lipton

Department of Computer Science
Princeton University
Princeton, NJ 08544
rjl@princeton.edu

Abstract. We show that if polynomials with many rational roots have polynomial length straight-line complexity, then integer factorization is "easy".

1 Introduction

We study the relationship between the straight-line complexity of polynomials and integer factorization. Our main result is the following: If for each m, there *exists* at least one polynomial of the form $(x - a_1)(x - a_2)...(x - a_m)g(x)$ with straight-line complexity less than $\log^{O(1)}(m)$ where each a_i is a distinct rational number, then integer factorization is not sufficiently hard for cryptographic proposes.

The importance of this result is that it demonstrates a connection between two different areas. It relates the complexity of evaluating polynomials to the complexity of integer factorization. The assumption that integer factorization is hard is central to modern cryptography. Recently, many beautiful results in cryptography and related areas have been obtained based on the assumption that factoring is hard [4, 6].

On the other hand, the straight-line complexity of polynomials has been studied for many years [12, 10, 15, 7, 8, 13]. However, two interesting questions are still open. The first question is: can one construct *explicit* polynomials that are hard to evaluate? A simple argument shows that all polynomials of degree n take at least $\log(n)$ steps: we will consider a polynomial *hard* if it takes at least $\log^d(n)$ for some $d > 1$. To date all known proofs that polynomials are hard fall into two categories: (i) existence proofs or (ii) growth arguments. Thus, the only examples of explicit polynomials that are hard to evaluate depend on very fast coefficient growth:

$$\sum_{i=1}^{n} 2^{2^i} x^i.$$

While these polynomials are explicit the main interest is in constructing ones that have "small" coefficients. The second question is an old question of Borodin and Cook [3] raised independently by Lipton and Stockmeyer [12]. It asks whether or not there are polynomials that are easy to evaluate and yet have many rational roots.

The key insight is that these two different areas, straight-line complexity and factorization, are closely related. If we assume that integer factorization is hard, then we can resolve these two old open problems concerning the complexity of polynomials. In particular, we can construct simple explicit polynomials that are hard to evaluate. Moreover, the reason these polynomials are hard to evaluate is precisely that they have *too many rational roots*. Another way to view these results is that they show us how difficult it will be to resolve certain open questions.

2 Straight Line Complexity

In order to state our results we need to define a complexity measure on polynomials. Let $f(x_1, \ldots, x_n)$ be a polynomial with rational coefficients. A *straight-line computation* for

$$f(x_1, \ldots, x_n)$$

is a sequence of rational functions $g_1, \ldots, g_m$ so that $f = g_m$ and each g_i satisfies one of the following: (i) g_i is the constant 1 or a variable x_j; (ii) $g_i = g_j \circ g_k$ where $\circ$ is either $+ - *$ and $j < i$ and $k < i$. The length of the computation is m. Then the straight-line complexity of $f(x_1, \ldots, x_n)$ denoted by $L(f)$ is the length of the shortest such computation.

While we restrict our computations to use just the constant 1, they can create any constant. In particular, an integer $s > 0$ can be created at the cost of at most $O(\log(s))$ steps of the computation. If a computation is "uniform", then the constants used are independent of the length of the computation; hence, there should be no loss of generality in using our model. Finally, we conjecture that our main results remain true for computations that allow any integer constants, but we cannot prove this at this time.

3 Factoring Assumption

In this section we will state our main assumptions about factoring. We need a number of standard notions from complexity and from number theory. As usual we measure the size of a boolean circuit by the number of its gates. A *polynomial family* of circuits $\{C_n\}$ consists of a sequence of boolean circuits so that C_n has n-inputs and n-outputs and has size $n^{O(1)}$. $C_n(x)$ denotes the output of the circuit at the input x.

As usual $|x|$ is the absolute value of x; $len(x)$ is the length of x. Also if $x = \frac{a}{b}$ is a rational number, then define $height(x)$ to be $|a| + |b|$. We will use $|A|$ to denote the cardinality of the set A. Let $x||y$ mean that x is a non-trivial factor of y, i.e. that $1 < x < y$ and x divides y. We also require the following concept that is important when studying the complexity of factoring:

$$\mathcal{Z}_2(k, l) = \{x : x = pq \land len(p) = k \land len(q) = l\}.$$

Here p and q always denote primes. Thus, this is just the set of integers that are the products of two primes with the given lengths.

The statement that it is difficult to factor is: Suppose that C_n is a polynomial family of circuits. Then for any $d > 0$ and n large enough,

$$Prob[C_n(\hat{x})||\hat{x} \in \mathcal{Z}_2(n/2, n/2)] \leq 1/n^d.$$

(Here $Prob[\ldots \hat{x} \ldots]$ is the probability that a random $\hat{x}$ satisfies the given property.) This is just the formal statement that it is hard to factor numbers that are products of two equal length primes [4]. Informally it states that for large enough n there is no circuit that can even factor "many" of the numbers of the form pq where p and q are both the same length. Here many means that the circuit cannot even factor a polynomial fraction of all the numbers of this form. It is essentially the standard assumption in much of modern cryptography [6]. For technical reasons it will be convenient to make our assumption about factoring slightly stronger. We believe, however, that it should be possible to use the weaker assumption to prove our results. Our assumption about factoring is:

Suppose that C_n is a polynomial family of circuits. Then for any $d > 0$ and $\epsilon > 0$ and n large enough,

$$Prob[C_n(\hat{x})||\hat{x} \in \mathcal{Z}_2(n^\epsilon, n - n^\epsilon)] \leq 1/n^d.$$

This assumption appears just as likely as the weaker one. As long as $x = pq$ with both p and q large primes there is no known efficient algorithm for factoring.

4 Main Results

In this section we will state our main results. A sequence of monic polynomials $\{f_m(x)\}$ is a $c - sequence$ provided,

(1). $deg(f_m(x)) \leq 2^{\log^c(m)}$;

(2). there are distinct rational roots $a_1, \ldots, a_m$ of $f_m(x)$ so that for all i and j, $height(a_i - a_j) \leq 2^{\log^c(m)}$.

Root Theorem: *Suppose that factoring is hard. Let $\{f_m(x)\}$ be a c-sequence of polynomials. Then for any constant $d > 0$,*

$$L(f_m(x)) \geq \log^d(m),$$

for m large enough.

This theorem states roughly that a polynomial that has many rational roots cannot be easy to evaluate.

5 Applications of the Root Theorem

In this section we will give two applications of the Root Theorem.

5.1 Examples of Hard Polynomials

As stated in the introduction, Borodin and Cook [3] and independently Lipton and Stockmeyer [12] both remark that there are rational polynomials with n distinct real roots that can be evaluated in $O(\log(n))$ steps. They leave open whether or not it is possible if the roots are additionally constrainted to be rational (see also [13]). This is resolved by the Root Theorem. We can also prove that many interesting polynomials are hard. For example, the following follows directly from the Root Theorem.

Theorem: *For any $c > 0$ the polynomial $(x - 1)(x - 2) \cdots (x - n)$ requires $\Omega(\log^c(n))$ steps to compute.*

Proof: This is immediate from the Root Theorem. q.e.d.

5.2 Chromatic Polynomials

As another application we turn to the *chromatic polynomial* of a graph G. Recall that this is a polynomial $P(G; x)$ that counts the number of possible ways to color the graph G with x colors [1]. We can prove the following theorem:

Theorem: *Let G_n be a sequence of n-vertex graphs. Suppose G_n requires at least $2^{\log^\epsilon(n)}$ colors for some $\epsilon > 0$. Then for any $d > 0$, $L(P(G_n; x)) \geq \Omega(\log(n)^d)$.*

Proof: Suppose that G_n is a graph a n-vertex graph with chromatic number k. Then, by definition, $P(G_n, i) = 0$ for all $i = 0, ..., k - 1$. Thus, $P(G_n, x)$ has at least k integer roots. The result then follows from the Root Theorem. q.e.d.

Note, that Read and Tutte [14] point out that chromatic polynomials are often "laborious" to compute. In a sense this theorem explains why: if they were easy to compute for graphs with large chromatic number, then factoring would be easy.

6 Proof of the Root Theorem

In this section we will prove the root theorem.

Lemma 6.1: *Let $x_1 + \cdots + x_k = m$ with each $x_i \geq 0$. Then*

$$\sum_{i=1}^{k} x_i^2 \geq m^2/k.$$

Proof: It is easy to see that the sum is minimized when each x_i is equal to m/k. Q.E.D.

Lemma 6.2: *Let $\varphi : A \to B$ with A and B finite sets. Then,*

$$|\varphi(A)| \geq |A|^2/\|\varphi\|$$

where $\|\varphi\| = |\{(x,y) : x \in A \wedge y \in A \wedge \varphi(x) = \varphi(y)\}|$.

Proof: For any $b \in B$ let A_b denote the number of $x \in A$ so that $\varphi(x) = b$. Clearly,

$$\sum_{b \in B} A_b = |A|.$$

Also it is clear that

$$\sum_{b \in B} A_b^2 = \|\varphi\|$$

since A_b^2 counts the number of (x, y) with $x \in A$ and $y \in A$ with $\varphi(x) = \varphi(y) = b$. Now assume that exactly k of the A_b's are non-zero. Then by lemma 6.1,

$$\|\varphi\| \geq |A|^2/k.$$

or

$$k \geq |A|^2/\|\varphi\|.$$

The proof now follows since k is just the size of the range of the function φ, i.e. $k = |\varphi(A)|$. Q.E.D.

Lemma 6.3: *Let A be a finite set of rationals so that height$(x - y)$ is at most L, for any x, y in A. Also let $\mathcal{P}$ be a finite set of primes. For each prime p in $\mathcal{P}$, let $\varphi_p(x) = x \bmod p$. Then,*

$$|\varphi_p(A)| \geq \frac{|\mathcal{P}|}{4\frac{|\mathcal{P}|}{|A|} + 4\log(L)}$$

for at least $1/2$ of the primes in $\mathcal{P}$.

Proof: Let us first make two definitions:

$$(x, y, p) \in C \Longleftrightarrow x \equiv y \bmod p$$

$$(x, y) \in C_p \Longleftrightarrow x \equiv y \bmod p.$$

We next make three claims:

   (1). $|C| = \sum_p |C_p|$.
   (2). $|C| \leq |A||\mathcal{P}| + |A|^2 \log(L)$.
   (3). For at least $1/2$ of the primes in $\mathcal{P}$,

$$|C_p| \leq \frac{4}{|\mathcal{P}|}(|A||\mathcal{P}| + |A|^2 \log(L)).$$

Claim (1) is immediate from the definitions. Claim (2) follows since there are at most $|A||\mathcal{P}|$ triples (x, y, p) with $x = y$ and at most $|A|^2 \log(L)$ triples with $x \neq y$. In order to see the latter, suppose that $x \neq y$ are in A. Then, $height(x - y) \leq L$: let $x - y = \frac{u}{v}$; thus, $|u| + |v| \leq L$. Now $x \equiv y \bmod p$ implies that $u \equiv 0 \bmod p$. Since $u \neq 0$ it can have at most $\log(L)$ prime factors, it follows that there are at most $\log(L)$ such primes.

Now we turn to proving the last claim. Suppose that claim (3) is false. Then,

$$|C_p| \geq \frac{4}{|\mathcal{P}|}(|A||\mathcal{P}| + |A|^2 \log(L))$$

for at least $1/2$ of the primes in $\mathcal{P}$. Thus, by claim (1),

$$|C| \geq 2(|A||\mathcal{P}| + |A|^2 \log(L)).$$

But this contradicts claim (2) so it follows that (3) is true.

Now let p be one of the primes that satisfy claim (3). Clearly, by the definition of C_p,

$$\|\varphi_p\| = |C_p| \leq \frac{4}{|\mathcal{P}|}(|A||\mathcal{P}| + |A|^2 \log(L))$$

Thus,

$$\|\varphi_p\| \leq \frac{4|A| + 4|A|^2 \log(L)}{|\mathcal{P}|}.$$

By lemma 6.2,

$$|\varphi_p(A)| \geq \frac{|A|^2}{\frac{4|A| + 4|A|^2 \log(L)}{|\mathcal{P}|}}$$

and so

$$|\varphi_p(A)| \geq \frac{|\mathcal{P}|}{4\frac{|\mathcal{P}|}{|A|} + 4\log(L)}.$$

Since this holds for at least $1/2$ of the primes in $\mathcal{P}$ the proof is complete. Q.E.D.

We now turn to proving the root theorem.

Proof: Assume that the theorem is false. Then for an infinite number of m,

$$L(f_m) \leq \log^d(m).$$

Choose a large m where m is large enough to satisfy certain inequalities and let

$$n = \lfloor \log^{\frac{1}{2}}(m) \rfloor.$$

We will show how to factor many $x = pq$ where p is a prime with $len(p) = \lfloor n^{\epsilon} \rfloor$ and q is a prime with $len(q) = n - \lfloor n^{\epsilon} \rfloor$. This will contradict our assumption about factoring.

The key is the the following program:

(1). $a \leftarrow \text{random}() \bmod x$.
(2). $\gcd(f_m(a), x) \| x$ then exit with the factor.
(3). goto (1).

Clearly, this program is correct, if it ever exits it does so with a correct factor of x. Recall that $w||x$ means that w is a non-trivial factor of x. Each iteration of the loop runs in polynomial time since $L(f_m) \leq \log^d(m)$. Thus, we need to show that at most a polynomial number of the loops are executed with a high probability. We do this by proving two claims:

(1). For a a random number modulo x, $f_m(a) \equiv 0 \bmod p$ with probability at least $1/\log^d(m)$ for some $d > 0$ and at least $1/2$ of the primes p.

(2). For a a random number modulo x, $f_m(a) \equiv 0 \bmod q$ with probability at most $2^{-n/2}$.

Let us consider claim (1) first. Since $\{f_m\}$ is a c-sequence there are $a_1, \ldots, a_m$ distinct rational roots of $f_m(z)$. So clearly, $f_m(a) \equiv 0 \bmod p$ if $a \equiv a_i \bmod p$ for some i. We therefore need to obtain a lower bound on the cardinality of the set $\{a_i \bmod p : i = 1, \ldots, m\}$. But by lemma 6.3, this is at least

$$\frac{|\mathcal{P}|}{4\frac{|\mathcal{P}|}{|A|} + 4\log(L)}$$

where $\mathcal{P} = \{p : len(p) = n^{\epsilon}\}$ and $L = 2^{\log^c(m)}$ for at least $1/2$ of the primes in $\mathcal{P}$. A simple calculation shows that $a_1, \ldots, a_m$ therefore map to at least $m/\log^{O(1)}(m)$ elements. So for at least $1/2$ of the primes p the probability that $f_m(a) \equiv 0 \bmod p$ is at least $1/\log^d(m)$ as claimed.

We now prove claim (2). Now $f_m(z)$ has at most $2^{\log^c(m)}$ roots modulo q since its degree is bounded by this quantity and it is monic. But $n^{\epsilon} \approx \log(m)$ and so the probability that a is a root is at most

$$\frac{2^{\log^c(m)}}{2^{n-n^{\epsilon}}} \leq \frac{1}{2^{n/2}}$$

This proves the second claim.

Now to complete the proof of the theorem we observe that (1) and (2) show that with probability at least $1/\log^{O(1)}(m)$, a random a satisfies

$$f_m(a) \equiv 0 \bmod p$$

and

$$f_m(a) \not\equiv 0 \bmod q.$$

But then the gcd calculated is a non-trivial factor of x. Note, since p and q are distinct primes the chinese remainder theorem shows that the behavior of $f_m(a)$ modulo p is independent of its behavior modulo q. Q.E.D.

7 Conclusion

We have uncovered a relationship between straight-line complexity and an important open question: the complexity of integer factoring. We believe that this relationship will help us to further our understanding of both these areas. We may be able to use the results here to resolve the complexity of integer factorization. Our intuition about the *existence* of combinatorial objects is often wrong. Thus, for example, it is entirely possible that there exists easy to compute polynomials with many rational roots. Since only the existence and not the explicit construction of these polynomials is required our intuition could be incorrect, i.e. integer factorization could be easy.

The connection between the difficulty of factoring and the cost of evaluation of certain polynomials has been made before [2]. For example, the rho method of Pollard [5] can be viewed as a special case of our method. In the rho method a fixed simple iteration is used to define a polynomial; in our method some arbitrary straight-line computation is used to define the polynomial. Perhaps our main contribution is the demonstration that any polynomial with many rational roots is enough to imply that factoring is easy. Indeed we further conjecture that rational can be replaced by roots from any low degree extension fields.

8   Acknowledgements

I would like to thank Dan Boneh for many conversations about this work. I especially would like to thank both Shafi Goldwasser and Avi Wigdeson without whose support this paper would never have been completed. Research supported in part by NSF Grant CCR-9304718.

References

1. N. Biggs. *Algebraic Graph Theory*. Cambridge University Press, 1974.
2. A. Borodin, private communication, 1994.
3. A. Borodin and S. Cook. On the Number of Additions to Compute Specific Polynomials. *SIAM Journal of Computing*, (5):7-15, 1976.
4. Manuel Blum, Paul Feldman, and Silvio Micali. Non-Interactive Zero-Knowledge and Its Applications. In *Proceedings of the 20th Annual ACM Symposium on the Theory of Computing*, pages 103-112, 1988.
5. D. Bressound. Factorization and Primality Testing. Springer-Verlag, 1989.
6. Joan Feigenbaum and Michael Merritt. *Distributed Computing and Cryptography*. DIMACS Workshop Series, 1989.
7. Erich Kaltofen. Computing with Polynomials Given by Straight-Line Programs II Sparse Factorization. In *Proceedings of the 26th IEEE Symposium on Foundations of Computer Science*, pages 451-458, 1985.
8. Erich Kaltofen. Uniform Closure Properties of P-Computable Functions. In *Proceedings of the 18th Annual ACM Symposium on the Theory of Computing*, pages 330-337, 1986.
9. Erich Kaltofen. Single-Factor Hensel Lifting and Its Application to the Straight-Line Complexity of Certain Polynomials. In *Proceedings of the 19th Annual ACM Symposium on the Theory of Computing*, pages 443-452, 1987.
10. Serge Lang. *Algebra*. Addison-Wesley, 1984.
11. Richard J. Lipton. Polynomials with 0-1 Coefficients that are Hard to Evaluate. *SIAM Journal of Computing*, 7(1):61-69, 1978.

12. Richard J. Lipton and Larry J. Stockmeyer. *Evaluation of Polynomials with Super Preconditioning.* Technical Report, IBM Thomas J. Watson Research Center, 1976.

13. Jean-Jacques Risler. Some Aspects of Complexity in Real Algebraic Geometry. *Journal of Symbolic Computation*, 5, 1988.

14. R.C. Read and W.T. Tutte. *Chromatic Polynomials,* in *Selected Topics in Graph Theory*, Academic Press, 1988.

15. Jean-Paul Van de Wiele. An Optimal Lower Bound on the Number of Total Operations to Compute 0-1 Polynomials Over the Field of Complex Numbers. In *Proceedings of the 19th IEEE Symposium on Foundations of Computer Science*, pages 159-165, 1978.

Decomposition of Algebraic Functions

Dexter Kozen[1]   Susan Landau[2]   Richard Zippel[1]

[1] Computer Science Department, Cornell University, Ithaca, NY 14853
[2] Computer Science Department, University of Massachusetts, Amherst, MA 01003

Abstract. Functional decomposition—whether a function $f(x)$ can be written as a composition of functions $g(h(x))$ in a nontrivial way—is an important primitive in symbolic computation systems. The problem of univariate polynomial decomposition was shown to have an efficient solution by Kozen and Landau [9]. Dickerson [5] and von zur Gathen [13] gave algorithms for certain multivariate cases. Zippel [15] showed how to decompose rational functions. In this paper, we address the issue of decomposition of algebraic functions. We show that the problem is related to univariate resultants in algebraic function fields, and in fact can be reformulated as a problem of *resultant decomposition*. We characterize all decompositions of a given algebraic function up to isomorphism, and give an exponential time algorithm for finding a nontrivial one if it exists. The algorithm involves genus calculations and constructing transcendental generators of fields of genus zero.

1 Introduction

Functional decomposition is the problem of representing a given function $f(x)$ as a composition of "smaller" functions $g(h(x))$. Decomposition of polynomials is useful in simplifying the representation of field extensions of high degree, and is provided as a primitive by many major symbolic algebra systems.

The first analyzed algorithms for decomposition of polynomials were provided in 1985 by Barton and Zippel [2] and Alagar and Thanh [1], who gave algorithms for the problem of decomposing univariate polynomials over fields of characteristic zero. Both solutions involved polynomial factorization and took exponential time. Kozen and Landau [9] discovered a simple and efficient polynomial time solution that does not require factorization. It works over fields of characteristic zero, and whenever the degree of h does not divide the characteristic of the underlying field, and provides *NC* algorithms for irreducible polynomials over finite fields and all polynomials over fields of characteristic zero. Dickerson [5] and von zur Gathen [13] gave algorithms for certain multivariate cases. In addition, von zur Gathen also found algorithms for the case in which the degree of h divides the characteristic of the field [14]. Zippel [15] showed how to decompose rational functions.

In this paper we address the decomposition problem for algebraic functions. We show that the problem bears an interesting and useful relationship to univariate resultants over algebraic function fields, and in fact can be reformulated as a certain resultant decomposition problem: whether some power of a given irreducible bivariate polynomial $f(x, z)$ can be expressed as the resultant with respect to y of two other bivariate polynomials $g(x, y)$, $h(y, z)$. We determine necessary and sufficient conditions for an algebraic function to have a nontrivial decomposition, and classify all such decompositions up to isomorphism. We give an exponential-time algorithm for finding a nontrivial decomposition of a given algebraic function if one exists. The algorithm involves calculating the genus of certain algebraic function fields and constructing transcendental generators of fields of genus zero.

The paper is organized as follows. In §2, we review the basic properties of univariate resultants, state the decomposition problem for algebraic functions, and describe the relationship between the two. In §3 we prove a general theorem that characterizes the set of all possible decompositions of an algebraic function. In §4 we give an exponential time algorithm for the decomposition problem. We conclude in §5 with an example.

2 Resultants and Algebraic Functions

2.1 The Univariate Resultant

Here we review some basic facts about the univariate resultant; see [8, 16] for a detailed introduction.

The *resultant* of two polynomials

$$g(y) = a \prod_{i=1}^{m}(y - \alpha_i) \qquad h(y) = b \prod_{j=1}^{\ell}(y - \beta_j)$$

with respect to y is the polynomial

$$\mathbf{res}_y(g, h) = a^\ell b^m \prod_{i,j}(\beta_j - \alpha_i) = b^m \prod_{h(\beta)=0} g(\beta) . \tag{1}$$

The resultant vanishes iff g and h have a common root. It can be calculated as the determinant of the *Sylvester matrix*, a certain $(m + \ell) \times (m + \ell)$ matrix containing the coefficients of g and h.

The following are some useful elementary properties.

$$\mathbf{res}_y(g, h) = (-1)^{m\ell} \mathbf{res}_y(h, g)$$
$$\mathbf{res}_y(g_1 g_2, h) = \mathbf{res}_y(g_1, h) \cdot \mathbf{res}_y(g_2, h)$$
$$\mathbf{res}_y(g, h_1 h_2) = \mathbf{res}_y(g, h_1) \cdot \mathbf{res}_y(g, h_2)$$
$$\mathbf{res}_y(c, h) = c^\ell$$
$$\mathbf{res}_y(g, 1) = \mathbf{res}_y(1, h) = 1$$
$$\mathbf{res}_y(g, y - \beta) = g(\beta)$$
$$\mathbf{res}_x\big(f(x), \mathbf{res}_y(g(x, y), h(y))\big) = \mathbf{res}_y\big(\mathbf{res}_x(f(x), g(x, y)), h(y)\big) \tag{2}$$

Property (2) is an associativity property. Because of this property, we can write

$$\mathbf{res}_{x,y}\big(f(x), g(x, y), h(y)\big)$$

unambiguously for the left or right hand side of (2).

We extend the definition to pairs of rational functions as follows. If neither g_1, h_2 nor g_2, h_1 have a common root, define

$$\mathbf{res}_y\left(\frac{g_1}{g_2}, \frac{h_1}{h_2}\right) = \frac{\mathbf{res}_y(g_1, h_1) \cdot \mathbf{res}_y(g_2, h_2)}{\mathbf{res}_y(g_1, h_2) \cdot \mathbf{res}_y(g_2, h_1)} .$$

This definition reduces to the previous one in the case of polynomials. All the properties listed above still hold, taking $m = \deg g_1 - \deg g_2$ and $n = \deg h_1 - \deg h_2$.

2.2 Resultants and Decomposition

Let K be an algebraically closed field, and let Ω be a *universal field* over K in the sense of van der Waerden [11]; *i.e.*, an algebraically closed field of infinite transcendence degree over K.

Algebraic functions of γ are usually defined as elements of some finite extension of $K(\gamma)$, the field of rational functions of γ. We can also view algebraic functions more concretely as multivalued functions $\Omega \to 2^\Omega$ or as binary relations on Ω defined by their minimum polynomials. In the latter view, the decomposition problem is naturally defined in terms of ordinary composition of binary relations:

$$R \circ S = \{(u, w) \mid \exists v \, (u, v) \in R \wedge (v, w) \in S\} .$$

Definition 1. For $f(x, z) \in K[x, z]$, let

$$V(f) = \{(\alpha, \gamma) \mid f(\alpha, \gamma) = 0\} \subseteq \Omega^2$$

be the affine variety generated by f. A *decomposition* of f is a pair of polynomials $g(x, y) \in K[x, y]$ and $h(y, z) \in K[y, z]$ such that

$$V(f) = \overline{V(g) \circ V(h)} ,$$

where the overbar denotes the Zariski closure in Ω^2 (see [6]). □

The Zariski closure is taken in order to account for points at infinity in a composition. An alternative approach would be to consider f as a binary relation on the projective line.

This notion of decomposition is strongly related to the univariate resultant:

$$V(g) \circ V(h) = \{(\alpha, \beta) \mid \exists \beta \, g(\alpha, \beta) = h(\beta, \gamma) = 0\}$$
$$= \{(\alpha, \beta) \mid \mathbf{res}_y(g(\alpha, y), h(y, \gamma)) = 0\}$$

by (1). The following results develop this relationship further.

Lemma 2. *Let $g(x, y) \in K[x, y]$ and $h(y, z) \in K[y, z]$. Considering $g(x, y)$ and $h(y, z)$ as polynomials in y, let $g_m(x)$ and $h_\ell(z)$ be their respective lead coefficients. Then*

$$V(\mathbf{res}_y(g, h)) = (V(g) \circ V(h)) \cup V(g_m, h_\ell) .$$

Proof. Consider the two expressions

$$\mathbf{res}_y(g(\alpha, y), h(y, \gamma)) \tag{3}$$
$$\mathbf{res}_y(g(x, y), h(y, z))[x := \alpha, z := \gamma] \ . \tag{4}$$

The difference is whether α and γ are substituted for x and z before or after the resultant is taken. We claim that for any α, γ,

(i) if $g_m(\alpha) = h_\ell(\gamma) = 0$, then (4) vanishes;
(ii) if either $g_m(\alpha) \neq 0$ or $h_\ell(\gamma) \neq 0$, then (3) and (4) vanish or do not vanish simultaneously.

In case (i), we have

$$\mathbf{res}_y(g(x, y), h(y, z)) = \det S(x, z) \ ,$$

where $S(x, z)$ is the Sylvester matrix of $g(x, y)$ and $h(y, z)$. Then

$$\mathbf{res}_y(g(x, y), h(y, z))[x := \alpha, z := \gamma] = \det S(\alpha, \gamma) = 0 \ ,$$

since the first row of $S(\alpha, \gamma)$ is the zero vector. In case (ii), say $h_\ell(\gamma) \neq 0$ (the other case is symmetric). Then

$$\mathbf{res}_y(g(x, y), h(y, z))[x := \alpha, z := \gamma] = \mathbf{res}_y(g(x, y), h(y, \gamma))[x := \alpha]$$
$$= h_\ell(\gamma)^{\deg_y g(x, y)} \prod_{h(\beta, \gamma) = 0} g(\alpha, \beta)$$
$$\mathbf{res}_y(g(\alpha, y), h(y, \gamma)) = h_\ell(\gamma)^{\deg_y g(\alpha, y)} \prod_{h(\beta, \gamma) = 0} g(\alpha, \beta)$$

thus both expressions are simultaneously zero or nonzero.

By (i) and (ii),

$$V(\mathbf{res}_y(g, h)) = \{(\alpha, \gamma) \mid \mathbf{res}_y(g(x, y), h(y, z))[x := \alpha, z := \gamma] = 0\}$$
$$= \{(\alpha, \gamma) \mid \mathbf{res}_y(g(\alpha, y), h(y, \gamma)) = 0 \vee g_m(\alpha) = h_\ell(\gamma) = 0\}$$
$$= (V(g) \circ V(h)) \cup V(g_m, h_\ell) \ .$$

$\square$

Theorem 3. *Let $g(x, y) \in K[x, y]$ and $h(y, z) \in K[y, z]$ be irreducible and non-degenerate (i.e., positive degree in each variable). Then*

$$V(\mathbf{res}_y(g, h)) = \overline{V(g) \circ V(h)} \ .$$

Proof. We have $\overline{V(g) \circ V(h)} \subseteq V(\mathbf{res}_y(g, h))$ by Lemma 2 and the fact that $V(\mathbf{res}_y(g, h))$ is Zariski-closed.

Conversely, it follows from the assumption that $g(x, y)$ and $h(y, z)$ are irreducible and nondegenerate that for all α, β, γ such that $g(\alpha, \beta) = h(\beta, \gamma) = 0$, either all $\alpha, \beta, \gamma \in K$ or all are transcendental over K. We use this to show that $\mathbf{res}_y(g, h)$ has no factor of the form $u(x)$. Suppose it did. Let $a \in K$ be a root of u (recall that K is algebraically closed). Then $\mathbf{res}_y(g, h)[x := a] = 0$. Let γ be transcendental over K. We have

$$
\begin{aligned}
0 &= \mathbf{res}_y(g(x, y), h(y, z))[x := a, z := \gamma] \\
&= \mathbf{res}_y(g(x, y), h(y, \gamma))[x := a] \\
&= h_\ell(\gamma)^m \prod_{h(\beta, \gamma)=0} g(x, \beta)[x := a] \\
&= h_\ell(\gamma)^m \prod_{h(\beta, \gamma)=0} g(a, \beta) \, ,
\end{aligned}
$$

thus $g(a, \beta) = h(\beta, \gamma) = 0$ for some β. But $a \in K$ and γ is transcendental over K, which contradicts our observation above.

By symmetry, $\mathbf{res}_y(g, h)$ has no factor $v(z)$.

Thus all irreducible factors of $\mathbf{res}_y(g, h)$ are nondegenerate. Let (α, γ) be a generic point of some irreducible component C of $V(\mathbf{res}_y(g, h))$. Then α and γ are transcendental over K. By Lemma 2, $(\alpha, \gamma) \in V(g) \circ V(h)$, so $C \subseteq \overline{V(g) \circ V(h)}$. Since C was arbitrary, $V(\mathbf{res}_y(g, h)) \subseteq \overline{V(g) \circ V(h)}$. □

Corollary 4. *Let* $f(x, z)$, $g(x, y)$, *and* $h(y, z)$ *be irreducible and nondegenerate. Then* g, h *give a decomposition of* f *iff* $f^k = \mathbf{res}_y(g, h)$ *for some* $k > 0$.

Proof. If $f^k = \mathbf{res}_y(g, h)$, then by Theorem 3,

$$
V(f) = V(f^k) = V(\mathbf{res}_y(g, h)) = \overline{V(g) \circ V(h)} \, .
$$

Conversely, if $V(f) = \overline{V(g) \circ V(h)}$, then by Theorem 3, $V(f) = V(\mathbf{res}_y(g, h))$, and $f^k = \mathbf{res}_y(g, h)$ follows immediately from the Nullstellensatz and the assumption that f is irreducible. □

In light of Corollary 4, the *decomposition problem* for algebraic functions becomes:

Given an irreducible polynomial $f(x, z)$, find polynomials $g(x, y)$ and $h(y, z)$ and a positive integer k such that $f^k = \mathbf{res}_y(g, h)$.

This formulation directly generalizes the definition for polynomials and rational functions: for polynomials $g(y)$ and $h(z)$,

$$
x - g(h(z)) = \mathbf{res}_y(x - g(y), y - h(z)) \, .
$$

Under this definition, every bivariate polynomial f is decomposable in infinitely many ways:

$$
\mathbf{res}_y(f(x, y^k), y^k - z) = \prod_{\beta^k = z} f(x, \beta^k) = \prod_{\beta^k = z} f(x, z) = f^k \, . \tag{5}
$$

However, these decompositions are not optimal in a sense to be made precise. In the next section we will define a notion of *minimality* for decompositions, and show that up to isomorphism there are only finitely many nontrivial minimal decompositions.

2.3 Irreducible Decompositions

A decomposition $f = \operatorname{res}_y(g, h)$ is called *irreducible* if both g and h are irreducible as polynomials in $K[x, y]$ and $K[y, z]$, respectively. By the multiplicativity of the resultant, every decomposition factors into a product of irreducible decompositions.

2.4 Monic Decompositions

A decomposition $f = \operatorname{res}_y(g, h)$ is called *monic* if $g \in K(y)[x]$ and $h \in K(z)[y]$ are monic. The next result says that we can restrict our attention to monic decompositions without loss of generality.

Lemma 5. *Let* $f \in K[x, z]$, $g \in K[x, y]$, $h \in K[y, z]$ *be nondegenerate,* g, h *irreducible,* f *a power of an irreducible polynomial. Let* $\widehat{f}$, $\widehat{g}$, *and* $\widehat{h}$ *be the monic associates of* f, g, h *in* $K(z)[x]$, $K(y)[x]$, *and* $K(z)[y]$ *respectively. Then* $f = \operatorname{res}_y(g, h)$ *iff* $\widehat{f} = \operatorname{res}_y(\widehat{g}, \widehat{h})$.

Proof. Let $f_n(z)$, $g_m(y)$, and $h_\ell(z)$ be the lead coefficients of f, g and h, respectively. Let

$$u(z) = \operatorname{res}_y(g_m(y), h(y, z)) \cdot h_\ell(z)^{\deg_y g - \deg_y g_m} .$$

Then

$$\operatorname{res}_y(g, h) = \operatorname{res}_y(g_m, h) \cdot \operatorname{res}_y(\widehat{g}, h_\ell) \cdot \operatorname{res}_y(\widehat{g}, \widehat{h}) = u \cdot \operatorname{res}_y(\widehat{g}, \widehat{h}) .$$

But since $\widehat{g}$ and $\widehat{h}$ are monic, so is $\operatorname{res}_y(\widehat{g}, \widehat{h})$, therefore if $f = \operatorname{res}_y(g, h) = u \cdot \operatorname{res}_y(\widehat{g}, \widehat{h})$, then $u = f_n$ and $\widehat{f} = \operatorname{res}_y(\widehat{g}, \widehat{h})$.

Conversely, if $\widehat{f} = \operatorname{res}_y(\widehat{g}, \widehat{h})$, then $uf = f_n \operatorname{res}_y(g, h)$. Remove common factors to get $vf = w \cdot \operatorname{res}_y(g, h)$, where $v, w \in K[z]$ are relatively prime. Now f has no factor in $K[z]$, so w is a unit. Likewise, as argued in the proof of Theorem 3, $\operatorname{res}_y(g, h)$ has no factor in $K[z]$, so v is a unit. $\square$

2.5 Inseparable Decompositions

In prime characteristic p, a decomposition $f(x, z)^k = \operatorname{res}_y(g(x, y), h(y, z))$ is *separable* if f is separable as a polynomial in $K(z)[x]$, g is separable as a polynomial in $K(y)[x]$, and h is separable as a polynomial in $K(z)[y]$. The following argument shows that we can restrict our attention to separable decompositions without loss of generality.

Any inseparable polynomial $f(x^q, z)$, $q = p^n$, has a nontrivial decomposition

$$f(x^q, z) = \operatorname{res}_y(x^q - y, f(y, z)) . \tag{6}$$

The polynomial $x^q - y$ decomposes into the composition of n copies of $x^p - y$. Also,

$$\begin{aligned}
\text{res}_y(g(x,y), y^q - z) &= \text{res}_y(g(x,y), (y - \sqrt[q]{z})^q) \\
&= \text{res}_y(g(x,y), y - \sqrt[q]{z})^q \\
&= g(x, \sqrt[q]{z})^q \\
&= g^{[q]}(x^q, z)
\end{aligned} \tag{7}$$

where $g^{[q]}(u,v)$ denotes the polynomial obtained from $g(u,v)$ by raising all the coefficients to the qth power.

Once we have decomposed $f(x^q, z)$ as in (6), we can attempt to decompose $f(y,z)$ further. The following argument shows that we can take this step without loss of generality.

Lemma 6. Let $f(x,z) = \text{res}_y(g(x,y), h(y,z))$, h monic in y. Let q, r be powers of p. Then

$$f^{[r]}(x^{qr}, z) = \text{res}_y(g(x^q, y), h(y^r, z)) .$$

Proof.

$$\begin{aligned}
\text{res}_y(g(x^q, y), h(y^r, z)) &= \text{res}_{w,y,u}(x^q - w, g(w,y), y^r - u, h(u,z)) \\
&= \text{res}_{w,u}(x^q - w, g^{[r]}(w^r, u), h(u,z)) \quad \text{by (7)} \\
&= \text{res}_{w,v,u}(x^q - w, w^r - v, g^{[r]}(v,u), h(u,z)) \\
&= \text{res}_v(x^{qr} - v, f^{[r]}(v,z)) \\
&= f^{[r]}(x^{qr}, z) .
\end{aligned}$$

$\square$

Lemma 7. If $f(x,z)^k = \text{res}_y(g(x,y), h(y,z))$ is a nondegenerate irreducible decomposition, g is separable in x, and h is separable in y, then f is separable in x.

Proof. Let γ be transcendental over K. Let β be a root of $h(y, \gamma)$ and let α be a root of $g(x, \beta)$. Then α is a root of $f(x, \gamma)$. Since h is separable in y, the extension $K(\beta, \gamma) : K(\gamma)$ is separable. Since g is separable in x, the extension $K(\alpha, \beta, \gamma) : K(\beta, \gamma)$ is separable. Combining these extensions, we have that the extension $K(\alpha, \beta, \gamma) : K(\gamma)$ is separable, hence $f(x, \gamma)$ is separable. $\square$

Theorem 8. Let q be a power of p and let $f(x^q, z)^k = \text{res}_y(g(x,y), h(y,z))$ be a monic nondegenerate irreducible decomposition, $f(x,z)$ separable. Then there exists a separable decomposition

$$f(x,z)^k = \text{res}_y(\hat{g}^{[s]}(x,y), \hat{h}(y,z))$$

where $g(x,y) = \hat{g}(x^r, y)$, $h(y,z) = \hat{h}(y^s, z)$, and $q = rs$.

Proof. Let r, s be powers of p such that g and h can be written $g(x, y) = \widehat{g}(x^r, y)$, $h(y, z) = \widehat{h}(y^s, z)$ with $\widehat{g}, \widehat{h}$ separable. Then $\widehat{g}, \widehat{h}$ are also irreducible. By Lemma 6,

$$\mathbf{res}_y(x^q - y, f(y, z)^k) = f(x^q, z)^k$$
$$= \mathbf{res}_y(g(x, y), h(y, z))$$
$$= \mathbf{res}_y(\widehat{g}(x^r, y), \widehat{h}(y^s, z))$$
$$= \mathbf{res}_{y,w}(x^{rs} - y, \widehat{g}^{[s]}(y, w), \widehat{h}(w, z))$$

and $\mathbf{res}_w(\widehat{g}^{[s]}(y, w), \widehat{h}(w, z))$ is separable by Lemma 7. Thus $q = rs$ and

$$f(y, z)^k = \mathbf{res}_w(\widehat{g}^{[s]}(y, w), \widehat{h}(w, z)) \ .$$

$\square$

This argument shows that in any irreducible decomposition of f, any inseparability of f must stem from the inseparability of one of the composition factors, and this inseparability ultimately emerges as a composition factor of the form $x^q - y$.

By Theorem 8, we can henceforth assume without loss of generality that all decompositions are separable.

3 A Characterization of All Decompositions

In this section we give a characterization of all possible irreducible decompositions of an algebraic function that can arise. As above, we assume that K is algebraically closed and that Ω is a universal field over K.

Let γ be transcendental over K and let α be a nonconstant algebraic function of γ with monic minimum polynomial $f(x, \gamma) \in K(\gamma)[x]$ of degree n. By results of the previous section, the functional decomposition problem reduces to the problem of finding all monic irreducible decompositions of the form

$$f(x, \gamma)^k = \mathbf{res}_y(g(x, y), h(y, \gamma)) = \prod_{h(\beta, \gamma)=0} g(x, \beta) \ .$$

Moreover, we can assume without loss of generality that $f(x, \gamma)$ is separable.

Let A be the set of conjugates of α over $K(\gamma)$, $|A| = n$. Let $\mathbf{Sym}\ A$ denote the field of symmetric functions of A. This is the smallest field containing all the coefficients of $f(x, \gamma)$. Note that $\mathbf{Sym}\ A$ properly contains K, for otherwise $f(x, \gamma)$ would factor into linear factors since K is algebraically closed, contradicting the assumption that α is nonconstant.

Now consider the following condition on algebraic functions β of γ:

Condition 9 *The monic minimum polynomial* $g(x, \beta)$ *of* α *over* $K(\beta)$ *divides* $f(x, \gamma)$.

If β is algebraic over $K(\gamma)$, then g exists, since α is algebraic over $K(\gamma)$ and γ is algebraic over $K(\beta)$. A subtle but important point to note is that Condition 9 does not imply that $f(x, \gamma)$ factors over $K(\beta)$. Indeed, $K(\beta)$ need not contain the coefficients of f or f/g. We give an example of this in Section 5. The following theorem states that any β satisfying Condition 9 uniquely determines a monic irreducible decomposition of α; moreover, all monic irreducible decompositions of α arise in this way.

Theorem 10. *Let α be an algebraic function of γ with monic minimum polynomial $f(x, \gamma) \in K(\gamma)[x]$ of degree n. Let β be algebraic over $K(\gamma)$ with monic minimum polynomial $h(y, \gamma) \in K(\gamma)[y]$ of degree ℓ. Let $g(x, \beta) \in K(\beta)[x]$ of degree m be the monic minimum polynomial of α over $K(\beta)$. If β satisfies Condition 9, i.e. if $g(x, \beta)$ divides $f(x, \gamma)$, then*

$$f(x, z)^{\frac{\ell m}{n}} = \mathrm{res}_y(g(x, y), h(y, z))$$

is a monic irreducible decomposition of α. Moreover, all monic irreducible decompositions of α arise in this way.

Proof. Let A be the set of roots of $f(x, \gamma)$ and let $B_\beta \subseteq A$ be the set of roots of $g(x, \beta)$. If η is a conjugate of β over $K(\gamma)$, let B_η be the set of roots of $g(x, \eta)$. The set B_η is the image of B_β under any Galois automorphism over $K(\gamma)$ mapping β to η. For any such conjugate η, $|B_\eta| = |B_\beta| = m$ and $B_\eta \subseteq A$, since the Galois group over $K(\gamma)$ preserves A setwise.

By the symmetry of the action of the Galois group on A, each $\delta \in A$ occurs in the same number of the B_η, say k. We determine k by counting in two ways the number of pairs (δ, η) such that $\delta \in B_\eta$. First, it is the number of conjugates η of β times the size of each B_η, or ℓm. Second, it is the number of $\delta \in A$ times the number of B_η containing δ, or nk. Equating these two values gives $k = \ell m/n$, the exponent in the statement of the theorem. Moreover, it follows from the same argument that

$$f(x, \gamma)^k = \prod_{\delta \in A}(x - \delta)^k = \prod_{h(\eta,\gamma)=0} \prod_{\delta \in B_\eta}(x - \delta)$$

$$= \prod_{h(\eta,\gamma)=0} g(x, \eta) = \mathrm{res}_y(g(x, y), h(y, \gamma)) \ .$$

Since γ is transcendental over K, we might as well replace it with the indeterminate z to get

$$f(x, z)^k = \mathrm{res}_y(g(x, y), h(y, z)) \ . \tag{8}$$

The decomposition is monic and irreducible by definition.

Now we show that every monic irreducible decomposition of α arises in this way. Suppose we have such a decomposition (8). Let β be a common root of $g(\alpha, y)$ and $h(y, \gamma)$. Such a β exists, since $f(\alpha, \gamma)$ vanishes, hence so does the resultant $\mathbf{res}_y(g(\alpha, y), h(y, \gamma))$. Then β is algebraic over $K(\gamma)$ with minimum polynomial $h(y, \gamma)$, $g(x, \beta)$ is the minimum polynomial of α over $K(\beta)$, and

$$f(x, \gamma)^k = \mathbf{res}_y(g(x, y), h(y, \gamma)) = \prod_{h(\eta, \gamma) = 0} g(x, \eta) .$$

Since $g(x, \beta)$ is one of the factors in the product, it divides $f(x, \gamma)$. □

At this juncture we make a few observations about minimal decompositions and uniqueness.

3.1 Minimal decompositions

There may exist β of arbitrarily high degree over $K(\gamma)$ satisfying Condition 9. For example, for any k, $\beta = \sqrt[k]{\gamma}$ gives the decomposition

$$(x - z)^k = \mathbf{res}_y(x - y^k, y^k - z) .$$

This is also the situation with (5) above. However, we can bound the search for a suitable β as follows. Observe that if there exists a β satisfying Condition 9 with factor $g(x, \beta)$ of f, say with roots $B \subseteq A$, then α will have the same degree over any subfield of $K(\beta)$ containing the coefficients of g. Furthermore, any such subfield is again a purely transcendental extension of K by Lüroth's Theorem (see [12, 16]), so a transcendental generator of that subfield would give a decomposition with the same g and smaller degree h and smaller k. For a given g, the degree of h and exponent k are minimized by taking the smallest subfield containing the coefficients of g, namely $\mathbf{Sym}\ B$.

3.2 Nontrivial decompositions

If the minimum polynomial $g(x, \beta)$ of α over $K(\beta)$ is f (as would occur in the case $\beta = \gamma$), then the minimal decomposition with this g occurs when β is a transcendental generator of $\mathbf{Sym}\ A$. Since $\mathbf{Sym}\ A \subseteq K(\gamma)$, β would be a rational function of γ and h would be linear of the form $y - u(\gamma)$, $u \in K(z)$, giving the decomposition

$$f(x, z) = \mathbf{res}_y(g(x, y), y - u(z)) = g(x, u(z)) .$$

In this case α is the composition of an algebraic function and a rational function.

In case $g(x, \beta)$ is linear, say $g = x - v(\beta)$, the smallest field containing the coefficients of g is $K(v(\beta))$, so by using $v(\beta)$ instead of β we would obtain the trivial decomposition

$$f(x, z) = \mathbf{res}_y(x - y, h(y, z)) = h(x, z) .$$

To find a nontrivial decomposition, we must find a β such that $K(\beta)$ does not contain α.

3.3 Uniqueness up to linear composition factors

The decomposition determined by β essentially depends only on the field $K(\beta)$, not on the choice of transcendental generator β. Any other transcendental generator of $K(\beta)$ is related to β by a nonsingular fractional linear transformation

$$\beta \mapsto \frac{a\beta + b}{c\beta + d}, \quad ad - bc \neq 0,$$

which extends to an automorphism of $K(\beta)$. Any two decompositions defined with respect to two transcendental generators of the same field are equivalent up to invertible composition factors of the form $(cz + d)y - (az + b)$.

4 An Algorithm

As determined in the previous section, up to fractional linear transformations there are only finitely many minimal irreducible monic decompositions of f, at most one for each factor g of f. We have thus reduced the decomposition problem to the problem of finding a subset $B \subseteq A$ (the roots of g) such that the field **Sym** B (the field generated by the coefficients of g) is a purely transcendental extension of K, and then finding a transcendental generator β of **Sym** B. Such a β is automatically algebraic over $K(\gamma)$, since **Sym** $B \subseteq K(A)$, the splitting field of f over $K(\gamma)$.

We must first determine whether f has a factor g whose coefficients lie in a purely transcendental extension of K. Equivalently, we want to know when the field **Sym** B of symmetric functions in the roots B of g is isomorphic to a rational function field over K. This is true iff **Sym** B is of genus zero. Thus the problem reduces to the problem of determining the genus of an algebraic function field.

The following is a synopsis of our algorithm.

Algorithm 11

1. Let g be a nontrivial factor of f. The coefficients of g lie in some finite extension $K(\gamma, \eta)$ of $K(\gamma)$ over which f has a nontrivial factorization. Then g can be written

$$g(x, \eta, \gamma) = x^m + u_{m-1}(\eta, \gamma)x^{m-1} + \cdots + u_0(\eta, \gamma).$$

For each such g, perform steps 2 and 3.

2. Construct the field $K(u_0, \ldots, u_{m-1})$. This is the field **Sym** B, where B is the set of roots of g. Pick one of the coefficients of g not in K, say u_0. We have two cases:

   (a) If $K(u_0, \ldots, u_{m-1}) = K(u_0)$, we are done: u is a transcendental generator of **Sym** B.

(b) If $K(u_0,\ldots,u_{m-1}) \neq K(u_0)$, construct a primitive element θ of the extension such that $K(u_0,\ldots,u_{m-1}) = K(u_0,\theta)$. Compute the genus of $K(u_0,\theta)$ by the Hurwitz genus formula or in some other fashion. An efficient algorithm is given in [3]. If the genus is nonzero, then no decomposition arises from this factor of f. If the genus is zero, compute a rational generator β of $K(u_0,\theta)$. Coates [4], Trager [10], and Huang and Ierardi [7] give efficient algorithms for computing rational generators. The coefficients of g can then be written as rational functions of β. $\quad\square$

3. Let $h(y,\gamma)$ be the minimum polynomial of β over $K(\gamma)$. Return $g(x,y)$ and $h(y,z)$ as the decomposition factors.

Under suitable assumptions about the complexity of operations in K, the complexity of the algorithm as given above is exponential in the worst case, since there are an exponentially many potential factors. For each such factor, the computation for that factor can be performed in polynomial time in the size of the representation of the algebraic numbers needed to express the result, or exponential time in the bit complexity model [7].

5 An Example

The following gives an example of a decomposition involving a β such that $g(x,\beta)$ divides $f(x,\gamma)$, but $f(x,\gamma)$ does not factor over $K(\beta)$. Consider the polynomial

$$f(x,z) = x^4 - zx^2(x+1) + z^3(x+1)^2 \ .$$

Let γ be transcendental over K, and let

$$\beta = \frac{\gamma(1 + \sqrt{1-4\gamma})}{2} \qquad \eta = \frac{\gamma(1 - \sqrt{1-4\gamma})}{2}$$

$$g(x,y) = x^2 - y(x+1) \qquad h(y,z) = y^2 - zy + z^3 \ .$$

Then β and η are conjugates over $K(\gamma)$ with minimum polynomial $h(y,\gamma)$, and

$$f(x,\gamma) = g(x,\beta) \cdot g(x,\eta) \ ,$$

thus Theorem 10 says that g and h should give a decomposition of f. Indeed,

$$\mathbf{res}_y(g(x,y), h(y,z)) = \begin{vmatrix} -(x+1) & 0 & 1 \\ x^2 & -(x+1) & -z \\ 0 & x^2 & z^3 \end{vmatrix} = f(x,z) \ .$$

To show $f(x,\gamma)$ does not factor over $K(\beta)$, it suffices to show that its trace γ is not in $K(\beta)$. But γ is a root of the irreducible polynomial $h(\beta,z)$, therefore is algebraic of degree three over $K(\beta)$.

Acknowledgements

We thank John Cremona, Ming-Deh Huang, John Little, Paul Pedersen, Moss Sweedler, Barry Trager, Emil Volcheck, Gary Walsh, and an anonymous reader for valuable comments. The support of the U.S. Army Research Office through the ACSyAM branch of the Mathematical Sciences Institute of Cornell University under contract DAAL03-91-C-0027, the National Science Foundation under grants CCR-9204630 and CCR-9317320, and the Advanced Research Projects Agency of the Department of Defense under Office of Naval Research grant N00014-92-J-1989 is gratefully acknowledged. This research was done while the second author was visiting the Cornell University Computer Science Department.

References

1. V. S. ALAGAR AND M. THANH, *Fast polynomial decomposition algorithms*, in Proc. EUROCAL85, Springer-Verlag Lect. Notes in Comput. Sci. 204, 1985, pp. 150–153.
2. D. R. BARTON AND R. E. ZIPPEL, *Polynomial decomposition algorithms*, J. Symb. Comp., 1 (1985), pp. 159–168.
3. G. A. BLISS, *Algebraic Functions*, Amer. Math. Soc., 1933.
4. J. COATES, *Construction of rational functions on a curve*, Proc. Camb. Phil. Soc., 68 (1970), pp. 105–123.
5. M. DICKERSON, *Polynomial decomposition algorithms for multivariate polynomials*, Tech. Rep. TR87-826, Comput. Sci., Cornell Univ., April 1987.
6. R. HARTSHORNE, *Algebraic Geometry*, vol. 52 of Graduate Texts in Mathematics, Springer, 1977.
7. M.-D. HUANG AND D. IERARDI, *Efficient algorithms for the effective Riemann-Roch problem and for addition in the Jacobian of a curve*, in Proc. 32nd Symp. Found. Comput. Sci., IEEE, November 1991, pp. 678–687
8. D. IERARDI AND D. KOZEN, *Parallel resultant computation*, in Synthesis of Parallel Algorithms, J. Reif, ed., Morgan Kaufmann, 1993, pp. 679–720.
9. D. KOZEN AND S. LANDAU, *Polynomial decomposition algorithms*, J. Symb. Comput., 7 (1989), pp. 445–456.
10. B. M. TRAGER, *Integration of Algebraic Functions*, PhD thesis, Massachusetts Institute of Technology, Cambridge, MA, September 1984.
11. B. L. VAN DER WAERDEN, *Algebra*, vol. 2, Frederick Ungar, fifth ed., 1970.
12. ———, *Algebra*, vol. 1, Frederick Ungar, fifth ed., 1970.
13. J. VON ZUR GATHEN, *Functional decomposition of polynomials: the tame case*, J. Symb. Comput., 9 (1990), pp. 281–299.
14. ———, *Functional decomposition of polynomials: the wild case*, J. Symb. Comput., 10 (1990), pp. 437–452.
15. R. E. ZIPPEL, *Rational function decomposition*, in International Symposium on Symbolic and Algebraic Computation, S. Watt, ed., New York, July 1991, ACM, pp. 1–6.
16. ———, *Effective Polynomial Computation*, Kluwer Academic Press, Boston, 1993.

A New Modular Interpolation Algorithm for Factoring Multivariate Polynomials (Extended Abstract)

Ronitt Rubinfeld* and Richard Zippel**

Dept. of Computer Science, Cornell University, Ithaca, NY 14853 USA

Abstract. In this paper we present a technique that uses a new interpolation scheme to reconstruct a multivariate polynomial factorization from a number of univariate factorizations. Whereas other interpolation algorithms for polynomial factorization depend on various extensions of the Hilbert irreducibility theorem, our approach is the first to depend only upon the classical formulation. The key to our technique is the interpolation scheme for multivalued black boxes originally developed by Ar et. al. [1]. We feel that this combination of the classical Hilbert irreducibility theorem and multivalued black boxes provides a particularly simple and intuitive approach to polynomial factorization.

Various versions of the problem of factoring polynomials, that is, writing a polynomial as the product of polynomials of smaller degree, have been studied for hundreds of years. In its earliest form it involved obtaining the zeroes of low degree univariate polynomials and was the subject of public competitions in the 15th and 16th centuries. In this case, the goal was to find linear factors with coefficients in a radical extension of the rational numbers, $\mathbb{Q}$.

The modern problem was first solved by Kronecker [25] and can be stated as follows. Let L be a field and $P(X, Y_1, \ldots, Y_n)$ a polynomial with coefficients in L. P is said to be *reducible* if there exists two polynomials $Q_1, Q_2 \in L[X, Y_1, \ldots, Y_n]$, neither of which are elements of L, such that $P = Q_1 \cdot Q_2$. Otherwise P is said to be *irreducible*. A *complete factorization* of P is a set of distinct, irreducible polynomials $P_1, \ldots, P_k$ such that

$$P(X, Y_1, \ldots, Y_n) = P_1^{e_1} \cdot P_2^{e_2} \cdots P_k^{e_k}. \tag{1}$$

The e_i are greater than or equal to 1 and not all of the P_i need have positive degree in X and the Y_i.

---

* Research supported by ONR Young Investigator Award N00014-93-1-0590 and United States—Israel Binational Science Foundation Grant 92-00226.

** Research supported in part by the Advanced Research Projects Agency of the Department of Defense under ONR Contract N00014-92-J-1989, by ONR Contract N00014-92-J-1839, United States—Israel Binational Science Foundation Grant 92-00234 and in part by the U.S. Army Research Office through the Mathematical Science Institute of Cornell University.

A univariate polynomial, *i.e.*, $n = 0$, can be factored in asymptotically polynomial time when the coefficient field L is either a finite field ($\mathbb{F}_p$ or $\mathbb{F}_{p^r}$) or the rational integers $\mathbb{Q}$. In the finite field case any of a number of algorithms can be used [3–5, 18, 19, 29]. When L is equal to $\mathbb{Q}$ various lattice reduction techniques can be used to factor P in polynomial time [26, 27, 31].

There are two main approaches to the multivariate factorization problem, the Hensel (or Newton) approach and the modular interpolation approach. Both approaches reduce the multivariate problem to a univariate or bivariate problem.

Intuitively the Hensel approach proceeds as follows: Choose a random n-tuple $(y_1, \ldots, y_n) \in L^n$ and factor the univariate polynomial $P(X, y_1, \ldots, y_n)$. This is viewed as factorization of $P(X, Y_1, \ldots, Y_n)$ modulo the ideal $\mathfrak{m} = (Y_1 - y_1, \ldots, Y_n - y_n)$. Using a constructive version of Hensel's lemma this factorization is lifted to one modulo $\mathfrak{m}^2$, $\mathfrak{m}^3$ and so on, yielding a factorization modulo $\mathfrak{m}^D$, where D is larger than the degree of any variable in $P(X, Y_1, \ldots, Y_n)$. This must then be a true factorization of $P(X, Y_1, \ldots, Y_n)$ [21, 35, 36, 38, 40].

The modular interpolation approach also chooses random values for the variables $Y_1, \ldots, Y_n$, but instead of using a single factorization of P, it uses interpolation techniques to combine many univariate factorizations to produce a multivariate factorization [22].

As described, both approaches reduce multivariate factorization to univariate factorization. The Hilbert irreducibility theorem guarantees that for most random choices the univariate factorizations have the same number of factors as the multivariate factorizations and thus the same structure. The authors of [20, 22] use Bertini's theorem, which is valid for reductions to bivariate factorizations. Their algorithms thus use an additional technique to reduce the required bivariate factorizations to univariate factorizations.

Our technique overcomes these earlier technical problems, allowing us to directly reduce multivariate factorization to univariate factorization and use the classical version of the Hilbert irreducibility theorem. Furthermore, given Conjecture 1, the computational complexity of our algorithm is comparable to the earlier algorithms. Briefly, we present a new, simple modular scheme for factoring a polynomial $P(X, Y_1, \ldots, Y_n)$ with integer coefficients:

$$P(X, Y_1, \ldots, Y_n) = P_1^{e_1}(X, Y_1, \ldots, Y_n) P_2^{e_2}(X, Y_1, \ldots, Y_n) \cdots P_k^{e_k}(X, Y_1, \ldots, Y_n).$$

The classical form of the Hilbert irreducibility theorem states that for almost all choices of $y_1, \ldots, y_n$, the factorization of $P(X, y_1, \ldots, y_n)$ has the same structure as the factorization of $P(X, Y_1, \ldots, Y_n)$. Our first step is to produce a black box $\mathscr{B}_{P_1, \ldots, P_k}$ that on input of $y_1, \ldots, y_n$ returns the set of factors of $P(X, y_1, \ldots, y_n)$. However, for different inputs, the factor corresponding to P_i may be returned in different positions. Nonetheless, using the techniques of Ar et. al. [1] we demonstrate how to construct k black boxes, each representing an individual factor P_i of P. These black boxes can then be interpolated using sparse polynomial interpolations schemes [2, 6, 37, 39].

The Hilbert irreducibility theorem is described in Section 1. In Section 2 we present the basic factoring algorithm. It relies on black box interpolation techniques discussed in Section 3 which in turn rely on well known Hensel techniques for solving equations that are described in Appendix A.

1 Hilbert Irreducibility Theorem

We make strong use of the Hilbert irreducibility theorem, which says that for almost all $\mathbf{y} = (y_1, \ldots, y_n)$, where $y_i \in \mathbb{Q}$, the univariate polynomial $P(X, y_1, \ldots, y_n)$ has the same number of irreducible factors as $P(X, Y_1, \ldots, Y_n)$. Thus the degree distribution of the multivariate polynomial is the same as that of the univariate polynomial.

If the factorization of $P(X, y_1, \ldots, y_n)$ has no more factors than that of $P(X, Y_1, \ldots, Y_n)$ then we call an n-tuple $(y_1, \ldots, y_n)$ *Hilbertian for* P. We need to quantify how often the factorization of $P(X, y_1, \ldots, y_n)$ corresponds to that of $P(X, Y_1, \ldots, Y_n)$. Let the number of non-Hilbertian n-tuples, $(y_1, \ldots, y_n)$ with $0 \leq y_i < N$, for an irreducible polynomial of degree d be denoted by $\bar{R}(d, n, N)$. More generally, the number of non-Hilbertian n-tuples $(y_1, \ldots, y_n)$ for a polynomial P, $R(d, n, N)$ is no greater than $k\bar{R}(d, n, N)$, where k is the number of irreducible factors of P.

The classical Hilbert irreducibility theorem asserts that $R(N)/N^n \to 0$, as N goes to infinity [9–14,17,23,24,30,33,34]. The sharpest result is due to Cohen [7]:

$$\bar{R}(d, n, N) < c(d) \cdot N^{n-\frac{1}{2}} \log N, \tag{2}$$

where c depends only on the degree of the irreducible polynomial. The distribution of non-Hilbertian points is invariant under translations of the Y_i, so we have the following proposition.

Proposition 1. *Let* $P(X, Y_1, \ldots, Y_n)$ *be a polynomial over* $\mathbb{Q}$ *and assume that the degree of* X *in* P *is less than* d. *Let* $a_1, \ldots, a_n$ *be any elements of* $\mathbb{Z}$. *For sufficiently large* N, *the number of* n-*tuples* $(y_1, \ldots, y_n) \in \mathbb{Z}^n$, $a_i \leq y_i < N + a_i$, *for which* $P(X, y_1, \ldots, y_n)$ *factors differently from* $P(X, Y_1, \ldots, Y_n)$ *is less than*

$$R(d, n, N) < c(d, n) \cdot N^{n-\frac{1}{2}} \log N.$$

As a consequence of this use of the Hilbert irreducibility theorem, the values used in the interpolation must be chosen randomly from a large enough domain.

Extensive experience has shown that n-tuples at which an irreducible polynomial is reducible are exceedingly rare. We believe the following conjecture to be true. This belief is reinforced by the implicit use of this conjecture by computer algebra systems over the last twenty years (*e.g.*, AXIOM, MACSYMA, MAPLE and REDUCE). As we do in this paper, these systems make extensive use of the classical version of Hilbert irreducibility theorem to factor multivariate polynomials over $\mathbb{Q}$.

Conjecture 1 *In the previous proposition, there exist absolute constants* c_1, c_2 *and* c_3 *such that* $c(d, n) < c_1 d^{c_2} n^{c_3}$, *i.e.*

$$R(d, n, N) < c_1 d^{c_2} n^{c_3} \cdot N^{n-\frac{1}{2}} \log N$$

Using this original form of the Hilbert irreducibility theorem and the new techniques presented in this paper we derive a simple multivariate polynomial factorization algorithm. Assuming Conjecture 1, our algorithm runs in random polynomial time.

2 Factoring Multivariate Polynomials

Assume that we want to factor a polynomial $P(X, Y_1, \ldots, Y_n)$ with coefficients in $\mathbb{Q}$. For clarity, assume also that the polynomial is square free and monic as a polynomial in X. Neither of these assumptions affect the complexity or correctness of the algorithm. The extension to non-square free polynomials is immediate. The extension to non-monic polynomials is less obvious and is outlined in Appendix B.

Assume the factorization of the monic square free polynomial $P(X, Y_1, \ldots, Y_n)$ is

$$P(X, Y_1, \ldots, Y_n) = P_1(X, Y_1, \ldots, Y_n) \cdot P_2(X, Y_1, \ldots, Y_n) \cdots P_k(X, Y_1, \ldots, Y_n).$$

For any Hilbertian point $\mathbf{y} = (y_1, \ldots, y_n)$ the factorization of $P(X, y_1, \ldots, y_n)$ will be

$$P(X, y_1, \ldots, y_n) = P_{1\mathbf{y}}(X) \cdot P_{2\mathbf{y}}(X) \cdots P_{k\mathbf{y}}(X),$$

where $P_{i\mathbf{y}}(X) = P_i(X, y_1, \ldots, y_n)$ is a univariate polynomial in X.

We construct a black box $\mathscr{B}_{P_1, \ldots, P_k}$ that represents the *set* of multivariate polynomials $\{P_1, \ldots, P_k\}$. When given a Hilbertian point $(y_1, \ldots, y_n)$ as input, the black box factors the univariate polynomial $P(X, y_1, \ldots, y_n)$ and returns the *unordered set* of factors

$$\mathscr{B}(y_1, \ldots, y_n) = \{P_{1\mathbf{y}}(X), P_{2\mathbf{y}}(X), \ldots, P_{k\mathbf{y}}(X)\}.$$

We call such a black box a *polynomial multivalued black box*, since it returns several unordered values on each call and each of these values is a univariate polynomial. By repeatedly querying this black box, we will recover the factors of P: $P_1, \ldots, P_k$. This process is called *interpolating* the black box. Thus, we have reduced the multivariate factorization problem to factoring univariate polynomials and interpolating "polynomial multivalued black boxes."

The black boxes we produce differ slightly from those studied by Ar et. al. [1], but in Section 3 we demonstrate how their techniques can be adapted to recover $P_1, \ldots, P_k$ in random polynomial time. Interpolating the black boxes requires factoring special bivariate polynomials, $Q(Z, \Theta)$, where we know the complete factorization of $Q(Z, \theta)$, for some integer θ. The bivariate polynomials $Q(Z, \Theta)$ have the special property that they have only linear factors of the form $Z - q_i(\Theta)$ where the q_i's are polynomials in Θ. With this additional information, factoring is not needed—instead, the zeroes of $Q(Z, \theta)$ can be refined to zeroes of $Q(Z, \Theta)$ using Newton's method (Appendix A).

The complete time complexity and probability of success of the factoring algorithm is given at the end of Section 3.3.

3 Interpolation Schemes

The simplest black box is one that represents a single polynomial $P(Y_1, \ldots, Y_n)$ with coefficients in $\mathbb{Q}$. We use the notation $\mathscr{B}_P$ to indicate such a black box. On input of $y_1, \ldots, y_n \in \mathbb{Q}$, such a black box returns an element of $\mathbb{Q}$, $P(y_1, \ldots, y_n)$. Given such a black box it is not difficult to determine P in time polynomial in the number of non-zero monomials in P [2, 6, 37, 39]. Extensions to black boxes representing rational functions are given in [22]. We note that the original randomized interpolation algorithm of [37] can use uniformly distributed evaluation points.

In [1], the concept of a black box is extended to a *multivalued black box* $\mathscr{B}_{P_1, \ldots, P_k}$, where the $P_i(\mathbf{Y})$ are distinct polynomials. On input $\mathbf{y}$, a multivalued black box returns an *unordered set* of values $\{P_1(\mathbf{y}), \ldots, P_k(\mathbf{y})\}$. Given a collection of t different k-tuples returned by such a black box,

$$\{P_1(\mathbf{y}_1), \ldots, P_k(\mathbf{y}_1)\}, \{P_1(\mathbf{y}_2), \ldots, P_k(\mathbf{y}_2)\}, \ldots, \{P_1(\mathbf{y}_t), \ldots, P_k(\mathbf{y}_t)\},$$

we want to determine the polynomials $P_1(\mathbf{Y}), \ldots, P_k(\mathbf{Y})$. We call this process *interpolating the black box*. Since the values are unordered, we don't know which values in each list correspond to which P_i. To find the correspondence by brute force requires time exponential in t, thus the standard interpolation techniques for black boxes are inadequate.

Denote the coefficient of X^j in the polynomial $P_i(X, Y_1, \ldots, Y_n)$ by a_{ij}, which is a polynomial in $Y_1, \ldots, Y_n$. Our approach is to reduce interpolation of the black box $\mathscr{B}_{P_1, \ldots, P_k}$ to independent interpolations of the black boxes $\mathscr{M}^{(j)}_{a_{1j}, \ldots, a_{kj}}$, $0 \le j \le d$, where d bounds the degree of X in the P_i. On input of integers $\mathbf{y} = (y_1, \ldots, y_n)$, $\mathscr{M}^{(j)}_{a_{1j}, \ldots, a_{kj}}$ returns the unordered set of values $a_{1j}(\mathbf{y}), \ldots, a_{kj}(\mathbf{y})$, which are elements of $\mathbb{Q}$. Each of the black boxes $\mathscr{M}^{(j)}_{a_{1j}, \ldots, a_{kj}}$ are then converted into several single valued black boxes $\mathscr{M}_{a_{ij}}$—in other words, the order of the values returned by $\mathscr{M}^{(j)}_{a_{1j}, \ldots, a_{kj}}$ is determined. This is accomplished using intermediate univariate multivalued black boxes of the form $\mathscr{U}_{q_1, \ldots, q_k}$ where the q_i are univariate polynomials in a new variable Θ.

In Section 3.1, we use the techniques of [1] to reduce the univariate case to the problem of refining linear factors of a bivariate polynomial. The case of $\mathbb{Q}$-valued black boxes representing multivariate polynomials is discussed in Section 3.2 and is reduced to the univariate problem. In Section 3.3, we deal with multivalued black boxes whose values are polynomials. This is again reduced to the univariate case discussed in Section 3.1. Furthermore, in all of the following sections calls will be made on randomly distributed points.

3.1 Black Boxes of Univariate Polynomials

Let $q_1(\Theta), \ldots, q_k(\Theta)$ be distinct univariate polynomials of degree no more than D. This section adapts the techniques developed in Ar et. al. [1] to our problem of interpolating a multivalued black box of univariate polynomials, $\mathscr{U}_{q_1, \ldots, q_k}$.

The symmetric functions in the $q_1(\theta), \ldots, q_k(\theta)$ are defined to be

$$\sigma_1(\theta) = q_1(\theta) + q_2(\theta) + \cdots + q_k(\theta),$$
$$\sigma_2(\theta) = q_1(\theta)q_2(\theta) + q_1(\theta)q_3(\theta) + \cdots + q_{k-1}(\theta)q_k(\theta),$$

$$\vdots$$

$$\sigma_k(\theta) = q_1(\theta)q_2(\theta)\cdots q_k(\theta).$$

Notice that the symmetric functions can be computed given the set of values $\{q_1(\theta), \ldots, q_k(\theta)\}$ without knowing which values correspond to which q_i. Therefore, we can construct k different single valued black boxes $\mathscr{B}_{\sigma_i}$, one for each symmetric function. On input of θ, $\mathscr{B}_{\sigma_i}$ returns the value $\sigma_i(\theta)$. For each black box $\mathscr{B}_{\sigma_i}$, the univariate polynomial $\sigma_i(\Theta)$ is of degree no more than iD and can be determined by Lagrangian interpolation in $O((iD)^2)$ time and $iD+1$ queries. This approach places no constraints on the points used in the interpolation process.

Using these univariate values as coefficients we can form the polynomial

$$Q(Z,\Theta) = Z^k - \sigma_1(\Theta)Z^{k-1} + \sigma_2(\Theta)Z^{k-2} + \cdots + (-1)^k\sigma_k(\Theta),$$

which by construction has only linear factors:

$$Q(Z,\Theta) = (Z - q_1(\Theta))(Z - q_2(\Theta))\cdots(Z - q_k(\Theta)).$$

The zeroes of $Q(Z,\theta)$ are $q_1(\theta), \ldots, q_k(\theta)$, which are the values of $\mathscr{B}_{q_1,\ldots,q_k}(\theta)$. This additional information allows us to use the Hensel techniques of Section A to quickly find the factorization of $Q(Z,\Theta)$ and thus determine the $q_i(\Theta)$.

Proposition 2. *Let $q_1, \ldots, q_k$ be polynomials in $\mathbb{Q}[\Theta]$ and the degree of q_i is bounded by D. Then all of the q_i's can be interpolated from a multivalued black box $\mathscr{U}_{q_1,\ldots,q_k}$ using $kD+1$ evaluations and time $O(k^3D^2)$. Furthermore, the evaluations can be made at arbitrary points.*

Note that Lagrangian interpolation does not depend upon the values chosen for θ. Thus we can use the same θ's when we need to interpolate several different black boxes at the same time.

3.2 Black Boxes of Multivariate Polynomials

Given a multivalued black box of multivariate polynomials $\mathscr{M}_{p_1,\ldots,p_k}$, we provide a technique for explicitly determining the p_i's. The approach we use converts the multivalued black box $\mathscr{M}$ into an *ordered* multivalued black box $\mathscr{M}'$ where the values are always returned in the same order. Standard interpolation techniques can then be used to recover the p_i's.

This method is a modification of the one given in Section 3 of [1]. The problem in [1] uses black boxes that with each call only return the value of an arbitrary P_i. The factorization problem yields black boxes that return the values of all of the P_i at each call. This additional information allows our technique to work over arbitrary fields, while that of [1] is only valid over finite fields.

The concept of a *reference point* is used to impose an ordering on the values returned by the $\mathscr{M}_{p_1,\ldots,p_k}$. We say that $\mathbf{y} = (y_1,\ldots,y_m)$ is a reference point if, for all i and j, $p_i(\mathbf{y}) \neq p_j(\mathbf{y})$. Given the range from which the y_i's are chosen (the interval $[0, N_P]$, where N_P is defined in Proposition 4), it can be shown that significantly more than half of the $\mathbf{y}$'s are reference points. Thus a reference point can be found by choosing a random point and verifying that it is a reference point. For each black box $\mathscr{M}$, we need only choose one reference point. Given the reference point $\mathbf{y}$, we compute the ordered sequence of reference values

$$\mathscr{M}(\mathbf{y}) = \langle p_1(\mathbf{y}), \ldots, p_k(\mathbf{y}) \rangle. \tag{3}$$

Define the multivalued black box $\mathscr{D}_{\mathbf{x},\mathbf{y}}(\theta)$, which on input θ calls the black box $\mathscr{M}(\mathbf{y}+\theta\cdot(\mathbf{x}-\mathbf{y}))$ and returns $\mathscr{M}$'s results. Fixing $\mathbf{x},\mathbf{y}$, $\mathscr{D}_{\mathbf{x},\mathbf{y}}$ is a black box of univariate polynomials in θ of degree nD. Applying the techniques described in Section 3.1 to $\mathscr{D}_{\mathbf{x},\mathbf{y}}$, we explicitly get the unordered set of univariate polynomials

$$S_{\mathbf{x}} = \{p_1(\mathbf{y} + \Theta \cdot (\mathbf{x} - \mathbf{y})), \ldots, p_k(\mathbf{y} + \Theta \cdot (\mathbf{x} - \mathbf{y}))\}$$

By substituting $\Theta = 0$ into each of the polynomials in $S_{\mathbf{x}}$, and comparing with the reference values (3), we can determine the correspondence between polynomials in $S_{\mathbf{x}}$ and $p_1,\ldots,p_k$. By substituting $\Theta = 1$ into the polynomials in $S_{\mathbf{x}}$, we can now determine $p_1(\mathbf{x}),\ldots,p_k(\mathbf{x})$.

This technique allows us to create a set of single valued black boxes (for multivariate polynomials) from a multivalued black box. It is a simple matter to use the randomized multivariate polynomial interpolation techniques of [37] to explicitly determine the p_i's.

Proposition 3. *Let θ be a fixed integer and $\mathbf{y}$ be a fixed n-tuple. For $\mathbf{x}$ such that x_i is uniformly distributed in the interval $0 \leq x_i < N$, there exists an n-dimensional box I of volume $(\theta N)^n$ such that $\mathbf{y} + \theta \cdot (\mathbf{x} - \mathbf{y})$ is uniformly distributed on a subset of I of volume N^n.*

Proposition 4. *Let $p_1,\ldots,p_k$ be polynomials in $\mathbb{Q}[Y_1,\ldots,Y_n]$ and let the total degree of the p_i be bounded by D. Furthermore, assume that no p_i has more than T non-zero terms. Then the p_i can be interpolated from a multivalued black box $\mathscr{M}_{p_1,\ldots,p_k}$ using time $O(knDT^3 + n(kD)^3T)$ and $O(knD^2T)$ black box evaluations, with likelihood of success $> 1/2$. The black box evaluations involves inputs whose magnitudes are less than $N_P = \bar{O}((kD)^nT)$ bits[3].*

Proof. Let T be the maximum number of non-zero terms in any p_i. Using the original probabilistic sparse interpolation algorithm in [37], which places no restrictions on the evaluation points, we can reconstruct the multivariate polynomials $p_1,\ldots,p_k$ from nDT the values of $p_1,\ldots,p_k$, in time $O(knDT^3)$ Each ordered set of values of $p_1,\ldots,p_k$ at $\mathbf{x}$ is determined by interpolating the black

---

[3] $\bar{O}$ indicates that additional log factors have been ignored.

box $\mathcal{D}_{\mathbf{x},\mathbf{y}}$, as described above. The black box $\mathcal{D}_{\mathbf{x},\mathbf{y}}$ represents univariate polynomials of degree at most D. So by Proposition 2, we can produce the k univariate polynomials of each $\mathcal{D}_{\mathbf{x},\mathbf{y}}$ with $kD+1$ evaluations of $\mathcal{M}$ and $O(k^3 D^2)$ operations.

The multivariate interpolation algorithm uses the values of $p_1(\mathbf{x}), \ldots, p_k(\mathbf{x})$ at nDT randomly chosen $\mathbf{x}$, at most. So the total number of values of $\mathcal{M}$ is $(nDT)(kD + 1) \approx knD^2T$. The computational cost of this approach in integer operations is the sparse interpolation cost plus the cost of the computing the nDT ordered values, or $O(knDT^3 + n(kD)^3T)$.

If $\mathbf{y} + \theta \cdot (\mathbf{x} - \mathbf{y})$ is a Hilbertian point, then each of these values can be computed by a univariate factorization. For success, each of these knD^2T points must be Hilbertian. Since kD values of θ are needed we can assume that all of the θ are less than kD. The points $\mathbf{y} + \theta \cdot (\mathbf{x} - \mathbf{y})$ lie in a box I of volume $(kDN_P)^n$. There are at most $R(d, n, kDN_P)$ non-Hilbertian points in this box. By Proposition 3 the points $\mathbf{y} + \theta(\mathbf{x} - \mathbf{y})$ are uniformly distributed in a region of I of volume N_P^n. Therefore the likelihood that a single one of these points is Hilbertian is at least

$$1 - \frac{c(d,n)(kDN_P)^{n-\frac{1}{2}}\log kDN_P}{N_P^n} = 1 - \frac{c(d,n)(kD)^{n-\frac{1}{2}}\log kDN_P}{\sqrt{N_P}}.$$

The likelihood that knD^2T points are all Hilbertian is at least

$$1 - \frac{c(d,n)n(kD)^{n+\frac{3}{2}}T\log kDN_P}{k}\frac{1}{\sqrt{N_P}},$$

for sufficiently large N_P. For some constant c_4, this expression is greater than $1 - \epsilon$ when $\log N_P > c_4(n\log kD + \log T)/\epsilon$.

3.3 Polynomial Multivalued Black Boxes

This section extends the results of the previous subsections to black boxes whose values are polynomials. That is, let $P_1, \ldots, P_k$ be polynomials in $\mathbb{Q}[Y_1, \ldots, Y_n][X]$:

$$P_1(X, Y_1, \ldots, Y_n) = a_{1d_1}(Y_1, \ldots, Y_n)X^{d_1} + \cdots + a_{10}(Y_1, \ldots, Y_n),$$

$$\vdots$$

$$P_k(X, Y_1, \ldots, Y_n) = a_{kd_k}(Y_1, \ldots, Y_n)X^{d_k} + \cdots + a_{k0}(Y_1, \ldots, Y_n),$$

where the a_{ij} are polynomials of total degree no more than D in $Y_1, \ldots, Y_n$. Let d denote the maximum of the d_i. A *polynomial multivalued black box* for $P_1, \ldots, P_k$ is a multivalued black box whose values are polynomials:

$$\mathcal{B}_{P_1,\ldots,P_k}(y_1, \ldots, y_n) = \{P_1(X, y_1, \ldots, y_n), \ldots, P_k(X, y_1, \ldots, y_n)\}.$$

Given $\mathcal{B}_{P_1,\ldots,P_k}$, we could use the techniques of Sections 3.1 and 3.2 to reconstruct the P_i since $\mathbb{Q}[X]$ is a ring. However, this is not particularly efficient. Consider the univariate case, where $n = 1$. As in Section 3.1, we would construct black boxes for the symmetric functions in P_i. Since $\mathscr{B}_{P_1,\ldots,P_k}$ has polynomial values, the $\mathscr{B}_{\sigma_i}$ will also. In particular, the degree of X in each value of $\mathscr{B}_{\sigma_k}$ will be dk. Furthermore, the Q polynomial will be trivariate.

While Hensel techniques can still be used to factor Q in polynomial time, the approach is somewhat complex. Here we propose an alternative, simpler approach. First, we determine d, the actual maximum degree of X that appears in any P_i. We then replace the polynomial valued black box $\mathscr{B}_{P_1,\ldots,P_k}$ by $d+1$, $\mathbb{Q}$-multivalued black boxes. The polynomials (in $Y_1,\ldots,Y_n$) that these black boxes represent can be reconstructed using the techniques of Sections 3.1 and 3.2.

In more detail: For two purposes, choose a random value $0 \le y_{i0} < N$ for each of the Y_i and compute:

$$\mathscr{B}_{P_1,\ldots,P_k}(y_{10},\ldots,y_{n0}) = \{Q_1(X),\ldots,Q_k(X)\},$$

where we number the Q_i so that Q_i corresponds with P_i. We claim that with high probability,

(1) the maximum degree of the Q_i will be d_i, and
(2) for every i, j and ℓ, if $a_{i\ell}(Y_1,\ldots,Y_n)$ and $a_{j\ell}(Y_1,\ldots,Y_n)$ are different then then $a_{i\ell}(y_{10},\ldots,y_{n0})$ and $a_{j\ell}(y_{10},\ldots,y_{n0})$ are also different.

Now construct $d+1$ black boxes, $\mathscr{M}^{(i)}_{P_1,\ldots,P_k}$, for $0 \le i \le d$, that represent polynomials in $Y_1,\ldots,Y_n$, as follows: the values returned by $\mathscr{M}^{(i)}_{P_1,\ldots,P_k}(y_1,\ldots,y_n)$ are the coefficients of X^i in the polynomials returned by $\mathscr{B}_{P_1,\ldots,P_k}(y_1,\ldots,y_n)$. Thus $\mathscr{M}^{(i)}_{P_1,\ldots,P_k}$ represents the polynomials

$$S_i = \{a_{1i}(Y_1,\ldots,Y_n),\ldots,a_{ki}(Y_1,\ldots,Y_n)\}.$$

The $\mathscr{M}^{(i)}_{P_1,\ldots,P_k}$ are $\mathbb{Q}$-multivalued black boxes, for which we can use the techniques of Sections 3.1 and 3.2 in order to determine the polynomials a_{ij}.

We reconstruct the $P_i(X,Y_1,\ldots,Y_n)$'s using the information in the $Q_i(X)$. Let the coefficient of X^j in P_i be a_{ij}. Further, assume $Q_i(X)$ has the form

$$Q_i(X) = q_{id_i}X^{d_i} + \cdots + q_{i0}.$$

Now, for each $0 \le j \le d_i$, a_{ij} is the entry in S_j whose value at $y_{10},\ldots,y_{n0}$ is q_{ij}. By property (2), if there is more than 1 such element then they are equal.

Proposition 5. Let $P(X,Y_1,\ldots,Y_n)$ be a polynomial over $\mathbb{Q}$, where the degree of X is not greater than d and the total degree in the Y_i is bounded by D. Using the interpolation schemes of Section 3.2 with evaluation points chosen with coordinates between 0 and $\bar{O}(n\log ndD)$ the number of operations used to determine the factors of P is $O(nd^2DT^3 + nd^3D^3T)$ and the number of univariate factorizations is $O(ndD^2T)$, with high likelihood of success.

Proof. The time complexity of interpolating polynomial multivalued black boxes is the same as that for Q-multivalued black boxes in Proposition 4 except that we may have as many as d Q-valued black boxes to interpolate. The number of values of each black box, k in Proposition 4 is also bounded by d.

The degree of X in $P_i(X, Y_1, \ldots, Y_n)$ will differ from the degree of X in $P_i(X, y_1, \ldots, y_n)$ if and only if $a_{id_i}(y_1, \ldots, y_n) = 0$. By the "DeMillo-Lipton-Schwartz-Zippel lemma" (Proposition 7) the fraction of the n-tuples in y_i that have this property is less than D/N. Thus, with high probability, d is the maximum degree of the polynomials returned by $\mathscr{B}_{P_1, \ldots, P_k}(y_{10}, \ldots, y_{n0})$.

To ensure property (2) the y_{i0} must be chosen such that $W(y_{10}, \ldots, y_{n0}) \neq 0$, where

$$W(Y_1, \ldots, Y_n) = \prod_{0 \leq \ell \leq d} \prod_{\substack{1 \leq i < j \leq k \\ a_{i\ell} \neq a_{j\ell}}} (a_{\ell i}(Y_1, \ldots, Y_n) - a_{\ell j}(Y_1, \ldots, Y_n)).$$

The maximum total degree of the Y_i in W is $D(D+1)\binom{k}{2}$, so again by using Proposition 7, the fraction of n-tuples that are accidental zeroes of W is bounded by $D(D+1)k(k-1)/2N$. The probability that a randomly chosen n-tuple, $(y_1, \ldots, y_n)$, will meet all of these conditions is thus

$$1 - \left(\frac{D}{N} + \frac{D(d+1)k(k-1)}{2N} \right) \geq 1 - \frac{dDk^2}{N} \geq 1 - \frac{d^3 D}{N},$$

or $N > d^3 D$. Proposition 4 requires that N be at least this large.

Combining this result with the univariate factoring algorithm of V. Miller [27] gives the following proposition.

Proposition 6. *Let $P(X_1, \ldots, X_n)$ be a multivariate polynomial over $\mathbb{Q}$, where the total degree of P is bounded by D, and the sum of the absolute value of the coefficients in P is bounded by H. Then P can be factored into irreducible components in $O(n^3 D^{8+\epsilon} HT + nD^3 T^3)$ arithmetic operations (for any $\epsilon > 0$). With classical integer multiplication $O(n^3 D^{9+\epsilon} HT + nD^3 T^3)$ arithmetic operations are required.*

References

1. Sigal Ar, Richard J. Lipton, Ronitt Rubinfeld, and Madhu Sudan. Reconstructing algebraic functions from mixed data. In *33rd Symposium on Foundations of Computer Science*, pages 503–512. ACM, 1992.
2. Michael Ben Or and Prasoon Tiwari. A deterministic algorithm for sparse multivariate polynomial interpolation. In *20th Symposium on Theory of Computing*, pages 301–309. ACM, 1988.
3. Elwyn Ralph Berlekamp. Factoring polynomials over finite fields. *Bell System technical Journal*, 46:1853, 1967.
4. Elwyn Ralph Berlekamp. Factoring polynomials over large finite fields. *Mathematics of Computation*, 24(111):713–735, July 1970.

5. David G. Cantor and Hans Zassenhaus. A new algorithm for factoring polynomials over finite fields. *Mathematics of Computation*, 36(154):587–592, April 1981.

6. Michael Clausen, A. Dress, Johannes Grabmeier, and Marek Karpinski. On zero-testing and interpolation of k-sparse multivariate polynomials over finite fields. Research Report 8522-CS, Universität Bonn, May 1988.

7. S. D. Cohen. The distribution of Galois groups and Hilbert's irreducibility theorem. *Proceedings of the London Mathematical Society (3)*, 43:227–250, 1981.

8. Richard A. Demillo and Richard J. Lipton. A probabilistic remark on algebraic program testing. *Information Processing Letters*, 7(4):193–195, June 1978.

9. K. Dörge. Über die Seltenheit der reduziblen Polynome und der Normalgleichungen. *Mathematische Annalen*, 95:247–256, 1926.

10. K. Dörge. Zum Hilbertschen Irreduzibilitätssatz. *Mathematische Annalen*, 95:84–97, 1926.

11. K. Dörge. Einfacher Beweis des Hilbertschen Irreduzibilitätssatzes. *Mathematische Annalen*, 96:176–182, 1927.

12. Torsten Ekedahl. An effective version of the Hilbert irreducibility theorem. In Catherine Goldstein, editor, *Séminaire de Théorie des Nombres, Paris 1988-1989*, volume 91 of *Progress in Mathematics*, pages 241–249, Boston, 1990.

13. W. Franz. Untersuchugen zum Hilbertschen Irreduzibilitätssatz. *Mathematische Zeitschrift*, 33:275–293, 1931.

14. Michael D. Fried. On Hilbert's irreducibility theorem. *Journal of Number Theory*, 6:211–231, 1974.

15. Dima Yu. Grigor'ev and Marek Karpinski. The matching problem for bipartite graphs with polynomial bounded perminants is NC. In *28th Symposium on Foundations of Computer Science*, pages 166–172. ACM, 1987.

16. Dima Yu. Grigor'ev, Marek Karpinski, and Michael F. Singer. Fast parallel algorithms for sparse multivariate polynomial interpolation over finite fields. *SIAM Journal of Computing*, 19(6):1059–1063, 1990.

17. David Hilbert. über die Irreduzibilität ganzer rationaler Funktionen mit ganzzahligen Koeffizienten. *Journal für reine und angewante Mathematik*, 110:104–129, 1892.

18. Ming-Deh A. Huang. Factorization of polynomials over finite fields and decomposition of primes in algebraic number fields. *Journal of Algorithms*, 12(3):482–489, 1991.

19. Ming-Deh A. Huang. Generalized Riemann hypothesis and factoring polynomials over finite fields. *Journal of Algorithms*, 12(3):464–481, 1991.

20. Erich Kaltofen. Computing with polynomials given by straight-line programs II: Sparse factorization. In *26th Symposium on Foundations of Computer Science*, pages 451–457. ACM, 1985.

21. Erich Kaltofen. A polynomial-time reduction from bivariate to univariate integral polynomial factorization. *SIAM Journal of Computing*, 14(2):469–489, May 1985.

22. Erich Kaltofen and Barry Marshall Trager. Computing with polynomials given by black boxes for their evaluations: Greatest common divisors, factorization, separation of numerators and denominators. *Journal of Symbolic Computation*, 9(3):301–320, March 1990.

23. Hans-Wilhelm Knobloch. Zum Hilbertschen Irreduzibilitätssatz. *Abhandlung Mathematische Seminar Univ. Hamburg*, 19:176–190, 1955.

24. Hans-Wilhelm Knobloch. Die Seltenheit der reduziblen Polynome. *Jarhesbericht der Deutsche Mathematische Vergeinung*, 59(1):12–19, 1956.

25. Leopold Kronecker. Grundzüge einer arithmetischen Theorie der algebraischen Größen. *Journal für reine und angewante Mathematik*, 92:1–122, 1882.

26. Arjen K. Lenstra, Hendrik W. Lenstra, Jr., and Laslo Lovász. Factoring polynomials with rational coefficients. *Mathematische Annalen*, 261:515–534, 1982.

27. Victor S. Miller. Factoring polynomials via relation-finding. In Danny Dolev, Zvi Galil, and Michael Rodeh, editors, *Theory of Computing and Systems*, volume 601 of *Lecture Notes in Computer Science*, pages 115–121, New York, 1992. Springer-Verlag.

28. E. Ng, editor. *EUROSAM '79*, volume 72 of *Lecture Notes in Computer Science*, Berlin-Heidelberg-New York, 1979. Springer-Verlag.

29. Lajos Rónyai. Galois groups and factoring polynomials over finite fields. In 30th *Symposium on Foundations of Computer Science*, pages 99–104. ACM, 1989.

30. Andrej Schinzel. On Hilbert's irreducibility theorem. *Annales Polinici Mathematici*, 16:333–340, 1965.

31. Arnold Schönhage. Factorization of univariate integer polynomials by diophantine approximation and an improved basis reduction algorithm. In Jan Paredaens, editor, *Automata, Languages and Programming*, volume 172 of *Lecture Notes in Computer Science*, pages 436–447, Berlin-Heidelberg-New York, 1984. Springer-Verlag.

32. Jacob T. Schwartz. Probabilistic algorithms for verification of polynomial identities. *Journal of the Association for Computing Machinery*, 27:701–717, 1980.

33. V. G. Sprindžuk. Reducibility of polynomials and rational points on algebraic curves. *Soviet Mathematics*, 21:331–334, 1980.

34. V. G. Sprindžuk. Arithmetic specializations in polynomials. *Journal für reine und angewante Mathematik*, 340:26–52, 1983.

35. Joachim von zur Gathen. Hensel and Newton methods in valuation rings. *Mathematics of Computation*, 42(166):637–661, April 1984.

36. Joachim von zur Gathen and Erich Kaltofen. Factoring sparse multivariate polynomials. *Journal of Computer and System Sciences*, 31:265–287, 1985.

37. Richard Eliot Zippel. Probabilistic algorithms for sparse polynomials. In Ng [28], pages 216–226.

38. Richard Eliot Zippel. Newton's iteration and the sparse Hensel algorithm. In Paul Wang, editor, *SYMSAC '81: Proceedings of the 1981 ACM Symposium on Symbolic and Algebraic Computation*, pages 68–72. Association for Computing Machinery, 1981.

39. Richard Eliot Zippel. Interpolating polynomials from their values. *Journal of Symbolic Computation*, 9:375–403, March 1990.

40. Richard Eliot Zippel. *Effective Polynomial Computation*. Kluwer Academic Press, Boston, 1993.

A   Finding Linear Factors of Bivariate Polynomials

In this section we demonstrate that, given the linear factors of $Q(Z, \theta)$, the linear factors of $Q(Z, \Theta)$ can be found in a very simple fashion.

A subcase of the work in [37, 38], uses Newton's method to determine the linear factors of a bivariate square free polynomial $Q(Z, \Theta)$ from the linear factors of $Q(Z, \theta)$. Here we give a complete description of this subcase. For clarity we assume that the leading coefficient of Z in $Q(Z, \Theta)$ is 1. In Appendix B we describe how this technique can be adapted to non-monic polynomials.

Let $Q(Z, \Theta)$ be a square free monic polynomial over a ring $L[\Theta]$, where L is a field. The linear factors of $Q(Z, \Theta)$ are of the form $Z - \alpha(\Theta)$ where α is a polynomial in Θ and $Q(\alpha(\Theta), \Theta) = 0$. Let $\alpha(\Theta)$ have the form

$$\alpha(\Theta) = a_0 + a_1(\Theta - \theta) + \cdots + a_\ell(\Theta - \theta)^\ell.$$

In the following paragraphs we develop an iteration formula based on Newton's method modulo powers of $(\Theta - \theta)$ that allows us to compute $\alpha(\Theta)$ efficiently from Q, θ and a_0, where $Q(a_0, \theta) = 0$.

For simplicity, translate θ to the origin. Let $\bar{Q}(Z, \Theta) = Q(Z, \Theta + \theta)$. Then $Z - a_0$ will be a factor of $\bar{Q}(Z, 0)$ and $Z - \bar{\alpha}(\Theta)$ will be a factor of $\bar{Q}(Z, \Theta)$, where

$$\bar{\alpha} = a_0 + a_1\Theta + a_2\Theta^2 + \cdots + a_\ell\Theta^\ell.$$

The image of $\bar{\alpha}$ modulo Θ^{k+1} is denoted by $\alpha^{(k)}$. Thus $\alpha^{(0)} = a_0$, $\alpha^{(1)} = a_0 + a_1\Theta$ and

$$\alpha^{(k)} = a_0 + a_1\Theta + \cdots + a_k\Theta^k. \tag{4}$$

Using the Taylor series expansion, $\bar{Q}(Z, \Theta)$ can be written as a polynomial in $Z - \alpha^{(k-1)}$:

$$\bar{Q}(Z, \Theta) = \bar{Q}(\alpha^{(k-1)}, \Theta) + \bar{Q}'(\alpha^{(k-1)}, \Theta)(Z - \alpha^{(k-1)}) + \cdots \tag{5}$$

In this and all following expressions, primes ($'$) refer to the partial derivative with respect to the first argument.

Since $Z = \bar{\alpha}$ is a zero of $\bar{Q}(Z, \Theta)$, substituting $Z = \bar{\alpha}$ into (5) gives

$$0 = \bar{Q}(\bar{\alpha}, \Theta) = \bar{Q}(\alpha^{(k-1)}, \Theta) + \bar{Q}'(\alpha^{(k-1)}, \Theta)(\bar{\alpha} - \alpha^{(k-1)}) + \cdots.$$

Since $\bar{\alpha} - \alpha^{(k-1)} = a_k\Theta^k + a_{k+1}\Theta^{k+1} + \cdots$, reducing the above equality modulo Θ^{k+1} gives

$$0 = \bar{Q}(\alpha^{(k-1)}, \Theta) + \bar{Q}'(\alpha^{(k-1)}, \Theta) \cdot a_k \cdot \Theta^k \pmod{\Theta^{k+1}}. \tag{6}$$

Since $\alpha^{(k-1)}$ is the image of $\bar{\alpha}$ modulo Θ^k, $\bar{Q}(\alpha^{(k-1)}, \Theta) = 0 \bmod \Theta^k$. Thus we can write $\bar{Q}(\alpha^{((k-1)}, \Theta) = Q_k\Theta^k \bmod \Theta^{k+1}$, where Q_k is a function of Q and $a_0, \ldots, a_{k-1}$. Q_k is an element of L. Dividing by Θ^k in (6) gives

$$0 = Q_k + \bar{Q}'(\alpha^{(k-1)}, \Theta)a_k \pmod{\Theta}.$$

Reducing the expression on the right hand side modulo Θ is equivalent to substituting $\Theta = 0$ into the right hand side, so solving for a_k gives

$$a_k = -\frac{Q_k}{\bar{Q}'(\alpha^{(k-1)}(0), 0)} = -\frac{Q_k}{\bar{Q}'(a_0, 0)} = -\frac{Q_k}{Q'(a_0, \theta)}. \tag{7}$$

If the characteristic of F is zero and $Q(Z, \theta)$ is square free then $Q'(a_0, \theta)$ does not vanish, since $Z = a_0$ is a simple zero of $Q(Z, \theta)$.

```
LinearNewton (Q(Z,Θ), θ, a₀, ℓ) := {
    w ← -[∂Q(Z,Θ)/∂Z |_{Z=a₀, Θ=θ}]⁻¹ ;
    α⁽⁰⁾ ← a₀;
    Q̄(Z,Θ) ← Q(Z,Θ + θ);
    for k = 1,...,ℓ do {
        set aₖ to w times the coefficient of Θᵏ in Q̄(α⁽ᵏ⁻¹⁾,Θ)
        α⁽ᵏ⁾ ← α⁽ᵏ⁻¹⁾ + aₖΘᵏ;
    }
    α ← α⁽ℓ⁾(Θ - θ);
    return(α);
}
```

Fig. 1. Procedure to obtain linear factors

Equation (4) allows us to compute a_k given $a_0, \ldots, a_{k-1}$. This allows us to compute $\alpha^{(k)}$ for any value of k efficiently. Since $\alpha(\Theta)$ does not contain any powers of Θ greater than ℓ, $\alpha(\Theta) = \alpha^{(\ell)}(\Theta - \theta)$.

The procedure in Figure 1 takes $Q(Z,\Theta)$, starting point (θ, a_0) and a degree bound ℓ as inputs and returns $\alpha(\Theta)$, a root of $Q(Z,\Theta)$ corresponding to $(Z,\Theta) = (\theta, a_0)$. LinearNewton is a linearly convergent iteration uses $O(\ell^3)$ operations and only requires one division. Quadratically convergent iterations can also be developed and are discussed in [35, 38].

B Non-Monic Polynomials

If we want to factor a non-monic square free polynomial $P(X, Y_1, \ldots, Y_n)$ over a field we proceed as follows. First assume that P is primitive, $i.e.$, no non-constant polynomial in $Y_1, \ldots, Y_n$ divides its coefficients. Assume the true factorization of $P(X, Y_1, \ldots, Y_n)$ is

$$P(X, Y_1, \ldots, Y_n) = P_1(X, Y_1, \ldots, Y_n) \cdots P_k(X, Y_1, \ldots, Y_n),$$

where the leading coefficient of P is $L(Y_1, \ldots, Y_n)$ and the leading coefficient of P_i is $\ell_i(Y_1, \ldots, Y_n)$. Construct a multivalued black box, $\mathscr{B}_P$ that proceeds as follows. Given $\mathbf{y} = (y_1, \ldots, y_n)$ it obtains the monic factorization:

$$\frac{P(X, y_1, \ldots, y_n)}{L(y_1, \ldots, y_n)} = \bar{P}_1(X) \cdots \bar{P}_k(X).$$

Note that the coefficients of the $\bar{P}(X)$ are the images of rational functions, not polynomials. $\mathscr{B}_P$ then returns the polynomials:

$$L(y_1, \ldots, y_n) \cdot \bar{P}_1(X), \ldots, L(y_1, \ldots, y_n) \cdot \bar{P}_k(X).$$

The coefficients of these polynomials are also polynomials, so we can use the techniques of this paper to compute the polynomials:

$$\tilde{P}_1(X, Y_1, \ldots, Y_n), \ldots, \tilde{P}_k(X, Y_1, \ldots, Y_n),$$

where

$$\tilde{P}_i(X, Y_1, \ldots, Y_n) = \frac{L(Y_1, \ldots, Y_n)}{\ell_i(Y_1, \ldots, Y_n)} P_i(X, Y_1, \ldots, Y_n).$$

Since the $P_i(X, Y_1, \ldots, Y_n)$ are primitive (by Gauss' lemma), P_i is the primitive part of $\tilde{P}_i$, i.e., the quotient of $\tilde{P}_i$ and the GCD of the coefficients of $\tilde{P}_i$ (with respect to X).

Since L has no more terms than $P(X, Y_1, \ldots, Y_n)$, ncne of the $\tilde{P}_i$ will have more than T^2 terms, where T is the maximum number of terms of any of the P_i. Thus we have shown that primitive sparse multivariate polynomials can be factored in polynomial time.

If the polynomial is not primitive then we can still follow this general approach but it must be done one variable at a time. Immediately removing the content of P may not be satisfactory because it has not known how much larger the primitive part of a polynomial can be than its factorization.

C   Probabilistic Zero Testing

Given a black box $\mathscr{B}_P$ for a polynomial $P(X_1, \ldots, X_n)$ over a field F, one can probabilistically determine if $P \equiv 0$, by examining $\mathscr{B}_P(x_1, \ldots, x_n)$ for randomly chosen x_i. If the value is ever different from zero then P cannot be the zero polynomial. An estimate of the number of times a non-zero polynomial can take on the value zero is given by the following proposition.

Proposition 7. *Let F be a field, $f \in F[X_1, \ldots, X_n]$ and the degree of f in each of X_i be bounded by d_i. Let $Z_n(B)$ be the number of zeroes of f, $\mathbf{x}$ such that x_i is chosen from a set with B elements, $B \gg d$. Then*

$$Z_n(B) \leq B^n - (B - d_1)(B - d_2) \cdots (B \cdot d_n)$$

$$\approx O\left((d_1 + d_2 + \cdots + d_n)B^{n-1}\right).$$

A proof of Proposition 7 is given in [39].

The idea of probabilistic zero testing was independently discovered by at least three different groups of researchers at almost the same time. DeMillo and Lipton presented essentially this result in the context of testing the correctness of algebraic programs [8] in 1978. Schwartz [32], used this result in the context of testing for polynomial identities. Zippel [37] used Proposition 7 not just to determine if a polynomial was identically zero, but as part of a complete randomized interpolation algorithm for sparse polynomials.

If a bound is known for the number of non-zero terms in P, then a deterministic algorithm for zero-testing and for sparse interpolation can be given. Results along these lines are given in [2, 6, 15, 16, 39].

The Function Field Sieve

Leonard M. Adleman

Department of Computer Science
University of Southern California
Los Angeles, California, 90089
Adleman@cs.usc.edu

Abstract. The fastest method known for factoring integers is the 'number field sieve'. An analogous method over function fields is developed, the 'function field sieve', and applied to calculating discrete logarithms over $GF(p^n)$. An heuristic analysis shows that there exists a $c \in \Re_{>0}$ such that the function field sieve computes discrete logarithms within random time:

$$L_{p^n}[1/3, c]$$

when $\log(p) \le n^{g(n)}$, where g is any function such that $g : N \to \Re_{>0}^{<.98}$ approaches zero as $n \to \infty$.

1 Introduction

It is well known that algebraic number fields and algebraic function fields with finite constant field - the so called global fields - share a rich theory. In this paper we generalize the 'number field sieve' [LL1, Ad1, Co1, Go1, Go2] from algebraic number fields to algebraic function fields.

In the number field sieve a positive integer n is given. One finds an irreducible polynomial $f \in Z[y]$ of degree d and an $m \in Z$ such that $f(m) \equiv 0 \bmod n$. Then working simultaneously in Q and the number field $L = Q[y]/(f)$, one seeks 'double smooth pairs' $< r, s > \in Z \times Z$ such that r and s are relatively prime and both $rm + s$ and $N_Q^L(ry + s) = r^d f(-s/r)$ are 'smooth' in Z (i.e. have no prime factors greater than some predetermined 'smoothness bound' B). In the simplest case where O_L, the ring of integers of L, is a PID then from such a double smooth pair we have:

$$rm + s = \prod_{i=1}^{z} p_i^{e_i}$$

where $p_1, p_2, ..., p_z$ are the primes less than B. And:

$$ry + s = \prod_{i=1}^{v} \epsilon_i^{f_i} \prod_{i=1}^{w} \wp_j^{g_j}$$

where $\epsilon_1, \epsilon_2, ...\epsilon_v$ is a basis for the units of O_L and $\wp_1, \wp_2, ..., \wp_w \in O_L$ are generators of the (residue class degree one) prime ideals of norm less than B.

Since there is a homomorphism ϕ from O_L to $Z/(n)$ induced by sending $y \mapsto m$, the above equations give rise to the congruence:

$$\prod_{i=1}^{z} p_i^{e_i} \equiv \prod_{i=1}^{v} \phi(\epsilon_i)^{f_i} \prod_{i=1}^{w} \phi(\wp_j)^{g_j}$$

In the case that n is composite and its factorization is sought, one collects sufficiently many such double smooth pairs and puts the corresponding congruences together using linear algebra to create a congruence of squares of the form:

$$(\prod_{i=1}^{z} p_i^{e_i'})^2 \equiv (\prod_{i=1}^{v} \phi(\epsilon_i)^{f_i'} \prod_{i=1}^{w} \phi(\wp_j)^{g_j'})^2$$

Creating such congruences is tantamount to factoring n.

In the case that n is prime and discrete logarithms in $Z/(n)$ are sought, one collects sufficiently many such double smooth pairs and uses linear algebra on the discrete logarithms of the corresponding congruences to solve for the discrete logarithms of the primes less than B [Gol].

Of course when O_L is not a PID complications arise. These complications can be overcome by viewing the factorization of $ry + s$ as an ideal factorization and introducing singular integers and character signatures [Ad1]. At a conceptual level this approach makes the divisor associated with $ry + s$ rather than $ry + s$ itself central. The character signature of $ry + s$ is introduced as a surrogate for the values of $ry + s$ 'at infinity'.

In the 'function field sieve' one starts with a finite field F and a monic irreducible polynomial $n \in F[x]$. One finds an absolutely irreducible polynomial $H \in F[x, y]$ of degree d in y and an $m \in F[x]$ such that $H(x, m) \equiv 0 \bmod n$. Then working simultaneously in $F(x)$ and the function field $L =$ Quotient$(F[x, y]/(H))$, one seeks 'double smooth pairs' $< r, s > \in F[x] \times F[x]$ such that r and s are relatively prime and both $rm + s$ and the intersection of $ry + s$ with H, $r^d H(x, -s/r)$ are 'smooth' in $F[x]$ (i.e. have no irreducible factors of degree greater than the 'smoothness bound' B).

Having collected sufficiently many double smooth pairs, the approach taken now becomes value theoretic and geometric. The divisors of the $ry+s$'s are calculated. As usual there are difficulties 'at infinity'. The value of $ry + s$ at the points at infinity on the smooth projective model of the affine curve associated with H are calculated geometrically using 'blow up'.

In the case that n is not irreducible, one could proceed to factor n in a manner analogous to that used in the number field sieve. However, since there exists a random polynomial time method to factor polynomials over finite fields [Be1] this approach is not needed. However, in the case n is irreducible and discrete logarithms in $F[x]/(n)$ are sought, one collects sufficiently many such double smooth pairs and uses linear algebra to solve for the discrete logarithms of the irreducible polynomials of degree less than B. A key fact that makes this approach possible is that the elements of L which have divisor 0 are precisely the elements of F^*. Using linear algebra on the divisors of the $ry + s$'s, a product $\prod_{i=1}^{z}(r_i y + s_i)^{e_i}$ can be created which has divisor 0 and hence:

$$\prod_{i=1}^{w}(r_i y + s_i)^{e_i} = a$$

for some $a \in F^*$

Using the homomorphism ϕ from $F[x,y]/(H)$ to $F[x]/(n)$ induced by sending $y \mapsto m$, this gives rise to a congruence:

$$\prod_{i=1}^{w}(r_i m + s_i)^{e_i} \equiv a$$

And since the $r_i m + s_i$ are smooth we finally achieve a congruence of the form:

$$\prod_{i=1}^{z} p_i^{e_i} \equiv a$$

where $p_1, p_2, ..., p_z$ are the irreducible monic polynomials of degree less than B.

When sufficiently many such congruences are created the discrete logarithms (up to the logarithm of an element of F^*) of the irreducible monic polynomials of degree less than B can be calculated using linear algebra. To improve the running time of this approach we would prefer to do linear algebra over a finite ring rather that Q. This can be accomplished if the class number of the field L meets a certain condition. The calculations of the discrete logarithms of elements which are not irreducible of degree less than B are carried out in an manner similar to that just described using a construction due to Gordon [Go1].

2 Preliminaries

In this sections some basic facts are presented [Ha1, CF1, Fu1].

A discrete valuation on a field k will be a map $\nu : k \to Z \cup \{\infty\}$ such that:

- The restriction of ν to k^* is a homomorphism onto Z.
- $\nu(0) = \infty$.
- For all $x, y \in k$: $\nu(x+y) \geq \min\{\nu(x), \nu(y)\}$ (where for all $a \in Z, \min\{a, \infty\} = \min\{\infty, a\} = a$ and $\min\{\infty, \infty\} = \infty$).

Let k be a field and ν a discrete valuation on k. Let k' be an extension field of k and ν' a valuation on k'. Then ν' is an extension of ν iff ν is the restriction of ν' to k.

Associate with a discrete valuation ν of k, the ring $R_\nu = \{x \in k | \nu(x) \geq 0\}$ and the maximal ideal $P_\nu = \{x \in k | \nu(x) > 0\}$.

Let T be a set of discrete valuations on k, then for all $\kappa \in k$, κ is T-smooth (also called a T unit) iff for all discrete valuations $\nu \notin T$, $\nu(\kappa) = 0$. Let T_k be the empty set of discrete valuations on k.

Let $p \in Z$ be prime and let $F = Z/pZ$. Let $K = F(x)$. K is the field of rational functions over F. The valuations on K are as follows:

- ν_f where $f \in F[x]$ is monic irreducible, and for all $\kappa \in K$ such that $\kappa = a/b$ with $a, b \in F[x]$, $\nu_f(\kappa) = i_a - i_b$ where $f^{i_a} || a$ and $f^{i_b} || b$.
- ν_∞ where for all $\kappa \in K$ such that $\kappa = a/b$ with $a, b \in F[x]$, $\nu_\infty = \deg(b) - \deg(a)$.

For all irreducible $f \in F[x]$, R_{ν_f}/P_{ν_f} is a finite extension of F of degree $f_\nu = dg(f)$. $R_{\nu_\infty}/P_{\nu_\infty}$ is an extension of F of degree $f_{\nu_\infty} = 1$.

Let $H \in F[x, y]$ be absolutely irreducible with $H = \sum_{i=1}^{d} a_i y^i$ where $a_i \in F[x]$, $i = 1, 2, ..., d$. Let L be the quotient field of $F[x, y]/(H)$. L is a function field over F.

The valuations on L are exactly the extensions of the valuations on K. Let $g \in F[x]$ be irreducible and monic. Let $g' \in (F[x]/(g))[y]$ be irreducible and monic and assume that $H \equiv 0 \bmod (g, g')$ but either $\partial H/\partial x \not\equiv 0 \bmod (g, g')$ or $\partial H/\partial y \not\equiv 0 \bmod (g, g')$. Let $\alpha \in \overline{F}$ be a root of g and let $\beta \in \overline{F}$ be a root of g'. Then $< \alpha, \beta >$ is a non-singular point on H and there exists a unique valuation ν of L corresponding to $< g, g' >$ such that for all $\lambda \in L$:

$$\nu(\lambda) = I(< \alpha, \beta >, H \cap \lambda)$$

where $I(< \alpha, \beta >, H \cap \lambda)$ denotes the intersection number of λ and H at $< \alpha, \beta >$. This valuation extends the valuation ν_g on K. It is independent of the choice of $< \alpha, \beta >$. Further, all but finitely many valuations of L are of this form.

For all discrete valuations ν of L, R_ν/P_ν is a finite extension of F. Let the degree of this extension be denoted f_ν. The set of T_L-smooth element is F^* [Hal].

Let k be either the field of rational functions over F or a function field over F. A divisor of k is a finite formal sum of discrete valuations $\sum a_i \nu_i$ where $a_i \in Z$. The degree of such a divisor is $\sum a_i f_{\nu_i}$. If $\kappa \in k$ then $[\kappa]$ denotes the 'principal' divisor $\sum \nu(\kappa)\nu$ where the sum is over all discrete valuations of k. Principal divisors are of degree 0. The set of degree zero divisors modulo the principal divisors is a finite group of cardinality h_k - the class number of k.

3   The Function Field Sieve Algorithm

Stage I: Calculating The Logarithms of Small Elements

Let $p \in Z_{>0}$ be prime and let $F = Z/pZ$. Let $f \in F[x]$, such that f is monic irreducible of degree n. Then $F[x]/(f)$ is a finite field with p^n elements. We will assume for convenience that x generates $(F[x]/(f))^*$ (i.e. that f is 'primitive'). Throughout, $F[x]/(f)$ will be identified with the complete set of representatives: $\{\alpha \in F[x] | \deg(\alpha) < n\}$. Assume that $p, f, \alpha \in (F[x]/(f))^*$ are given and $l \in Z_{\geq 0}^{<p^n - 1}$ such that $\alpha \equiv x^l \bmod f$ is sought.

For elements $g \in (F[x]/(f))^*$, the goal is to calculate $ind(g) \in Z_{\geq 0}^{<p^n - 1}$ such that $x^{ind(g)} \equiv g \bmod f$. However, in the following routines we will work in $(F[x]/(f))^*$ modulo F^* and calculate the 'restricted discrete logarithm': $\overset{*}{ind}$ $(g) \in Z_{\geq 0}^{<(p^n - 1)/(p-1)}$ such that there exists a $u \in F^*$ with $x^{\overset{*}{ind}(g)} \equiv ug \bmod f$. It will be convenient to write $\alpha \overset{*}{\equiv} \beta \bmod f$ to denote that there exists a $u \in F^*$ such that $\alpha \equiv u\beta \bmod f$.

Begin by setting $d = \lceil \frac{1}{c_2}(n^{1/3}/\log(n)^{1/3}) \rceil$ (c_2 will be determined later). Next find an $m \in F[x]$ of degree $d' = \lceil (n+1)/(d+1) \rceil$ and an $H = H(x, y) = \sum_{i=0}^{d} \sum_{j=0}^{d'-1} h_{i,j} y^i x^j \in F[x, y]$ satisfying the following conditions:

- (1) H is absolutely irreducible.
- (2) $H(x, m) \equiv 0 \bmod f$.
- (3) $h_{d,d'-1} = 1$.
- (4) $h_x = \sum_{i=0}^{d} h_{i,d'-1} y^i$ is square free in $\overline{F}[y]$.
- (5) $h_{0,d'-1} \neq 0$.
- (6) $h_y = \sum_{j=0}^{d'-1} h_{d,j} x^j$ is square free in $\overline{F}[x]$.
- (7) $h_{d,0} \neq 0$.
- (8) $(h_L, (p^n - 1)/(p - 1)) = 1$ where L is the quotient field of $F[x, y]/(H)$.

Basically this is accomplished by choosing a monic m of degree d' at random and writing f base m which is possible because $F[x]$ is a Euclidean domain. However, to meet the condition (3) that $h_{d,d'-1} = 1$ it is necessary to make a slight adjustment first. Choose a random monic $r \in F[x]$ of degree $(d' - 1) + dd' - n$ and write fr base m, this will insure that $h_{d,d'-1} = 1$ as required by condition (3). Let the resulting polynomial be $H(x, y)$. Then $H(x, m) = rf \equiv 0 \bmod f$ as required in condition (2). Since F is a perfect field h_x and h_y are square free over $\overline{F}$ iff they are square free over F which can be easily checked by factoring them over F (or by taking their GCDs with their respective derivatives). Hence conditions (4),(5),(6) and (7) can be readily checked. Conditions (1) can be checked in deterministic polynomial time in the total degree of H and $\log(p)$ [Ka1]. There appears to be no algorithm known for efficiently checking whether condition (8) holds, hence the algorithm should be tried and if the correct restricted discrete logarithms are not produced (which can be efficiently checked since the elements of $F[x]/(f)$ which are in F are precisely those fixed by the Frobenius automorphism) then a different m should be chosen and the process tried again.

Next collect double smooth pairs. This is done by considering all relatively prime $r, s \in F[x]$ where r and s have degree less than or equal to $d'' = c_3 n^{1/3} \log(n)^{2/3}$ (c_3 will be determined later). For each such pair check if both $rm + s$ and the intersection of $ry + s$ with H, $I(ry + s) = I(ry + s, H) = r^d H(x, -s/r)$ are 'smooth' in $F[x]$. Essentially 'smooth' means all irreducible factors are of degree less than B (B will be determined later); however, certain irreducible elements of degree less than B are best avoided. With the number field sieve small primes which divide the discriminant of the polynomial used to generate the number field are avoided. In the function field sieve irreducible polynomials of low degree whose roots are the x coordinates of singular points on the affine curve defined by H are avoided. Hence $ry+s$ will be considered 'smooth' iff all of its irreducible factors belong to the set S_1 defined below. $I(ry + s)$ will be considered 'smooth' iff all of its irreducible factors belong to the set S_2 defined below.

$$S_1 = \{g | g \in F[x] \ \& \ g \text{ monic irreducible } \& \ \mathrm{dg}(g) \leq B\}$$

$$S_2' = \{< g, y - w > | g \in S_1 \ \& \ w \in F[x]/(g) \ \& \ H \in F[x, y]/(g, y - w)$$

$$\& \ (\partial H/\partial x \notin F[x, y]/(g, y - w) \text{ or } \partial H/\partial y \notin F[x, y]/(g, y - w))\}$$

$$S_2 = \{g | (\exists w \in F[x]/(g))[< g, y - w > \in S_2']\}$$

Notice that because there exists a random polynomial time algorithm to factor polynomials over finite fields, testing a pair $< r, s >$ for double smoothness requires a negligible amount of time (this should be compared with the analogous problem in the number field sieve).

For each double smooth pair $< r, s >$ calculate the divisor $[ry + s]$. This is done in two steps. Step one: for each g which divides $I(ry + s)$ there is a unique $w = -s/r \in F[x]/(g)$ such that $< g, y - w > \in S'_2$. As described in the preliminaries section each such element of S'_2 gives rise to a unique valuation ν on L such that:

$$\nu(ry + s) = I(< \alpha, \beta >, H \cap ry + s)$$

where $\alpha \in \overline{F}$ is a root of g, $\beta \in \overline{F}$ is a root of $y - w$ and $I(< \alpha, \beta >, H \cap ry + s)$ denotes the intersection number of $ry + s$ and H at $< \alpha, \beta >$.

Since α and $\beta \in F[x]/g$ and intersection numbers are efficiently computed (see for example [Fu1]), $\nu(ry + s)$ is easily calculated. All other valuations ν on L for which $\nu(ry + s) \neq 0$ lie at infinity. These are handled in Step two:

Homogenize H with respect to a new variable z. By condition (3) on H the only points at infinity on the projective closure are $< 1, 0, 0 >$ and $< 0, 1, 0 >$. Consider the point $< 1, 0, 0 >$. Dehomogenizing with respect to x results in the polynomial $H_x = H_x(z, y) \in F[z, y]$. Writing H_x as a sum of forms (and using condition (3)) gives $\sum_{k=d}^{d+d'-1} H_{x,k}(z, y)$ where $H_{x,k} \in F[z, y]$ is a form of degree k. It follows that $< 0, 0 >$ (the image of $< 1, 0, 0 >$ on the affine curve defined by H_x) is a singularity of multiplicity d. Further, $H_{x,d} = \sum_{i=0}^{d} h_{i,d'-1} z^{d-i} y^i$. By conditions (4) and (5) above, $H_{x,d}(1, y) = h_x = \prod_{i=1}^{d}(y - \alpha_i)$ where the α_i are distinct and non zero. It follows that $< 0, 0 >$ is an ordinary singularity with distinct tangent lines $y - \alpha_1 z, y - \alpha_2 z, ..., y - \alpha_d z$ none of which equals y. By blowing up H_x we obtain a new curve $H_x^b(v, w) \in F[v, w]$ which has d simple points $< 0, \alpha_1 >, < 0, \alpha_2 >, ..., < 0, \alpha_d >$ each of which gives rises to a valuation ν_{α_i} which can be evaluated using intersection numbers. These are precisely the valuations whose rings dominate the local ring at $< 1, 0, 0 >$. The details are as follows:

$$H_x^b(v, w) = \sum_{k=d}^{d+d'-1} v^{k-d} H_{x,k}(1, w)$$

The image of $r(x)y + s(x)$ in the quotient field of $F[v, w]/(H_x^b)$ is $\gamma = r(1/v)w + s(1/v)$ and:

$$\nu_{\alpha_i}(\gamma) = I(< 0, \alpha_i >, \gamma \cap H_x^b)$$

Whenever α_i and α_j are conjugate over F then $\nu_{\alpha_i}(\gamma) = \nu_{\alpha_j}(\gamma)$. Hence, though it is inconsequential when considering the asymptotic running time of the function field sieve, only one valuation need be calculated for each set of conjugate αs.

The point $< 1, 0, 0 >$ is handled in a similar fashion. Dehomogenizing with respect to y results in the polynomial $H_y = H_y(z, x) \in F[z, x]$. Writing H_y as a sum of forms (and using condition (3)) gives $\sum_{k=d'-1}^{d+d'-1} H_{y,k}(z, x)$ where $H_{x,k} \in F[z, x]$ is a form of degree k. It follows that $< 0, 0 >$ (the image of $< 1, 0, 0 >$ on the affine curve defined by H_y) is a singularity of multiplicity $d' - 1$. Further, $H_{y,d'-1} = \sum_{j=0}^{d'-1} h_{d,j} z^{d'-1-j} x^j$. By conditions (6) and (7) above, $H_{y,d'-1}(1, x) = h_y = \prod_{j=1}^{d'-1}(x - \beta_j)$ where the β_j are distinct and non zero. It follows that $< 0, 0 >$ is an ordinary singularity with distinct tangent lines $x - \beta_1 z, x - \beta_2 z, ..., x - \beta_{d'-1} z$ none of which equals x. After blowing up H_y we obtain a new curve $H_y^b(v, w) \in F[v, w]$ which has $d' - 1$ simple points $< 0, \beta_1 >$, $< 0, \beta_2 >, ..., < 0, \beta_{d'-1} >$ each of which gives rises to a valuation ν_{β_j} which can be evaluated using intersection numbers. These are precisely the valuations whose rings dominate the local ring at $< 0, 1, 0 >$. The details are as follows:

$$H_y^b(v, w) = \sum_{k=d'-1}^{d+d'-1} v^{k-(d'-1)} H_{y,k}(1, w)$$

The image of $r(x)y + s(x)$ in the quotient field of $F[v, w]/(H_y^b)$ is $\gamma = r(w)(1/v) + s(w)$ and:

$$\nu_{\beta_j}(\gamma) = I(< 0, \beta_j >, \gamma \cap H_y^b)$$

Where as above only one valuation need be calculated for each set of conjugate βs.

In the number field sieve, the analog of the function field L is a number field. When this number field has class number greater than 1 problems arise. When the number field sieve is used to factor integers, then these problems can be overcome efficiently with the use of singular integers and character signatures [Ad1]. When the class number is not 1 in the function field case similar problems arise. However, if $(h, (p^n - 1)/(p - 1)) = 1$, then these problems also can be efficiently overcome. The basic idea is to 'pretend' that $h = 1$ and that all divisors are principle.

By condition (2) $H(x, m) \equiv 0 \mod f$. It follows that there exists a homomorphism $\phi : F[x, y]/(H) \to F[x]/(f)$ is induced by sending $y \mapsto m$.

Let $< r, s >$ be a double smooth pair. Let:

(0) $rm + s = \prod_{j=1}^{w} g_j^{b_j}$

where $g_1, g_2, ..., g_w$ is an ordering of S_1.

(1) $[ry + s] = \sum_{i=1}^{z} \nu_i(ry + s)\nu_i$

where $\nu_1, \nu_2, ..., \nu_z$ is an ordering of the valuations associated with S_2 and the points $< 1, 0, 0 >$ and $< 0, 1, 0 >$.

Let ν' be a valuation on L such that $\nu' \neq \nu_i$, $i = 1, 2, ..., z$ and $f_{\nu'} = 1$ (such exists by condition (1) [Ha1]). Then $[ry + s] = \sum_{i=1}^{z} \nu_i(ry + s)(\nu_i - f_{\nu_i}\nu')$ (since $[ry + s]$ has degree zero) and $[(ry + s)^{h_L}] = \sum_{i=1}^{z} \nu_i(ry + s)(h_L\nu_i - h_L f_{\nu_i}\nu')$. But for $i = 1, 2, ..., z$, $h_L\nu_i - h_L f_{\nu_i}\nu' = [\lambda_i]$ where $\lambda_i \in L$ is unique up to an element in F^* (i.e. an element which is T_L-smooth). Hence $[(ry + s)^{h_L}] = \prod_{i=1}^{z} \lambda_i^{\nu_i(ry+s)}$ and thus there exists a $u \in F^*$ such that:

(1) $\qquad (ry + s)^{h_L} = u \prod_{i=1}^{z} \lambda_i^{\nu_i(ry+s)}$.

Applying ϕ gives:

(2) $\qquad (rm + s)^{h_L} \stackrel{*}{\equiv} \prod_{i=1}^{z} \phi(\lambda_i^{\nu_i(ry+s)})$.

Taking h_L^{th}-roots in $(F[x]/(f))/F^*$ (which exist and are unique by condition (8) since $(p^n - 1)/(p - 1)$ is the order of $(F[x]/(f))/F^*$) and letting κ_i denote the h_L^{th}-root of $\phi(\lambda_i)$ gives:

(3) $\qquad (rm + s) \stackrel{*}{\equiv} \prod_{i=1}^{z} (\kappa_i^{\nu_i(ry+s)})$.

From (0) and (3):

(4) $\qquad \prod_{i=1}^{w} g_j^{b_j} \stackrel{*}{\equiv} \prod_{i=1}^{z} \kappa_i^{\nu_i(ry+s)}$

Taking discrete logarithms with respect to x mod F^* gives

(5) $\qquad \sum_{i=1}^{w} b_j \stackrel{*}{\mathrm{ind}} (g_j) = \sum_{i=1}^{z} \nu_i(ry + s) \stackrel{*}{\mathrm{ind}} (\kappa_i)$

Hence double smooth pairs give rise to $\stackrel{*}{\mathrm{ind}}$ equations. If sufficiently many independent $\stackrel{*}{\mathrm{ind}}$-equations are obtained from these double smooth pairs, then the $\stackrel{*}{\mathrm{ind}}$ (g) for all $g \in S_1$ can be determined by linear algebra. Notice that the matrix involved is sparse and that we need only do the calculation modulo $(p^n - 1)/(p - 1)$. It follows that fast matrix methods may be applied [Wi1]. Notice also that we need not know the κ_is nor h_L (however, these techniques may be useful for calculating h_L if that is desired).

Stage II: Calculating The Logarithms of Arbitrary Elements

We wish to calculate the $\overset{*}{ind}(\alpha)$ where $\alpha \in (F[x]/(f))^*$ is arbitrary. Our method is due to Gordon and the reader is referred [Go1] for the details.

First we will choose random $l \in Z$ with $1 \le l \le (p^n - 1)$ until an l is found such that

$$x^l \alpha \equiv q_1 q_2 ... q_t \bmod f$$

where the q_i are of degree less than or equal to $d' = c_2 n^{2/3} \log(n)^{1/3}$ (see Stage I). For each $i = 1, 2, ..., t$ calculate $m_i = x^{l_i} q_i$ where l_i is chosen so that the degree of m_i is d'. Then to calculate $\overset{*}{ind}(\alpha)$ it is enough to calculate $\overset{*}{ind}(m_i)$ for $i = 1, 2, ..., t$.

For each i calculate an H_i as in Stage I meeting conditions (2)-(8) with m_i taking the role of m. Define for $j \in F[x]$:

$$H_{i,j} = H_i + j(m_i - y)$$

Try random js of degree less than or equal to $2c_3 n^{1/3} \log(n)^{2/3}$ (actually the degree of j required is much less, but this choice is convenient in the running time analysis) until an $H_{i,j} = \sum_{i=0}^d h_i y^i$ is found which still meets conditions (1)-(8) of Stage I and has the further property that h_0 is B smooth. Let L denote the quotient field of $F[x, y]/(H_{i,j})$. Notice $y \in L$ is such that $I(y, H_{i,j}) = h_0$ so $I(y, H_{i,j})$ is smooth. Also that $\phi(y) = m_i$. It follows that by collecting double smooth elements $< r, s >$ as in Stage I that we can use linear algebra to obtain a congruence of the following form:

$$m_i \overset{*}{\equiv} \prod_{i=1}^v g_i$$

where the g_i have degree less than B. Now the results of Stage I can be used to readily obtain $\overset{*}{ind}(m_i)$ for $i = 1, 2, ..., t$ and from these $\overset{*}{ind}(\alpha)$ can be quickly obtained.

Stage III: Calculating The $ind(\alpha)$ from $\overset{*}{ind}(\alpha)$

We have that

$$x^{\overset{*}{ind}(\alpha)} \alpha^{-1} \equiv u \bmod (f)$$

where $u \in F^*$. Hence to obtain $ind(\alpha)$ it is enough to calculate the index of u with respect to the base $x^{(p^n-1)/(p-1)}$ in F^*. This is a discrete logarithm problem over $GF(p)$ and can be done using Gordon's method [Go1].

Running Time Analysis

The following analysis assumes that the polynomial H satisfies conditions (1)-(8). These conditions present no serious problems except perhaps (1) and (8). For (8) we assume that h_L 'behaves' like a random integer with regard to being relatively prime to $(p^n - 1)/(p - 1)$ and hence the probability condition (8) is satisfied is adequately high. At this time condition (1) is more problematic and for now it will simply be assumed that our method of generating H produces absolutely irreducible polynomials with adequate probability.

A detailed, optimal analysis of the function field sieve would be quite lengthy. For the sake of brevity we will make the non-optimal assumption that:

$$(*) \log(p) \le n^{g(n)}$$

Where g can be any function such that $g : N \to \Re_{>0}^{<.98}$ approaches zero as $n \to \infty$. It is likely that a finer analysis would allow g to be replaced by a positive constant.

Our analysis will make use of the results of Lovorn [Lo1] on the number of smooth polynomials of bounded degree.

First consider Stage I:

For pairs $< r, s >$ each of degree less than d'', the degree of $rm + s \le (c_2 + o(1))n^{2/3} \log(n)^{1/3}$ and the degree of $I(ry + s) \le (c_2 + \frac{c_3}{c_2} + o(1))n^{2/3} \log(n)^{1/3}$. Hence the degree D of $(rm + s)(I(ry + s))$ is less than or equal to $(\frac{2c_2^2+c_3}{c_2} + o(1))n^{2/3} \log(n)^{1/3}$. Let $Q = p^D$. Let the smoothness bound $B = \log_p(L_Q[1/2, 1/\sqrt{2}])$. Then by Lovorn's Theorem 3.2.48 and the assumption (*) the probability that a pair $< r, s >$ is B smooth is at least:

$$1/L_Q[1/2, 1/\sqrt{2}]$$

The number of elements in the factor base is less than or equal to $p^B = L_Q[1/2, 1/\sqrt{2}]$. We will assume that this many double smooth pairs will be sufficient to generate the independent $\overset{*}{ind}$-equations needed to determine the $\overset{*}{ind}$ of all elements in S_1. This assumption is similar those made in many factoring and discrete logarithm algorithms based on smoothness. This assumption is similar to assuming that the $\overset{*}{ind}$-equations which arise from double smooth pairs 'behave' like 'random' relations on the multiplicative group. Hence $L_Q[1/2, \sqrt{2}]$ $< r, s >$ pairs must be tried. There are $p^{2d''} = p^{(2c_3+o(1))n^{1/3}\log(n)^{2/3}}$ pairs available. Therefore we have the constraint that:

$$L_Q[1/2, \sqrt{2}] \leq p^{(2c_3+o(1))n^{1/3}\log(n)^{2/3}}$$

Which, again using assumption (*), simplifies to

$$\frac{2c_2^2 + c_3}{3c_2} \leq c_3^2 \log(p)$$

Solving gives the solution $c_2 = 3/9^{2/3}\log(p)^{1/3}$, $c_3 = 2/9^{1/3}\log(p)^{2/3}$.

The running time assuming sparse matrix methods is then:

$$e^{(2c_3\log(p)+o(1))n^{1/3}\log(n)^{2/3}} =$$

$$e^{((4\log(p)^{1/3})/(9^{1/3})+o(1))n^{1/3}\log(n)^{2/3}} =$$

$$L_{p^n}[1/3, 4/9^{1/3}] \approx.$$

$$L_{p^n}[1/3, 1.92].$$

For Stage II:

The probability that a random element of degree n will have all factors of degree less that $d' = c_2 n^{2/3}\log(n)^{1/3}$ is by Lovorn

$$1/e^{((\frac{1}{3c_2}+o(1))n^{1/3}\log(n)^{2/3})} =$$

$$L_{p^n}[1/3, -9^{1/3}]$$

subject to the constraint that $n^{1/100} \leq c_2 n^{2/3}\log(n)^{1/3} \leq n^{99/100}$. These constraints are satisfied for all but finitely many p^n using assumption (*).

hence finding an l such that

$$x^l \alpha \equiv q_1 q_2 ... q_t \bmod f$$

where the q_i have degree less than or equal to d' requires at most $L_{p^n}[1/3, 9^{1/3}] \approx L_{p^n}[1/3, 2.08]$ steps.

For each possible $H_{i,j}$, h_0 is of degree at most $(c_2 + o(1))n^{2/3}\log(n)^{1/3}$. Hence the probability that it will have all factors of degree less than B is the same as the probability that a single $rm + s$ will be B-smooth in Stage I. Since we have allowed $p^{(2c_3+o(1))n^{1/3}\log(n)^{2/3}}$ possible j it follows from the analysis of Stage I that we will be find the desired $H_{i,j}$ within the time required for Stage I.

The remaining parts of Stage II are essentially the same as Stage I and require the same amount of time.

Stage III requires a negligible amount of time.

4 Discussion

The similarity of the function field sieve and the number field sieve should make it possible to study them simultaneously. This may lead to improved methods for both integer factoring and discrete logarithms. There are several natural open problems:

1. Are the integer factoring problem and the discrete logarithm problem over finite fields (random) polynomial time equivalent?
2. Can the function field sieve be optimized and specialized to give an algorithm over $GF(2^n)$ which has a better worst case running time than the algorithm of Coppersmith?
3. Can an $L[1/3]$ algorithm be found for discrete logarithms over all finite fields?

It seems probable that the answer to all three of the above is yes. The first problem seems difficult and currently there are no obvious approaches to pursue. For the second, it is possible that the tools needed to solve it are currently available. The version of the function field sieve presented here was designed to facilitate the exposition rather than optimize the running time. It is likely that the running time can be substantially improved. For the third, it may suffice to create a 'global field sieve' by combining Gordon's version of the number field sieve (for n very small) with the function field sieve (for n larger).

5 Acknowledgments

The author was greatly helped by discussions with Ming-Deh Huang and Jonathan DeMarrais. The author's research was supported by the National Science Foundation (Grant number CCR-9214671).

References

[Ad1] Adleman L.M., Factoring numbers using singular integers, *Proc. 23rd Annual ACM Symposium on Theory of Computing*, 1991, pp. 64-71.

[Be1] Berlekamp E., Factoring polynomials over large finite fields. *Math. Comp.* 24, 1970. pp. 713-735.

[CF1] Cassels J.W.S. and Fröhlich A., *Algebraic Number Theory*, Thompson Book Company, Washington, D.C. 1967.

[Co1] Coppersmith D., Modifications to the Number Field Sieve. IBM Research Report #RC 16264, November, 1990.

[Co2] Coppersmith D., Fast Evaluation of Logarithms in Fields of Characteristic Two. *IEEE Trans on Information Theory*, vol IT-30, No 4, July 1984, pp. 587-594.

[Fu1] Fulton W., *Algebraic Curves*, The Benjamin/Cummings Publishing Company, Menlo Park, 1969.

[Go1] Gordon D.M., Discrete logarithms in $GF(p)$ using the number field sieve, manuscript, April 4, 1990.

[Go2] Gordon D.M., Discrete logarithms in $GF(p^n)$ using the number field sieve (preliminary version), manuscript, November 29, 1990.

[Ha1] Hasse H., *Number Theory*. English Translation by H. Zimmer. Springer-Verlag, Berlin. 1980.

[Ka1] Kalofen E., Fast parallel absolute irreducibility testing. *J. Symbolic Computation* 1, 1985, pp. 57-67.

[LL1] A.K. Lenstra and H.W. Lenstra Jr. (Eds.), *The development of the number field sieve*, Lecture NOtes in Mathematics 1554, Springer-Verlag, Berlin. 1993.

[Lo1] Lovorn R., Rigorous Subexponential Algorithms For Discrete Logarithms Overr Finite Fields. Ph.D. Thesis. Universiy of Georgia, Athens, Georgia. 1992.

[Wi1] Wiedermann D. Solving sparse linear equations over finite fields. *IEEE Trans. Inform. Theory*. IT-32, pp. 54-62

Heegner Point Computations

Noam D. Elkies

Department of Mathematics, Harvard University, Cambridge, MA 02138 USA
elkies@zariski.harvard.edu

Abstract. We discuss the computational application of Heegner points
to the study of elliptic curves over $\mathbf{Q}$, concentrating on the curves

$$E_D : Dy^2 = x^3 - x$$

arising in the "congruent number" problem.[1] We begin by briefly review-
ing the cyclotomic construction of units in real quadratic number fields,
which is analogous in many ways to the Heegner-point approach to the
arithmetic of elliptic curves, and allows us to introduce some of the key
ideas in a more familiar setting. We then quote the key results and conjectures that we shall need concerning elliptic curves and Heegner points,
and show how they yield practical algorithms for finding rational points
on E_D and other properties of such curves. We conclude with a report on
more recent work along similar lines on the elliptic curves $x^3 + y^3 = A$.

1 Review: the cyclotomic approach to "Pell's equation"

This classical Diophantine equation[2] asks in effect for nontrivial units u in an
order $O = O_D$ in a real quadratic number field K. Here D is the discriminant
of O, so $O_D = \mathbf{Z} + \mathbf{Z}w$ where $w = \sqrt{D}/2$ or $(1 + \sqrt{D})/2$ according as D is
even or odd. It is known that for all D (subject to the usual condition that D
is positive, congruent to 0 or 1 mod 4, and not a perfect square), there exists
a "fundamental unit" $u_1 = a_1 - b_1 w$ with nonnegative $a, b \in \mathbf{Z}$ such that the
group of units in O_D is $\{\pm 1\} \times u_0^{\mathbf{Z}}$. For any unit $u = a - bw$ with $a \geq 0, b > 0$
the rational number a/b approximates w well; indeed $|w - a/b|$ is small enough
that all such a/b, and in particular the first one, a_1/b_1, are among the rational
approximations of w obtained by expanding w in a continued fraction (a.k.a.
applying the Euclidean algorithm to 1 and w). Thus the fundamental unit can
be readily computed from the continued-fraction expansion of w. At least for

---

[1] The problem is for which D does E_D have nontrivial rational points, or equivalently
positive rank. Such D are called "congruent", because they are precisely the numbers that arise as the common difference ("congruum") of a three-term arithmetic
progression of rational squares, namely the squares of $(x^2 - 2x - 1)/2y$, $(x^2 + 1)/2y$,
and $(x^2 + 2x - 1)/2y$. See the Preface and Chapter XVI of [Di] for the early history
of this problem, and [Kob] for a contemporary treatment of the curves E_D.

[2] "...the equation $x^2 - Ny^2 = 1$ to which [Pell's] name, because of Euler's mistaken at-
tribution, has remained attached; since its traditional designation as 'Pell's equation'
is unambiguous and convenient, we will go on using it, even though it is historically
wrong." [We, p.174]

even D, this was first explained by Euler and Lagrange in the late 1760's, though algorithms implementing this method of computing the fundamental unit were known since the age of Fermat in Europe, and centuries earlier in India [We].

Gauss found an alternative construction of units in K. Let ω be a primitive D-th root of unity. Then he showed that K is contained in $\mathbf{Q}(\omega)$, and observed that $(1 - \omega^a)/(1 - \omega^b)$ is a unit whenever a, b are integers relatively prime to D. A suitable product of these "cyclotomic units" then yields a unit $U \in K$. Such a unit will not in general be fundamental, and indeed it is not a priori obvious that U is not one of the trivial units ± 1; that U is indeed nontrivial was first shown for all D by Dirichlet, who further showed that $U = \pm u_1^{\pm h}$ where $h = h(K)$ is the class number of K by expressing both $h \big| \log |u_1| \big|$ and $\big| \log |U| \big|$ in terms of the residue of the zeta function $\zeta_K(s)$ at $s = 1$.

This cyclotomic approach, though certainly more circuitous and cumbersome than the usual continued-fraction method, still yields a polynomial-time algorithm for computing both U and u_1. Since the conjugate of U is $\pm U^{-1}$, both $U \pm U^{-1}$ and $(U \mp U^{-1})/\sqrt{D}$ are integers. Thus by merely calculating U *as a real number* to more than $\big| \log_{10} |U| \big|$ digits we can recognize U as a $\mathbf{Z}$-linear combination of 1 and w. Dirichlet's formula, together with easy upper bounds on $L(1, \chi)$, shows that $\big| \log |U| \big| \ll D^{1/2+\epsilon}$, so the computation time is dominated by the evaluation of $O(D)$ trigonometric functions to $O(D^{1/2+\epsilon})$ digits and is thus clearly polynomial in D. For instance, if D is prime then we may take

$$U = D^{-1/2} \prod_a (2 \sin \frac{a\pi}{D}),$$

the product extending over all positive integers $a < D/2$ that are quadratic residues mod D; e.g. when $D = 1093$ we calculate (in a few CPU seconds on a SUN-4) that

$$
\begin{aligned}
U &= \qquad\qquad .0000000254354271700067\ldots, \\
U^{-1} &= 39315243.0000000254354271700067\ldots,
\end{aligned}
$$

from which U is the unit $(-39315243 + 1189189\sqrt{1093})/2$ of norm -1. The class number h is then the largest integer such that $|U|^{1/h}$ is a unit in K, which again we detect by the integrality of $|U|^{1/h} \pm |U|^{-1/h}$. It quickly follows that having found U we can also determine h and recover u_1 in polynomial time. Returning to our example of $D = 1093$, we see that h must be odd since U has negative norm (this always happens when D is a prime); for $h = 3$ we find

$$U^{1/3} = .002940896\ldots, \qquad U^{-1/3} = 340.032416171\ldots$$

whence $\sqrt[3]{U} \notin K$, but for $h = 5$

$$U^{1/5} = .030275255\ldots, \qquad U^{-1/5} = 33.030275255\ldots$$

indicates that $U = [(-33 + \sqrt{1093})/2]^5$, and we soon confirm that K has fundamental unit $(33 - \sqrt{1093})/2$ and thus that $h(K) = 5$. Indeed it turns out that the prime ideal $(3, (1 + \sqrt{1093})/2)$ generates the class group, and its fifth power is the principal ideal $((11 - \sqrt{1093})/2)$.

Our use of "polynomial time" here and later requires some comment, since we mean some power of the input D rather than its size $\log(D)$, so that the running time is actually exponential in the size of the input. But the size $|\log|U||$ of the *output* is heuristically expected, and experimentally observed, to be $D^{1/2 \pm \epsilon}$. Since it takes at least $|\log|U||$ time just to write U out in full, no algorithm can be expected to run in time less than polynomial in D, so our designation of $D^{O(1)}$ time as "polynomial" appears reasonable. Moreover it is expected that for most D the class number $h(K)$ is small enough that even the fundamental unit u_1 has at least $D^{1/2 - \epsilon}$ digits.[3] Admittedly, when (as with $D = 1093$) the fundamental unit is unusually small, the continued-fraction algorithm will find it quickly, taking time polynomial in its size, whereas the cyclotomic approach cannot exploit a small fundamental unit well enough to reduce the running time below a power of D. More generally, even if U is a fundamental unit and thus is not an exact power of a much smaller unit, it may be possible to write U more compactly as a product than in its explicit $a - bw$ form; for instance the fundamental unit

$$9482010153904475351 + 2763835460065778460\sqrt{1177}$$

of the quadratic field $\mathbf{Q}(\sqrt{1177})$ of class number 1 equals

$$2^{-50}3^{-13}(31 + \sqrt{1177})^2(35 + \sqrt{1177})^{13}(37 + \sqrt{1177})^7$$

(cf. [Co, 281–282]). If we accept such representations it becomes reasonable to ask for an algorithm faster than $D^{O(1)}$, and indeed Buchmann [Co, 283–289], adapting ideas of Hafner and McCurley ([HMcC], see also [Co, 247–256]) for imaginary quadratic fields, has a factor-base algorithm that finds both the class number and a product form of the fundamental unit in expected time $D^{o(1)}$ subexponential in the input size. When we later consider elliptic curves depending on a parameter D or A, we again expect that the output often has size proportional to a power of that parameter, and thus use "polynomial time" to mean $|D|^{O(1)}$ or $|A|^{O(1)}$. Our method, analogous to the cyclotomic approach to Pell's equation, will require such running times even when the fundamental solutions are unusually small. There is as yet no elliptic-curve analogue known for the continued-fraction approach to Pell's equation, nor a subexponential factor-base method, nor even a more compact representation of large generators. The analogy between Pell's equation and the congruent number problem suggests that progress in these directions may yield not only faster algorithms but also a better theoretical understanding of the arithmetic of elliptic curves.

---

[3] See [CL] and [Co, 292–293] for these heuristics on the size of $U, h(K), u_1$.

2 Elliptic curves and Heegner points

Let E be an elliptic curve over $\mathbf{Q}$, with L-series (see [Si, App.C, §16])

$$L(E, s) = \sum_{n=1}^{\infty} a_n n^{-s}.$$

We assume that E is *modular*, i.e. that there exists a nonconstant map from $X_0(N)$ to E defined over $\mathbf{Q}$, or equivalently that

$$\varphi(\tau) := \sum_{n=1}^{\infty} a_n q^n$$

(with $q = \exp(2\pi i\tau)$ as usual) is a modular form of weight 2 for $\Gamma_0(N)$ for some integer N. A modular parametrization $X_0(N) \to E$ can be constructed from φ as follows: regard $X_0(N)$ over $\mathbf{C}$ as the quotient of the upper half-plane by the fractional linear transformations in $\Gamma_0(N)$; then the function[4]

$$I_\varphi(\tau) := 2\pi i \int_{i\infty}^{\tau} \varphi(z)\, dz$$

on the upper half-plane descends to an analytic map from $X_0(N)$ to $\mathbf{C}/L$ for some lattice $L \subset \mathbf{C}$ such that $\mathbf{C}/L$ is isomorphic with E as a Riemann surface.[5] Elliptic curves with complex multiplication, such as our "congruent number" curves

$$E_D : Dy^2 = x^3 - x,$$

are known to be modular, and it is expected but not yet proved that every $E/\mathbf{Q}$ is modular — a seminal conjecture variously credited to Shimura, Taniyama, and/or Weil.

The L-series of a modular elliptic curve can be expressed as a Mellin transform

$$L(E, s) = \frac{(2\pi)^s}{\Gamma(s)} \int_0^{\infty} \varphi(it) t^s \frac{dt}{t}$$

of φ, giving an analytic continuation of $L(E, s)$ to an entire function of s with a functional equation relating $L(E, s)$ and $L(E, 2-s)$. The *Birch and Swinnerton-Dyer (BSD) Conjecture* relates the behavior of $L(E, s)$ at the fixed point $s=1$ of the involution $s \leftrightarrow 2-s$ with the arithmetic of $E/\mathbf{Q}$. Let $r = r(E)$ be the rank of the abelian group $E(\mathbf{Q})$ of rational points of E; Mordell showed that this group is finitely generated, so r, the *(arithmetic) rank* of E, is a finite nonnegative integer. The BSD conjecture asserts that r equals the order of vanishing of $L(E, s)$

---

[4] We insert the factor $2\pi i$ because $2\pi i\, dz = dq/q$.

[5] Note that L contains, but might be strictly larger than, the period lattice of ϕ; thus it is not claimed that E is the strong Weil curve associated with ϕ, only that it is isogenous to that curve. This requires the isogeny theorem, which for general E is a difficult theorem (due to Faltings) even if we assume the Eichler-Shimura modular theory, but is much easier for curves with complex multiplication such as our E_D.

at $s = 1$; that order of vanishing is thus known as the *analytic rank* $r_{an} = r_{an}(E)$ of E. The conjecture also expresses the "leading term" $\lim_{s \to 1}(s-1)^{-r_{an}}L(E, s)$ as a product of invariants of the curve, including notably the *regulator* of $E(\mathbf{Q})$ and the order of the conjecturally finite *Tate-Šafarevič group* $\text{III} = \text{III}(E)$, analogous to the factors of the regulator and class number in Dirichlet's formula for $L(1, \chi)$. (Unlike the class group, though, III is known to have square order if it is finite, because then it carries a nondegenerate alternating pairing.) The parity of r_{an}, and thus conjecturally also of r, depends only on the sign of the functional equation for $L(E, s)$. For the curves E_D, where we may assume that D is a squarefree integer because square factors may be absorbed into y^2, it is known that r_{an} is odd if and only if $|D|$ mod 8 is 5, 6, or 7. Thus for such D, the arithmetic rank of E_D is conjecturally odd, so in particular positive, whence the Diophantine equation $Dy^2 = x^3 - x$ is expected to have infinitely many nontrivial solutions in rational x, y. The BSD conjecture has almost been proved for modular elliptic curves of analytic rank 0 or 1: it is now known [Kol] that for such curves the arithmetic and analytic ranks are equal, III is finite, and the conjectured leading term of $L(E, s)$ at $s = 1$ is correct except perhaps for "small" prime factors.

A crucial ingredient in the proof is the construction of a non-torsion point in $E(\mathbf{Q})$ when $r_{an} = 1$. Recall that $X_0(N) - \{\text{cusps}\}$ parametrizes elliptic curves C with a cyclic subgroup G of order N; over $\mathbf{C}$, the image in $X_0(N)$ of any τ in the upper half-plane corresponds to $C = \mathbf{C}/(\mathbf{Z} + \mathbf{Z}\tau)$ with the cyclic subgroup G generated by $1/N$. A *Heegner point* is a point of $X_0(N)$ parametrizing (C, G) such that C and the quotient elliptic curve C/G are curves with complex multiplication *of the same discriminant*. Let the "CM field" of such a point be the fraction field of the endomorphism ring of C. A Heegner point, and thus also its image $P \in E(\mathbf{C})$, must be defined over some number field (described in detail by the arithmetic theory of complex multiplication); the sum in the group law of E of all the conjugates[6] of P is then a rational point ΣP on E. If $X_0(N_1) \to E_1$ is a modular parametrization of some elliptic curve E_1 of analytic rank 0 which becomes isomorphic with E over some imaginary quadratic field K (so E is a quadratic twist of E_1), then it is sometimes possible to construct a rational point on E from the image $P_1 \in E_1$ of a Heegner point of $X_0(N_1)$ with CM field K: the sum of the $\text{Gal}(\bar{\mathbf{Q}}/\mathbf{Q})$-conjugates of P_1 will be a torsion point on E_1, but the conjugates *over* K sum to a K-rational point on E_1 which yields a $\mathbf{Q}$-rational point $\Sigma_1 P_1$ on E. For our curves E_D of odd analytic rank this happens when D is odd using $N_1 = 32$ and

$$E_1 : y^2 = x^3 - x$$

and $K = \mathbf{Q}(\sqrt{-|D|})$ [note that in our family E_D and E_{-D} are isomorphic elliptic curves over $\mathbf{Q}$, since $(x, y) \leftrightarrow (-x, y)$ maps one to the other — in general any CM

---

[6] More precisely, the sum of the images of conjugates of the Heegner point: it is conceivable that P might be defined over a smaller number field, and thus have fewer conjugates, but then each one is counted several times in the sum. Likewise for the variant construction using quadratic twists, and the computations described in the next section.

elliptic curve over **Q** is isogenous with its twist by the CM field]; this indeed was the original setting for Heegner's construction [He], later amplified and extended by Birch [Bi1, Bi2]. When $D \equiv 6 \bmod 8$ a similar construction [Mo] works with E_1 replaced by E_2 and $N_1 = 64$ instead of 32.

Our rational point ΣP or $\Sigma_1 P_1$ on E need not in general be a generator of $E(\mathbf{Q})$ mod torsion, and indeed our construction does not preclude the possibility of finding only a trivial (torsion) point. Occasionally (e.g. for $\Sigma_1 P_1$ on E_D with $D \equiv 7 \bmod 8$ prime) triviality may be ruled out by a parity argument, analogous to the observation that when a cyclotomic unit has norm -1 it cannot be a root of unity. A more powerful tool is Gross and Zagier's result [GrZ], analogous to but much harder than Dirichlet's class number formula, which shows that the canonical height of this point is proportional to $L'(E, 1)$. Thus the Heegner-Birch construction, which can be applied to any modular elliptic curve of odd analytic rank, yields a nontorsion point only when $L'(E, 1) \neq 0$, i.e. when $r_{an} = 1$. Comparing the Gross-Zagier formula with the BSD conjecture one further finds[7] that when $r_{an} = 1$ the height of that point should be $|\text{III}|$ times the height of a generator of $E(\mathbf{Q})$ mod torsion, which means that the Heegner-point construction yields the $\pm\sqrt{|\text{III}|}$ multiple of the generator up to torsion.

3 Heegner-point computations for E_D

As was true for the cyclotomic and fundamental units of a real quadratic field, we can compute, in time polynomial in $|D|$, the point $\Sigma_1 P_1$ on E_D, and (provided $\Sigma_1 P_1$ is not a torsion point) also a fundamental solution and thus the conjectural size of III. Here "computing" a rational point means exhibiting its coordinates as rational numbers. Nevertheless, as in the computation of U, it will be enough to compute those coordinates as real numbers to sufficient accuracy: $O(H + 1)$ digits, where H is either the naïve or the canonical logarithmic height of $\Sigma_1 P_1$, determine the numerators and denominators via the Euclidean algorithm. By the Gross-Zagier formula, the canonical height of $\Sigma_1 P_1$ is proportional to $\sqrt{|D|} \cdot L'(E_D, 1)$; one readily shows that $L'(E_D, 1)$, and thus also H, is bounded by a power of D, so we need only compute $\Sigma_1 P_1$ to $|D|^{O(1)}$ digits. One expects that $L'(E_D, 1)$ (and indeed the r-th derivative of $L(E_D, s)$ at $s = 1$ for any fixed r) is actually bounded by $O_\epsilon(|D|^\epsilon)$ for all $\epsilon > 0$, as was true for the residue at $s = 1$ of $\zeta_K(s)$ with $K = \mathbf{Q}(\sqrt{+|D|})$, and thus that $H \ll_\epsilon |D|^{1/2+\epsilon}$, but this has not yet been proved.

Let $K = \mathbf{Q}(\sqrt{-|D|})$ as before, with D odd. The prime 2 splits or ramifies in K according as $|D| \equiv 7$ or $|D| \equiv 5 \bmod 8$. We shall henceforth assume that $|D| \equiv 7 \bmod 8$, the other case being similar but somewhat more delicate. Then the number of conjugates of P_1 over K is the class number of K, and is thus

---

[7] At least for the curves E_D with D prime; for general E, the factor $\sqrt{|\text{III}(E)|}$ must be multiplied by $\sqrt{|\text{III}(E_1)|}$ and perhaps by products of small primes. With our $E_1 : y^2 = x^3 - x$ the BSD conjecture predicts that $|\text{III}(E_1)| = 1$; this was proved by Rubin [Ru].

$O(|D|^{1/2} \log|D|)$. These conjugates may be exhibited explicitly as follows: let O be the ring of integers of K, and $I \subset O$ the fifth power of one of the two ideals of O lying above 2; thus $O/I \cong \mathbf{Z}/32$. For each ideal class of K choose a representative ideal J. Fix an imbedding of K in $\mathbf{C}$; then each J is a lattice in $\mathbf{C}$ determined up to homothecy, and IJ is a sublattice with $J/IJ \cong O/I \cong \mathbf{Z}/32$. For each J we regard $\mathbf{C}/J$ and $\mathbf{C}/IJ$ as elliptic curves with complex multiplication of discriminant $-|D|$, and J/IJ as a cyclic subgroup of $\mathbf{C}/IJ$, whence $(\mathbf{C}/IJ, J/IJ)$ yields a Heegner point on $X_0(32)$. The images of these points under the modular parametrization $X_0(32) \to E_1$ are then the conjugates of P_1. It is easy to compute in polynomial time the ideal I and representatives J of all ideal classes, and thus also the Heegner points as points $\tau \in K$ in the upper half-plane. Thus it remains only to compute for each such τ the image on E_1 of the corresponding point on $X_0(32)$.

We may compute these images either as complex solutions (x,y) of the cubic equation $y^2 = x^3 - x$, or as points on the complex torus $\mathbf{C}/L$. The latter approach is more convenient since we will have to sum many such points and adding complex numbers mod L is easier than the addition law on E_1; also the $\mathbf{C}/L$ approach directly generalizes to quadratic twists of other modular elliptic curves. At any rate the generators of L and the elliptic functions giving (x,y) as functions on $\mathbf{C}/L$ can be easily computed to arbitrary precision in polynomial time, so we may compute on $\mathbf{C}/L$ and recover (x,y) at the end. We noted already that the image of τ on $\mathbf{C}/L$ is $I_\varphi(\tau)$ mod L. We integrate φ termwise to find

$$ I_\varphi(\tau) = \sum_{n=1}^\infty \frac{a_n}{n} q^n = q - \frac{2}{5} q^5 - \frac{3}{9} q^9 + \frac{6}{13} q^{13} \cdots . $$

For our curve E_1 the coefficients a_n can be rapidly computed from either a product or a sum expansion of φ:

$$ \varphi = \eta_4^2 \eta_8^2 = q \prod_{m=1}^\infty \left[(1 - q^{4m})(1 - q^{8m}) \right]^2 = \sum_{m_1, m_2} m_1 q^{m_1^2 + m_2^2}, $$

where the sum runs over all $m_1, m_2 \in \mathbf{Z}$ such that m_1 is odd, m_2 is even, and $m_1 \equiv m_2 + 1 \bmod 4$. Each of the coefficients a_n/n in the expansion of $I_\varphi(\tau)$ is of absolute value at most 1 (indeed $a_n \ll_\epsilon n^{1/2+\epsilon}$), so the sum of the first $O(H/\log(|q|^{-1})) = O(H/\mathrm{Im}(\tau))$ terms $(a_n/n)q^n$ already gives $I_\varphi(\tau)$ to the required precision $\exp(-O(H))$. Thus we conclude that the entire computation can be done in polynomial time provided we have a polynomial upper bound on $1/\mathrm{Im}(\tau)$.

Now τ is only defined up to a fractional linear transformation in $\Gamma_0(32)$, corresponding to a choice of generators for the lattice IJ. It is not hard to choose these generators to make $\mathrm{Im}(\tau) \gg |D|^{-1/2}$, which is enough to compute $\Sigma_1 P_1$ in polynomial time; but in fact we can always choose an equivalent τ with imaginary part $> 1/8$. This is because in addition to the invariance of $\varphi(z)\,dz$ under $\Gamma_0(32)$ the modular form φ also satisfies

$$ \varphi\left(z + \frac{1}{4}\right) = i\varphi(z), \qquad \varphi\left(-\frac{1}{32z}\right) = -\frac{1}{32z^2}\varphi(z), $$

whence I_φ satisfies the identities

$$I_\varphi\left(\tau + \frac{1}{4}\right) = iI_\varphi(\tau), \qquad I_\varphi\left(-\frac{1}{32\tau}\right) = 2I_\varphi(\frac{i}{\sqrt{32}}) - I_\varphi(\tau),$$

and the transformations $\tau \mapsto \tau + 1/4$ and $\tau \mapsto -1/32\tau$ generate an arithmetic subgroup of $\mathrm{PSL}_2(\mathbf{Q})$ with fundamental domain

$$\{z : |\mathrm{Re}(z)| \leq \frac{1}{8}, |z|^2 \geq \frac{1}{32}\}.$$

(For a general modular elliptic curve of conductor N_1 such a trick is not available, but could still find an image of any τ under $\Gamma_0(N_1)$ that is close enough to some cusp to compute $I_\varphi(\tau)$ by one of a finite list of q-expansions with $\log|1/q| \gg 1/N_1$, the implied constant not depending on D.) So the computation of $\Sigma_1 P_1$ takes time polynomial in $|D|$ as claimed.

We further find that the most time-consuming part of the computation is the calculation of $O(|D|^{1/2}\log|D|)$ values of I_φ, each of which requires on the order of H arithmetic operations to as many digits of precision; if $H \ll_\epsilon |D|^{1/2+\epsilon}$ as expected then the entire computation takes time $O_\epsilon|D|^{2+\epsilon}$, or $O_\epsilon|D|^{3/2+\epsilon}$ with fast multiplication and division. Moreover, the space requirement is only a constant multiple of the $\max(H, \log|D|)$ needed just to write down D and the digits of $\Sigma_1 P_1$. It may seem that we must know H in advance to achieve this, but we can get around this as follows: perform the computation assuming $H < 2$; if it fails (as it most likely will unless $r_{\mathrm{an}} > 1$), start anew with the assumption $H < 4$; then try $H < 8$, $H < 16$, etc. until the first success. In practice we initially assume that H is less than a small multiple of $\sqrt{|D|}$, set the precision accordingly, and increase it (or check for bugs!) if no suitable rational approximation to the coordinates of $\Sigma_1 P_1$ is found.

If $r_{\mathrm{an}} = 1$, i.e. if $\Sigma_1 P_1$ is not a torsion point, then we can find a fundamental solution, and thus the conjectural $|\mathrm{III}|$, by trying to divide $\Sigma_1 P_1$ by $S = 2, 3, \cdots$ in $E_D(\mathbf{Q})$. For each S there are at most $2S$ points $(\Sigma_1 P_1)/S$ on $E_D(\mathbf{R})$, which we can conveniently calculate from our $\mathbf{C}/L$ representation of $\Sigma_1 P_1$. If one of them is rational it has height $\ll H/S^2$; thus it is enough to try all $S \ll \sqrt{H}$, and we find that the recovery of the generator takes less time than the computation of $\Sigma_1 P_1$. For instance when $|D|$ is the prime 751 we compute $\Sigma_1 P_1$ in less than one SUN-4 CPU minute, and in a few more seconds find that $S = 3$ and that $751y^2 = x^3 - x$ has fundamental solution

$$(x, y) = \left(\frac{323761}{17^2 751}, \frac{136696560}{17^3 751^2}\right).$$

It is actually quite rare to find D such that $\Sigma_1 P_1$ is not the fundamental solution or its 2^s-th multiple for small s: usually (assuming the BSD conjecture) III is either trivial or a 2-group accounted for entirely by the 2-descent. For instance, of the 565 curves E_D with $|D| < 20000$ a prime congruent to 7 mod 8 (such for such D it is known that the 2-part of $\mathrm{III}(E_D)$ is trivial), only 24 have $S = 3$, of which 751 is the first example; in that range there are no instances of $S = 5$ or greater, the first one occurring at $|D| = 26423$. (For $|D| \equiv 5 \bmod 8$, the smallest primes with $S = 3$, $S = 5$ are 1493, 12269.) When $r_{an} = 1$, the generator typically has about $\sqrt{|D|}$ digits, and then the algorithm outlined above is the fastest known for computing it: as noted already, there is as yet no known analogue for E_D to the continued-fraction approach to Pell's equation (for which again the class number is usually 1 or a small power of 2, and fundamental units often have about $\sqrt{D}$ digits), and the traditional approach of making a few 2-descents and then searching exhaustively takes time exponential in $\sqrt{|D|}$. In practice the 2-descents also significantly streamline the computation of $\Sigma_1 P_1$ by reducing the precision required; e.g. if $|D| \equiv 7 \bmod 8$ is prime then the x-coordinate of $\Sigma_1 P_1$ must be of the form $x = (D^2 X^4 + 1)/(2DX^2)$ for some rational X which has only about 1/4 as many digits as x. This allowed us to compute that the x-coordinate of a generator of $E_{1063}(\mathbf{Q})$ mod torsion is $(1063^2 X^4 + 1)/(2 \cdot 1063 X^2)$ where

$$X = \frac{11091863741829769675047021635712281767382339667434645}{31734265754477218073520797732090001252280793677887},$$

a solution so large that to date it could not have been computed except by Heegner-point methods.

If $r_{an} > 1$ then $H = 0$ and $\Sigma_1 P_1$ is a torsion point. This can be detected by computing $\Sigma_1 P_1$ to low precision (in practice 19 digits, and a threshold of 10^{-12}, are more than enough) because there are only four rational torsion points on E_D. This is a very fast computation, requiring time only $O_\epsilon(|D|^{1/2+\epsilon})$, much less than the $O_\epsilon(|D|^{1+\epsilon})$ required to calculate $L'(E_D, 1)$ directly as in [BGZ]. We were thus able screen all squarefree $|D| < 2 \cdot 10^5$ congruent to 7 mod 8, and found that only 75 had $r_{an} \geq 3$, of which the smallest are

$$4199 = 13 \cdot 17 \cdot 19, \quad 4895 = 5 \cdot 11 \cdot 89, \quad \text{and} \quad 6671 = 7 \cdot 953.$$

For $|D| \equiv 5 \bmod 8$ we find 89 examples in the same range, beginning with

$$2605 = 5 \cdot 521, \quad 4669 = 7 \cdot 23 \cdot 29, \quad \text{and} \quad 8005 = 5 \cdot 1601;$$

and there are 173 examples with $|D| \equiv 6 \bmod 8$, starting with

$$1254 = 2 \cdot 3 \cdot 11 \cdot 19, \quad 2774 = 2 \cdot 19 \cdot 73, \quad 3502 = 2 \cdot 17 \cdot 13.$$

We have checked that in the first few cases the curve E_D has arithmetic rank 3. (See also [GeZ, p. 81].) The heuristics used above to surmise $H \ll_\epsilon |D|^{1/2+\epsilon}$ when $r(E_D) = 1$ indicate that if $r(E_D) \geq 3$ then E_D should have a nontorsion point of height $O_\epsilon(|D|^{1/6+\epsilon})$, considerably smaller than in the rank-1 case. Indeed on each of our $75 + 89 + 173$ curves E_D with analytic rank at least 3 we were able to find at least one nontorsion point by a rather simple-minded numerical search. Having already computed a nontorsion point such as $\Sigma_1 P_1$ on E_D for all other $|D| < 2 \cdot 10^5$ congruent to 5, 6, or 7 mod 8, we deduce that E_D has positive rank for all $|D| < 2 \cdot 10^5$ such that the functional equation of $L(E_D, \cdot)$ has sign -1.

4   The twisted Fermat cubics $x^3 + y^3 = A$

The question of which numbers can be written as the sum of two rational cubes leads to the family of elliptic curves

$$E_A : x^3 + y^3 = A.$$

We take the point at infinity as the origin of the curve. By absorbing cubic factors in A into x, y we may assume that A is a positive cube-free integer; we also assume that $A \geq 3$, since it is well known that other than the point at infinity $(1,1)$ is the only rational point on E_2, and $(1,0)$, $(0,1)$ the only rational points on E_1, the latter being of course the exponent-3 case of Fermat's "Last Theorem". For $A \geq 3$ the torsion group of E_A is trivial ([Se, p.212], attributed to A. Hurwitz), so A is the sum of two rational cubes if and only if E_A has positive arithmetic rank.

The family of curves E_A recalls in many ways the family of curves E_D of the last two sections; each E_A is an elliptic curve with complex multiplication and a twist (cubic, not quadratic) of a curve E_1 admitting a simple modular parametrization, this time by $X_0(27)$. The parity of $r_{an}(E_A)$ was computed by Birch and Stephens [BS]; in particular if $A > 3$ is prime, $r_{an}(E_A)$ is odd precisely when A is congruent to 4, 7, or 8 mod 9, and the same is true of $r_{an}(E_{A^2})$. Thus Birch and Stephens conjectured that every prime $p \equiv 4$, 7, or 8 mod 9 can be written as a sum of two rational cubes, a conjecture prefigured by Selmer [Se] and traditionally ascribed to Sylvester. We have proved this conjecture for the first two cases $p \equiv 4, 7$ mod 9, and also shown that p^2 is a sum of two cubes for any such p; e.g. for $p = 337$ the smallest solution is

$$337^2 = \frac{77739716821190096^3 - 5751462787384249^3}{1350225936494447^3}.$$

One might try to construct points on E_p by mimicking the Heegner-Birch approach to E_D: find a combination of CM points on $X_0(27)$ that sum to a point on E_1 defined over $\mathbf{Q}(\sqrt[3]{p})$. Indeed a similar construction starting from the parametrization $X_0(36) \to E_2$ yields non-torsion points on E_{2p} for certain primes p [Sa]. But all attempts to construct rational points on E_p along similar lines have produced only the zero point, a fact Gross can now predict from a conjectured generalization of the Gross-Zagier formula. This generalization also suggested how to modify the CM construction to obtain nontrivial points on E_p. Let ϖ be one of the prime factors of p in $\mathbf{Z}[e^{2\pi i/3}]$, chosen so that $\varpi \equiv 1$ mod 3. Then the curve E_ϖ has a modular parametrization by $X_1(9p)$ or $X_1(27p)$ defined over $\mathbf{Q}(e^{2\pi i/3})$. We find a point on the modular curve parametrizing a cyclic isogeny between elliptic curves with complex multiplication of discriminant -27, and show that the image of this point on E_ϖ is a nontorsion point defined over $\mathbf{Q}(e^{2\pi i/3}, \sqrt[3]{\varpi})$ which yields a nontrivial point of $E_{\varpi\bar\varpi} = E_p$ defined over $\mathbf{Q}$. Likewise we construct a nontrivial point of $E_{p^2}(\mathbf{Q})$ from a modular parametrization of E_{ϖ^2}. As in the previous section we find that this yields a practical algorithm with running time polynomial in p, dominated this time by

$O(pH)$ arithmetic operations to $O(H)$-digit precision, where H is the height of the constructed point, expected to be $O_\epsilon(p^{1/3+\epsilon})$ or $O_\epsilon(p^{2/3+\epsilon})$ for E_p, E_{p^2} respectively. This point should again be the $S = \pm\sqrt{|\text{III}|}$ multiple of the generator of $E_p(\mathbf{Q})$ or $E_{p^2}(\mathbf{Q})$, where S is usually 1 but occasionally larger: on E_{547} and E_{229^2} (the smallest instances) our construction gives not the generators

$$547 = 14^3 - 13^3, \quad 229^2 = \frac{154699^3 - 69515^3}{4004^3},$$

but twice those points, so we guess that $|\text{III}| = 4$ in these cases. So far no higher multiples have been observed, though presumably S and $|\text{III}|$ can be arbitrarily large.

Acknowledgements

Thanks to Benedict Gross for introducing me to the arithmetic of elliptic and modular curves and for many enlightening conversations on the Heegner-point construction and its variations. This work was made possible in part by funding from the Harvard Society of Fellows, the National Science Foundation, and the Packard Foundation; some of the computations were carried out using the GP/PARI and MACSYMA packages.

References

[Bi1] Birch, B.J.: Elliptic curves and modular functions. *Symp. Math.* **4** (1970), 27–37.

[Bi2] Birch, B.J.: Heegner points of elliptic curves. *Symp. Math.* **15** (1975), 441–445.

[BS] Birch, J., Stephens, N.M.: The parity of the rank of the Mordell-Weil group. *Topology* **5**, 295–299 (1966).

[BGZ] Buhler, J.P., Gross, B.H., Zagier, D.: On the conjecture of Birch and Swinnerton-Dyer for an elliptic curve of rank 3. *Math. of Computation* **44** (1985) #175, 473–481.

[Co] Cohen, H.: *A course in computational algebraic number theory.* Berlin: Springer 1993.

[CL] Cohen, H., Lenstra, H.W.: Heuristics on class groups of number fields. Lect. Notes in Math. **1068** (1984: Proceedings of Journées Arithmétiques 1983 at Noordwijkerhout), 33–62.

[Di] Dickson, L.E.: *History of the Theory of Numbers, Vol. II: Diophantine Analysis.* New York: Stechert 1934.

[GeZ] Gebel, J., Zimmer, H.G.: Computing the Mordell-Weil Group of an Elliptic Curve over **Q**. *CRM Proc. & Lect. Notes* **4** (1994), 61–83.

[GrZ] Gross, B.H., Zagier, D.: Heegner points and derivatives of L-series. *Invent. Math.* **84** (1986), 225–320.

[HMcC] Hafner, J., McCurley, K.: A rigorous subexponential algorithm for computation of class groups. *J. AMS* **2** (1989), 837–850.

[He] Heegner, K.: Diophantische Analysis und Modulfunktionen. *Math. Z.* **56** (1952), 227–253.

[Kob]   Koblitz, N.: *Introduction to elliptic curves and modular forms.* New York: Springer 1984.

[Kol]   Kolyvagin, V.A.: Euler systems. Pages 435–483 of *The Grothendieck Festscrhift* Vol. II, Birkhäuser: Boston 1990.

[Mo]   Monsky, P.: Mock Heegner points and congruent numbers. *Math. Z.* **204** (1990) #1, 45–67.

[Ru]   Rubin, K.: Tate-Shafarevich groups and L-functions of elliptic curves with complex multiplication. *Invent. Math.* **89** (1987), 527–560

[Sa]   Satgé, Ph.: Un analogue du calcul de Heegner. *Invent. Math.* **87**, 425–439 (1987).

[Se]   Selmer, E.S.: The Diophantine equation $ax^3 + by^3 + cz^3 = 0$. *Acta Math.* **85**, 203–362 (1951).

[Si]   Silverman, J.H.: *The Arithmetic of Elliptic Curves.* New York: Springer 1986.

[We]   Weil, A.: *Number theory: An approach through history; From Hammurapi to Legendre.* Boston: Birkhäuser, 1983.

Computing the Degree of a Modular Parametrization

J. E. Cremona

University of Exeter

Abstract. The Weil–Taniyama conjecture states that every elliptic curve $E/\mathbb{Q}$ of conductor N can be parametrized by modular functions for the congruence subgroup $\Gamma_0(N)$ of the modular group $\Gamma = PSL(2, \mathbb{Z})$. Equivalently, there is a non-constant map φ from the modular curve $X_0(N)$ to E. We present here a method of computing the degree of such a map φ for arbitrary N. Our method, which works for all subgroups of finite index in Γ and not just $\Gamma_0(N)$, is derived from a method of Zagier in [2]; by using those ideas, together with techniques which have been used by the author to compute large tables of modular elliptic curves (see [1]), we are able to derive an explicit and general formula which is simpler to implement than Zagier's. We discuss the results obtained, including several examples.

1   Introduction

The Weil–Taniyama conjecture states that every elliptic curve $E/\mathbb{Q}$ of conductor N can be parametrized by modular functions for the congruence subgroup $\Gamma_0(N)$ of the modular group $\Gamma = PSL(2, \mathbb{Z})$. Equivalently, there is a non-constant map φ from the modular curve $X_0(N)$ to E. We present here a method of computing the degree of such a map φ for arbitrary N. Our method is derived from a method of Zagier in [2]; by using those ideas, together with techniques which have been used by the author to compute large tables of modular elliptic curves (see [1]), we are able to derive an explicit formula which is in general much simpler to implement than Zagier's, for arbitrary subgroups of finite index in Γ. To implement this formula one needs to have explicit coset representatives for the subgroup, but it is not necessary to determine an explicit fundamental domain for its action on the upper half-plane $\mathcal{H}$. In particular, it is simple to implement for $\Gamma_0(N)$ for arbitrary N, in contrast with Zagier's formula which is only completely explicit for N prime.

In the following section, we review the necessary background on modular parametrizations of elliptic curves. In Section 3 we introduce some machinery concerning coset representatives and fundamental regions, and state the main result (Theorem 5). In Section 4 we discuss the implementation of the method for the case of $\Gamma_0(N)$, and the results of a systematic computation of the degree of the parametrization of all modular elliptic curves of conductor less than 2000.

2   Modular Parametrizations of Elliptic Curves

Let $\Gamma = PSL(2, \mathbb{Z})$ be the modular group, and Γ_0 a subgroup of Γ of finite index. Both act discretely on the upper half-plane $\mathcal{H}$ and the extended upper half-plane $\mathcal{H}^* = \mathcal{H} \cup \mathbb{Q} \cup \{\infty\}$ obtained by adjoining the cusps $\mathbb{Q} \cup \{\infty\}$, which form a single Γ-orbit. The quotient $X = X_{\Gamma_0} = \Gamma_0 \backslash \mathcal{H}^*$ can be given the structure of a Riemann surface; in the case we are most interested in, where Γ_0 is a congruence subgroup, X is also an algebraic curve defined over a number field, and is called a modular curve.

An elliptic curve E defined over $\mathbb{Q}$ is called a modular elliptic curve if there is a non-constant map $\varphi: X \to E$ for some modular curve X. The pull-back of the unique (up to scalar multiplication) holomorphic differential on E is then of the form $2\pi i f(\tau)d\tau$ where $f(\tau)$ is a holomorphic cusp form of weight 2 for Γ_0. According to the Weil–Taniyama conjecture, this should be the case for every elliptic curve defined over $\mathbb{Q}$, with $\Gamma_0 = \Gamma_0(N)$, where N is the conductor of E. Moreover, the cusp form $f(\tau)$ should be a newform in the usual sense.

We will suppose that we are given a cusp form $f(\tau)$ of weight 2 for Γ_0. Since the differential $f(\tau)d\tau$ is holomorphic, the function

$$\varphi_1(\tau) = 2\pi i \int_{\infty}^{\tau} f(\zeta)d\zeta \quad (\tau \in \mathcal{H}^*)$$

is well-defined (independent of the path from ∞ to τ). Also, for $\gamma \in \Gamma_0$, the function

$$\omega(\gamma) = \varphi_1(\gamma(\tau)) - \varphi_1(\tau) = 2\pi i \int_{\tau}^{\gamma(\tau)} f(\zeta)d\zeta$$

is independent of τ, and defines a function $\omega: \Gamma_0 \to \mathbb{C}$ which is a group homomorphism. The image Λ_f of this map will, under suitable hypotheses on f which we will assume to hold, be a lattice of rank 2 in $\mathbb{C}$, so that $E_f = \mathbb{C}/\Lambda_f$ is an elliptic curve. Hence φ_1 induces a map

$$\varphi: X = \Gamma_0 \backslash \mathcal{H}^* \to E_f = \mathbb{C}/\Lambda_f$$

via $\varphi(\tau \bmod \Gamma_0) = \varphi_1(\tau) \bmod \Lambda_f$.

The period map $\omega: \Gamma_0 \to \Lambda_f$ is surjective (by definition) and its kernel contains all elliptic and parabolic elements of Γ_0. We may write $\Lambda_f = \mathbb{Z}\omega_1 + \mathbb{Z}\omega_2$ with $\mathrm{Im}(\omega_2/\omega_1) > 0$. Then

$$\omega(\gamma) = n_1(\gamma)\omega_1 + n_2(\gamma)\omega_2$$

where $n_1, n_2: \Gamma_0 \to \mathbb{Z}$ are homomorphisms. It is important to observe here that these functions are explicitly and easily computable in terms of modular symbols: for the case $\Gamma_0 = \Gamma_0(N)$, see [1] for details. Alternatively, given sufficiently many Fourier coefficients of the cusp form $f(\tau)$ we may evaluate the period integrals $\varphi_1(\tau)$ to sufficient precision that (assuming that the fundamental periods ω_1 and ω_2 are also known to some precision) one can determine the values of $n_1(\gamma)$ and $n_2(\gamma)$ for all $\gamma \in \Gamma_0$. The latter approach is used in [2]. The advantage of the modular symbol approach here is that exact values are obtained directly, and that it is not necessary to compute (or even know) any Fourier coefficients of $f(\tau)$. On the other hand, it becomes computationally infeasible to carry out the modular symbol computations when the index of Γ_0 in Γ is too large, whereas the approximate approach can still be used, provided that one has an explicit equation for the curve E to hand, from which one can compute the periods and the Fourier coefficients in terms of traces of Frobenius (assuming that E is modular and defined over $\mathbb{Q}$.) This method was used, for example, to compute $\deg(\varphi)$ for the curve of rank 3 with conductor 5077, in [2].

The special case we are particularly interested in is where $\Gamma_0 = \Gamma_0(N)$ and $f(\tau)$ is a normalised newform for $\Gamma_0(N)$. Then $f(\tau)$ is a Hecke eigenform with rational integer eigenvalues and therefore rational integer Fourier coefficients. The periods of $2\pi i f(\tau)$ do in this case form a lattice Λ_f, and the modular elliptic curve $E_f = \mathbb{C}/\Lambda_f$ is defined over $\mathbb{Q}$ and has conductor N.

In order to compute the degree of the map $\varphi: X \to E_f$, the idea used in [2] is to compute the Petersson norm $||f||$ in two ways. The first way involves $\deg(\varphi)$ explicitly, while the second expresses it as a sum of terms involving periods, which can be evaluated as above.

Proposition 1. *Let $f(\tau)$ be a cusp form for Γ_0 as above, and $\varphi: X \to E_f$ the associated modular parametrization. Then*

$$4\pi^2||f||^2 = \deg(\varphi)\mathrm{Vol}(E_f) \ .$$

Remark. In terms of the fundamental periods ω_1, ω_2 of E_f, the volume is given by $\mathrm{Vol}(E_f) = |\mathrm{Im}\,(\overline{\omega_1}\omega_2)|$. More generally, if ω, $\omega' \in \Lambda_f$, with $\omega = n_1(\omega)\omega_1 + n_2(\omega)\omega_2$ and $\omega' = n_1(\omega')\omega_1 + n_2(\omega')\omega_2$, then (up to sign) we have

$$\mathrm{Im}\,(\overline{\omega}\omega') = \mathrm{Vol}(E_f) \cdot \begin{vmatrix} n_1(\omega) & n_1(\omega') \\ n_2(\omega) & n_2(\omega') \end{vmatrix} \ .$$

3 Coset Representatives and Fundamental Domains

Let $S = \begin{pmatrix} 0 & -1 \\ 1 & 0 \end{pmatrix}$ and $T = \begin{pmatrix} 1 & 1 \\ 0 & 1 \end{pmatrix}$ be the usual generators for Γ, so that S has order 2 and TS has order 3.

As fundamental domain for Γ we may take the triangular region $\mathcal{F}$ with vertices at 0, $\rho = (1 + i\sqrt{3})/2$, and ∞. Since TS fixes ρ and permutes 0, ∞ and 1 cyclically, the three transforms of $\mathcal{F}$ by I, TS and $(TS)^2$ fit together around ρ to form an "ideal triangle" $\mathcal{T}$ with vertices at 0, 1 and ∞. Let $\langle \gamma \rangle$ denote the transform of $\mathcal{T}$ by γ for $\gamma \in \Gamma$. Then these triangles $\langle \gamma \rangle$ form a triangulation of the upper half-plane $\mathcal{H}$, whose vertices are precisely the cusps: the vertices of $\langle \gamma \rangle$ are the cusps $\gamma(0)$, $\gamma(1)$ and $\gamma(\infty)$. Note that

$$\langle \gamma \rangle = \langle \gamma TS \rangle = \langle \gamma(TS)^2 \rangle$$

but that otherwise the triangles are distinct. The triangle $\langle\gamma\rangle$ has three (oriented) edges; in the modular symbol notation of [1], these are

$$\langle\gamma\rangle = \{\gamma(0), \gamma(\infty)\}\ ,$$
$$\langle\gamma TS\rangle = \{\gamma TS(0), \gamma TS(\infty)\} = \{\gamma(\infty), \gamma(1)\}\ ,$$

and

$$\langle\gamma(TS)^2\rangle = \{\gamma(TS)^2(0), \gamma(TS)^2(\infty)\} = \{\gamma(1), \gamma(0)\}\ .$$

Here the modular symbol $\{\alpha, \beta\}$ denotes a geodesic path in $\mathcal{H}^*$ from α to β.

Assume, for simplicity, that Γ_0 has no non-trivial elements of finite order, i.e., no conjugates of either S or TS. (This assumption is merely for ease of exposition; in fact, it is easy to see that elliptic elements of Γ_0 contribute nothing to the formulas in Theorems 4 and 5 below in any case.) Choose, once and for all, a set $\mathcal{S}$ of right coset representatives for Γ_0 in Γ, such that $\gamma \in \mathcal{S} \Rightarrow \gamma TS \in \mathcal{S}$; this is possible since, by hypothesis, Γ_0 contains no conjugates of TS.

Let $\mathcal{S}'$ be a subset of $\mathcal{S}$ which contains exactly one of each triple γ, γTS, $\gamma(TS)^2$, so that $\mathcal{S} = \mathcal{S}' \cup \mathcal{S}'TS \cup \mathcal{S}'(TS)^2$. Then a fundamental "domain" for the action of Γ_0 on $\mathcal{H}$ is given by

$$\mathcal{F}_{\Gamma_0} = \bigcup_{\gamma \in \mathcal{S}'} \langle\gamma\rangle\ .$$

In general, this set need not be connected, but this does not matter for our purposes: it can be treated as a disjoint union of triangles, whose total boundary is the sum of the oriented edges $\langle\gamma\rangle$ for $\gamma \in \mathcal{S}$.

The key idea in our algebraic reformulation of Zagier's method is to make use of the coset action of Γ on the set $\mathcal{S}$. We introduce notation for the actions of the generators S and T of Γ.

Action of S. For each $\gamma \in \mathcal{S}$ we set $\gamma S = s(\gamma)\sigma(\gamma)$, where $s\colon \mathcal{S} \to \Gamma_0$ is a function and $\sigma\colon \mathcal{S} \to \mathcal{S}$ is a permutation. Since S^2 is the identity, the same is true of σ, and $s(\sigma(\gamma)) = s(\gamma)^{-1}$. For brevity we will write $\gamma^* = \sigma(\gamma)$, so that $\gamma^{**} = \gamma$ for all $\gamma \in \mathcal{S}$.

Note that the triangles $\langle\gamma\rangle$ and $\langle\gamma S\rangle$ are adjacent in the triangulation of $\mathcal{H}$, since they share the common side $\langle\gamma\rangle = \{\gamma(0), \gamma(\infty)\} = -\langle\gamma S\rangle$. (Here the minus sign denotes reverse orientation.) However, since in general we do not have $\gamma S \in \mathcal{S}$, in the fundamental domain $\mathcal{F}_{\Gamma_0}$ for Γ_0 it is the triangles $\langle\gamma\rangle$ and $\langle\gamma^*\rangle$ which are glued together by the element $s(\gamma) \in \Gamma_0$ which takes $\langle\gamma^*\rangle$ to $-\langle\gamma\rangle$ (the orientation is reversed).

Action of T. Similarly, for $\gamma \in \mathcal{S}$ we set $\gamma T = t(\gamma)\tau(\gamma)$ with $t(\gamma) \in \Gamma_0$ and $\tau(\gamma) \in \mathcal{S}$. The permutation τ of $\mathcal{S}$ plays a vital part in what follows. Lemma 2 will not be used later, but is included for its own interest as it explains the geometric significance of this algebraic permutation.

Lemma 2. *(a) Two elements γ and γ' of $\mathcal{S}$ are in the same τ-orbit if and only if the cusps $\gamma(\infty)$ and $\gamma'(\infty)$ are Γ_0-equivalent.*

(b) The length of the τ-orbit of an element $\gamma \in \mathcal{S}$ is the width of the cusp $\gamma(\infty)$ of Γ_0.

Thus there is a one–one correspondence between the orbits of τ on $\mathcal{S}$ and the classes of Γ_0-inequivalent cusps, with the length of each orbit being the width of the corresponding cusp.

In each τ-orbit in $\mathcal{S}$, we choose an arbitrary base-point γ_1, and set $\gamma_{j+1} = \tau(\gamma_j)$ for $1 \leq j \leq k$, where k is the length of the orbit and $\gamma_{k+1} = \gamma_1$. Thus $\gamma_j T = t(\gamma_j)\gamma_{j+1}$, so that

$$\gamma_1 T^j = t(\gamma_1)t(\gamma_2)\ldots t(\gamma_j)\gamma_{j+1} .$$

In particular, $\gamma_1 T^k = \gamma_0 \gamma_1$, where

$$\gamma_0 = t(\gamma_1)t(\gamma_2)\ldots t(\gamma_k) \in \Gamma_0 .$$

Since γ_0 is parabolic, we obtain the following.

Lemma 3.

$$\sum_{j=1}^{k} \omega(t(\gamma_j)) = 0 .$$

Write $\gamma \prec \gamma'$ if γ and γ' are in the same τ-orbit in $\mathcal{S}$, and γ precedes γ' in the fixed ordering determined by choosing a base-point for each orbit. In the notation above, $\gamma \prec \gamma'$ if and only if $\gamma = \gamma_i$ and $\gamma' = \gamma_j$ where $1 \leq i < j \leq k$.

We can now state our main results.

Theorem 4. *Let f be a cusp form of weight 2 for Γ_0 with associated period function $\omega : \Gamma_0 \to \mathbb{C}$. Then (the square of) the Petersson norm of f is given by*

$$\|f\|^2 = \frac{1}{8\pi^2} \sum_{\gamma \prec \gamma'} \mathrm{Im}(\omega(t(\gamma))\overline{\omega(t(\gamma'))}) .$$

Here the sum is over all ordered pairs $\gamma \prec \gamma'$ in $\mathcal{S}$ which are in the same orbit of the permutation τ of $\mathcal{S}$ induced by right multiplication by T.

Combining this result with Proposition 1 of the previous section, we immediately obtain our explicit formula for the degree of the modular parametrization φ.

Theorem 5. *With the above notation,*

$$\deg(\varphi) = \frac{1}{2\mathrm{Vol}(E_f)} \sum_{\gamma \prec \gamma'} \mathrm{Im}(\omega(t(\gamma))\overline{\omega(t(\gamma'))}) = \frac{1}{2} \sum_{\gamma \prec \gamma'} \begin{vmatrix} n_1(t(\gamma)) & n_1(t(\gamma')) \\ n_2(t(\gamma)) & n_2(t(\gamma')) \end{vmatrix} .$$

Hence to compute $\deg(\varphi)$, we only have to compute the right coset action of T on an explicit set $\mathcal{S}$ of coset representatives for Γ_0 in Γ, and evaluate the integer-valued functions n_1 and n_2 on each of the matrices $t(\gamma)$ for $\gamma \in \mathcal{S}$. In the case of $\Gamma_0(N)$, these steps can easily be carried out within the framework described in [1], and we will give some further details in Section 4 below.

Remarks. 1. The formula given in Theorem 5 expresses $\deg(\varphi)$ explicitly as a sum which can be grouped as a sum of terms, one term for each cusp, by collecting together the terms for each τ-orbit. It is not at all clear what significance, if any, can be given to the individual contributions of each cusp to the total. In Section 4 we give an example in the case $\Gamma_0 = \Gamma_0(210)$.

2. The form of our formula is identical to the one in [2]. However, we should stress that in [2], the analogue of our coset action τ is defined not algebraically, as here, but geometrically, as a permutation of the edges of a fundamental polygonal domain for Γ_0 (and dependent on the particular fundamental domain used). Then it becomes necessary to have an explicit picture of such a fundamental domain, including explicit matrices which identify the edges of the domain in pairs. This is only carried out explicitly in [2] in the case $\Gamma_0 = \Gamma_0(N)$ where N is a prime. In our formulation, the details are all algebraic rather than geometric, which makes the evaluation of the formula more practical to implement. Also, we have the possibility of evaluating the functions n_1 and n_2 exactly using modular symbols, instead of using numerical evaluation of the periods, which reduces the computation of $\deg(\varphi)$ entirely to linear algebra and integer arithmetic.

4   The Case of $\Gamma_0(N)$: Implementation and Results

In this section we discuss the case $\Gamma_0 = \Gamma_0(N)$ in greater detail. We have implemented the algorithm in this case as part of our suite of modular elliptic curves programs which were described in [1]; to date (December 1993) we have computed all modular elliptic curves of conductors up to $N = 2000$, together with the degrees of their modular parametrizations (in all but a very small number of cases). It is not practical to give complete tables of these results here, as there are approximately 6000 curves (up to isogeny) with conductor up to 2000. Instead, we give results in a selection of specific cases. A complete table of results is available from the author, from which phenomena of interest (such as the growth of $\deg(\varphi)$ in terms of N, or the set of primes dividing $\deg(\varphi)$) can be obtained.

Let N be an arbitrary positive integer. The right coset representatives of $\Gamma_0(N)$ in Γ are in bijective correspondence with the set $P^1(N) = P^1(\mathbb{Z}/N\mathbb{Z})$ of "M-symbols" $(c : d)$, where $c, d \in \mathbb{Z}$, $\gcd(c, d) = 1$, and

$$(c : d) = (c' : d') \iff cd' \equiv c'd \pmod{N} .$$

We will also write $(c, d) \equiv (c', d')$ for this equivalence relation on $\mathbb{Z}^2$. The correspondence with right cosets is given by

$$(c : d) \leftrightarrow \Gamma_0(N) \begin{pmatrix} a & b \\ c & d \end{pmatrix}$$

where $a, b \in \mathbb{Z}$ are chosen so that $ad - bc = 1$, different choices of a, b giving the same right coset. The right coset action of Γ on $P^1(N)$ is given simply by

$$(c : d) \begin{pmatrix} p & q \\ r & s \end{pmatrix} = (cp + dr : cq + ds) ;$$

in particular, we have $\sigma(c:d) = (c:d)S = (d:-c)$ and $\tau(c:d) = (c:d)T = (c:c+d)$.

Lemma 6. *The length of the τ-orbit containing $(c:d) \in P^1(N)$ is $N/\gcd(N,c^2)$.*

In our earlier work [1], where we used M-symbols to compute modular elliptic curves, it was immaterial exactly which coset representatives were used, or in practice which pair (c,d) was used to represent the M-symbol $(c:d)$. For the application of Theorem 5, however, we must ensure that our set is closed under right multiplication by TS, where $(c:d)TS = (c+d:-c)$, unless $(c:d)$ is fixed by TS, which is if and only if $c^2 + cd + d^2 \equiv 0 \pmod{N}$. Thus each M-symbol $(c:d)$ will be represented by a specific pair $(c,d) \in \mathbf{Z}^2$ with $\gcd(c,d) = 1$, in such a way that our set S of representatives contains the pairs $(c+d,-c)$ and $(-d,c+d)$ whenever it contains (c,d), unless $(c:d)$ is fixed by TS. (Even when working with pairs $(c,d) \in \mathbf{Z}^2$ we will identify (c,d) and $(-c,-d)$.)

Fixing these triples of pairs (c,d) corresponds to fixing the triangles $\langle\gamma\rangle$ which form a (possibly disconnected) fundamental domain for $\Gamma_0(N)$. If $\gamma = \begin{pmatrix} a & b \\ c & d \end{pmatrix}$, the pair (c,d) corresponds to the directed edge $\{\gamma(0),\gamma(\infty)\} = \{b/d, a/c\}$. For this reason, we will refer to the pairs (c,d) as edges, and the triples of pairs as triangles. Right multiplication by TS corresponds geometrically to moving round to the next edge of the triangle, while right multiplication by S corresponds to moving across to the next triangle $\langle\gamma^*\rangle$ adjacent to the current one. The τ-action is given by composing these, taking $(c:d)$ (or edge $\{b/d, a/c\}$) to the symbol $(c:d)T = (c:c+d)$ with corresponding edge $\{(a+b)/(c+d), a/c\}$, up to translation by an element of $\Gamma_0(N)$. Note how in this operation the endpoint at the cusp a/c is fixed, as in Lemma 2 above.

We may therefore proceed as follows. For each orbit, start with a standard pair (c,d), chosen in an M-symbol class $(c:d)$ not yet handled. Apply T to obtain the pair $(c,c+d)$. If this pair is the standard representative for the class $(c:c+d)$, we need take no action and may continue with the orbit. But if $(c,c+d) \equiv (r,s)$, say, with $(r,s) \in S$, then we must record the "glueing matrix" δ, where

$$\delta = \begin{pmatrix} a & a+b \\ c & c+d \end{pmatrix} \begin{pmatrix} p & q \\ r & s \end{pmatrix}^{-1} \in \Gamma_0(N) ,$$

where $ad - bc = ps - qr = 1$, whose period $\omega(\delta)$ will contribute to the partial sum for this orbit. When this happens, we say that the orbit has a "jump" at this point. Different choices for a, b, p and q only change δ by parabolic elements, and so do not affect the period $\omega(\delta)$. We continue until we return to the starting pair, and then move to another orbit, until all M-symbols have been used. As checks on the computation we may used Lemmas 2 and 6: the length of the orbit starting at (c,d) can be precomputed as $N/\gcd(N,c^2)$, and the number of orbits is the number of $\Gamma_0(N)$-inequivalent cusps.

Example 1: $N = 11$. The 12 symbols form 4 triangles which we choose as follows:

$$(1,0), (-1,1), (0,1); \qquad (1,1), (-2,1), (-1,2) ;$$
$$(1,2), (-3,1), (-2,3); \qquad (1,3), (-4,1), (-3,4) .$$

There are two τ-orbits, corresponding to the two cusps at ∞ (of width 1) and at 0 (of width 11). The first contributes nothing. The second is as follows:

$$(1,0) \mapsto (1,1) \mapsto (1,2) \mapsto (1,3) \mapsto (1,4) \equiv (-2,3) \mapsto (-2,1)$$
$$\mapsto (-2,-1) \equiv (-3,4) \mapsto (-3,1) \mapsto (-3,-2) \equiv (-4,1)$$
$$\mapsto (-4,-3) \equiv (-1,2) \mapsto (-1,1) \mapsto (1,0) .$$

There are four jump matrices coming from the above sequence. From $(1,4) \equiv (-2,3)$ we obtain

$$\delta_1 = \begin{pmatrix} 0 & -1 \\ 1 & 4 \end{pmatrix} \begin{pmatrix} 1 & -2 \\ -2 & 3 \end{pmatrix}^{-1} = \begin{pmatrix} -2 & -1 \\ 11 & 5 \end{pmatrix} ;$$

the others are $\delta_2 = \begin{pmatrix} 4 & 1 \\ 11 & 3 \end{pmatrix}$, $\delta_3 = \begin{pmatrix} -5 & -1 \\ 11 & 2 \end{pmatrix}$ and $\delta_4 = \begin{pmatrix} -3 & 1 \\ 11 & -4 \end{pmatrix}$. Hence

$$\deg(\varphi) = \frac{1}{2\mathrm{Vol}(E)} \sum_{1 \le i < j \le 4} \mathrm{Im}(\omega(\delta_i)\overline{\omega(\delta_j)}) .$$

Now by using modular symbols, we can compute the coefficients of $\omega(\delta_i)$ with respect to a period basis ω_1, ω_2, to obtain

$$\omega(\delta_1) = -\omega_1; \qquad \omega(\delta_2) = -\omega_2 ;$$
$$\omega(\delta_3) = \omega_1; \qquad \omega(\delta_4) = \omega_2 .$$

Hence we obtain $\deg(\varphi) = \frac{1}{2}(+1 + 0 + (-1) + 1 + 0 + 1) = 1$. Of course, this answer was obvious a priori, since the modular curve $X_0(11)$ has genus 1, so that φ is the identity map in this case. The curve (11A1 in [1]) has coefficients $[a_1, a_2, a_3, a_4, a_6] = [0, -1, 1, -10, -20]$.

Example 2: $N = 26 = 2 \cdot 13$. Here the genus is 2 and there are two newforms. Of the four cusps, only $\frac{1}{2}$ (of width 13) contributes to $\deg(\varphi)$, which is 2 in both cases. The curves are 26A1= $[1,0,1,-5,-8]$ and 26B1= $[1,-1,1,-3,3]$.

Example 3: $N = 30 = 2 \cdot 3 \cdot 5$. Here the genus is 3, there are two oldforms from level 15 and a newform. The cusps $\frac{1}{2}, \frac{1}{5}$ and $\frac{1}{6}$ contribute respectively 1, $\frac{1}{2}$, and $\frac{1}{2}$ to $\deg(\varphi)$ which equals 2. The curve is 30A1= $[1,0,1,1,2]$.

Example 4: $N = 210 = 2 \cdot 3 \cdot 5 \cdot 7$. There are five newforms here giving five curves, A—E. There are 16 cusps, namely $\frac{1}{d}$ (of width $210/d$) for $d \mid 210$. The contributions to $\deg(\varphi)$ are as follows:

d	A	B	C	D	E
1	0	0	0	0	0
2	10	12	6	2	6
3	2	27	2	0	16
5	3	−5	5/2	4	−5/2
6	14	21	19/2	1	89/2
7	10	−13	4	0	49
10	9	8	5/2	4	3/2
14	−1	3	−5/2	0	21/2
15	2	19	7/2	3	27/2
21	3	4	5/2	0	3/2
30	2	8	2	2	0
35	−6	12	0	0	−12
42	0	0	0	0	0
70	0	0	0	0	0
105	0	0	0	0	0
210	0	0	0	0	0
Total = $\deg(\varphi)$	48	96	32	16	128

The curves are A=210A1= $[1, 0, 0, -41, -39]$, B=210B1= $[1, 0, 1, -498, 4228]$, C=210C1= $[1, 1, 1, 10, -13]$, D = 210D1 = $[1, 1, 0, -3, -3]$ and E = 210E1 = $[1, 0, 0, 210, 900]$.

References

1. J. E. Cremona, Algorithms for Modular Elliptic Curves, Cambridge University Press, 1992.
2. D. Zagier, Modular Parametrizations of Elliptic Curves, Canadian. Math. Bull. (1985) **28**, 372–384.

Galois Representations from the Cohomology of SL(3,$\mathbb{Z}$)

Mark McConnell

Dept. of Mathematics, Oklahoma State University, Stillwater, OK 74078-0613

Conjecturally, by the Langlands philosophy, a cuspidal cohomology class of level N for SL($n, \mathbb{Z}$) should have an attached Galois representation. This is a finite-dimensional l-adic representation π of Gal($\bar{\mathbb{Q}}/\mathbb{Q}$), unramified for primes p not dividing l and N, for which the image of a Frobenius element for p is related to the p-th Hecke eigenvalues of the class. By a conjecture of Ash [1], the same should hold for l-torsion cohomology classes α.

Given α, finding the representation π is equivalent to finding π's fixed field K, a finite Galois extension of $\mathbb{Q}$ with group G. To find K, one must solve the inverse Galois problem for G, with the extra constraints that the primes p must split in K in ways determined by the Hecke data.

Computational number theory gives interesting techniques for solving this problem. When G is solvable, as in [2] [3], one constructs a solvable tower $K = K_m \supset K_{m-1} \supset \cdots \supset K_1 \supset \mathbb{Q}$ from the bottom up. This involves knowledge of the rings of integers in some of the K_j (including the class group and generators of principal ideals), plus work with the integral theory of quadratic forms [4] [5]. When G is not solvable, one can use rigidity or other methods to look for K.

Ash and I [2] have found non-trivial α of level N for SL(3), for a range of N and l, and have computed their Hecke eigenvalues for $p \leq 97$. In the cases we could check, there did seem to be a Galois representation π attached to α. All our π so far have small G, such as S_3, $\hat{A}_4$, or A_5. These call for a mixture of computational and non-computational techniques. We predict other π will have non-solvable G equal to SL(3, q) for $q = 3, 7, 17$ and 23.

References

1. Ash, A.: Galois representations attached to mod p cohomology of GL($n, \mathbb{Z}$). Duke J. **65** (1992) 235–255
2. Ash, A., McConnell, M.: Experimental indications of three-dimensional Galois representations from the cohomology of SL(3, $\mathbb{Z}$). Journal of Experimental Math. **1** (1992) 209–223
3. Ash, A., Pinch, R., Taylor, R.: An $\hat{A}_4$ extension of $\mathbb{Q}$ attached to a non-selfdual automorphic form on GL(3). Math. Annalen **291** (1991) 753–766
4. Crespo, T.: Explicit construction of $\tilde{A}_n$ type fields. J. Algebra **127** (1989) 452–461
5. Serre, J.-P.: L'invariant de Witt de la forme Tr(x^2). Comm. Math. Helv. **59** (1984) 651–676

AN ANALYSIS OF THE GAUSSIAN ALGORITHM
FOR LATTICE REDUCTION

Hervé Daudé[1], Philippe Flajolet[2], and Brigitte Vallée[3]

[1] Département de Mathématiques, Université de Provence,
Case 96, 3 Place Victor Hugo F-13331 Marseille Cedex 3 (France)
[daude@gyptis.univ-mrs.fr].
[2] INRIA-Rocquencourt, F-78153 Le Chesnay (France),
[Philippe.Flajolet@inria.fr].
[3] Département d'Informatique,Université de Caen, F-14032 Caen (France),
[Brigitte.Vallee@univ-caen.fr].

Abstract. The Gaussian algorithm for lattice reduction in dimension 2 (under both the standard version and the centered version) is analysed. It is found that, when applied to random inputs, the complexity is asymptotically constant, the probability distribution decays geometrically, and the dynamics is characterized by a conditional invariant measure. The proofs make use of connections between lattice reduction, continued fractions, continuants, and functional operators. Detailed numerical data are also presented.

1 Introduction

The lattice reduction problem consists in finding a short basis of a lattice of Euclidean space given a (usually skew) basis. This reduction problem is well-known to be central to many areas of approximation and optimization with deep consequences in computational number theory, cryptography, and symbolic computation.

In dimension $d = 1$, lattice reduction may be viewed as a mere avatar of the Euclidean GCD algorithm and of continued fraction expansions. Lattice reduction *per se* really started with Gauss who gave an algorithm that solves the problem exactly using what resembles a lifting of the Euclidean algorithm to 2–dimensional lattices. In recent times, an important discovery was made by Lenstra, Lenstra and Lovász in 1982 [4, 12, 25]; their algorithm, called the LLL algorithm, is able to find reduced bases in all dimensions $d \geq 3$. The LLL algorithm itself proceeds by stages based on the Gaussian algorithm as the main reduction step.

The Euclidean algorithm and the continued fraction algorithm are by now reasonably well understood as regards complexity questions. Knuth's book [10] provides a detailed account till 1981. From results of Lamé, Dupré, Heilbronn, Dixon, Wirsing, Babenko, and Hensley, the following facts are known. The worst case complexity of the Euclidean algorithm is $\mathcal{O}(\log N)$ when applied to integers at most N (Lamé and Dupré); the average case on random inputs is also logarithmic (Dixon, Heilbronn); the distribution of the number of iterations obeys in the asymptotic limit a normal law with a variance that is logarithmic (a recent result of Hensley).

There are some deep connections between these properties and an invariant measure for the continued fraction transformation whose existence was first conjectured by Gauss and proved in this century by Lévy and Kuzmin. Most of these results are obtained by means of functional operators related to continued fractions and continuants of which extensive use will be made here. We refer in particular to the works of Wirsing [24], Babenko [1], Mayer [14, 15, 16, 17], and Hensley [8].

This paper provides a detailed analysis of the Gaussian algorithm, both in the average case and in probability. Like its one–dimensional counterpart, the algorithm is known to be of worst–case logarithmic complexity, a result due to Lagarias [11], with best possible bounds being provided by Vallée [21] and Kaib-Schnorr [9]. The probabilistic behaviour of the Gaussian algorithm turns out to be appreciably different however. The main results of the paper are as follows.

— The average–case complexity of the Gaussian algorithm (measured in the number of iterations performed) is asymptotically constant, and thus essentially independent on the size of the input vectors.
— The distribution of the number of iterations is closely approximated by a geometric law.
— The dynamics of the algorithm is governed by a conditional invariant measure that constitutes the analogue of the invariant measure first observed by Gauss for continued fractions.

Precise characterizations of the behaviour of the algorithm are given here. In particular the geometric rate of decrease of the distribution of costs and the conditional invariant measure are expressed simply in terms of spectral properties of an operator that generalizes the operator associated with Euclid's algorithm.

In this paper, we mostly focus on the analysis of what we call the "standard" version of the Gaussian reduction algorithm, which generalizes the standard Euclidean algorithm. Another often encountered version, called here the "centered" version, is analogous to the centered Euclidean algorithm and is amenable to a similar treatment as we briefly explain at the end of the paper.

Our analytic results are naturally expressed as multiple infinite sums involving the continuants of continued fraction theory. As such sums tend to be rather slowly convergent, some attention is also paid to obtaining precise estimates by means of simple convergence acceleration techniques. For instance, we establish that the average case complexity of the "inner part" of the algorithm is asymptotic to the constant $\mu = 1.35113\,15744\ldots$.

On average, the Gaussian algorithm is thus of complexity $\mathcal{O}(1)$, which is of an order different from the worst-case. The case of dimension $d = 2$ therefore departs significantly from its 1–dimensional analogue, and it would be of interest to determine to which extent such a phenomenon propagates to higher dimensions. Our analytic knowledge of the LLL algorithm in higher dimensions is of course less advanced, but Daudé and Vallée [6] already succeeded in proving that the LLL algorithm, when applied to d–dimensional lattices, has an

 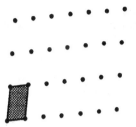

Fig. 1. A lattice and two of its bases represented by the parallelogram they span. The first basis is skew, the second one is minimal (reduced).

---

average–case complexity that is bounded from above by a constant K_d, where $K_d = \mathcal{O}(d^2 \log d)$. The present work thus fits as a component of a more global enterprise whose aim is to understand theoretically why the LLL algorithm performs in practice much better than worst–case bounds predict, and to quantify precisely the probabilistic behaviour of lattice reduction in higher dimensions.

2 Lattice reduction in dimension 2

*Lattices and bases.*This paper addresses specifically the reduction of 2–dimensional lattices. A *lattice* of rank 2 in the complex plane $\mathbb{C}$ is the set $\mathcal{L}$ of elements of $\mathbb{C}$ ("vectors") defined by

$$\mathcal{L} = \mathbb{Z}u \oplus \mathbb{Z}v = \{\lambda u + \mu v \mid \lambda, \mu \in \mathbb{Z}\},$$

where (u, v), called a *basis*, is a pair of $\mathbb{R}$–linearly independent elements of $\mathbb{C}$.

A lattice is generated by infinitely many bases that are related to each other by integer matrices of determinant ± 1. Amongst all the bases of $\mathcal{L}$, some, called *minimal*, enjoy the property of being formed with a shortest vector u of the lattice and another vector v which is shortest in the set of all vectors independent of u. Minimality is the specialization to dimension 2 of the general notion of reduced basis in arbitrary dimensions. A minimal basis (u, v), when it is in addition *acute*, is characterized by the two simultaneous conditions:

$$(I_1): \ \left|\frac{v}{u}\right| \geq 1 \quad \text{and} \quad (I_2): \ 0 \leq \Re\left(\frac{v}{u}\right) \leq \frac{1}{2}. \tag{1}$$

The angle between the two vectors of a minimal acute basis thus lies between $\frac{\pi}{3}$ and $\frac{\pi}{2}$.

The Gaussian reduction schema. A reduction algorithm takes an arbitrary basis of a lattice and determines another basis that is minimal. The Gaussian algorithm is a reduction algorithm whose principle consists in satisfying the two simultaneous conditions of (1). Condition I_1 is satisfied by exchanges between vectors, then condition I_2 is satisfied by an integral translation of the longer vector v parallel to the shorter vector u. The schema underlying this reduction process is then the following.

Input: an acute basis (u, v) of $\mathcal{L}$.
Output: a minimal acute basis (u, v) of $\mathcal{L}$.
    repeat
        (i). If $|u| > |v|$, then exchange u and v so as to satisfy condition I_1;
        (ii). Translate v parallel to u: $v := v - m\,u$ for some $m \in \mathbb{N}$, so as
            to satisfy condition I_2; if (u, v) is not acute then change v to $-v$;
    until $|v| \geq |u|$.

The complex framework. Many structural characteristics of lattices and bases are invariant under linear transformations —similarity transformations in geometric terms— of the form $S_\lambda \ : \ z \ \mapsto \ \lambda z$ with $\lambda \in \mathbb{C}$. An instance is the characterization of minimal acute bases that only depends on the ratio v/u. It is thus natural to consider lattices and bases taken up to equivalence under linear transformation (similarity). For such similarity invariant properties, it is sufficient to restrict attention to lattices generated by a basis of the form $(1, z)$. In that case, the property for a basis to be minimal and acute corresponds to the fact that z belongs to the so–called *fundamental domain* $\mathcal{F} = \{z \mid |z| \geq 1 \text{ and } 0 \leq \Re(z) \leq \frac{1}{2}\}$. Such a domain is familiar from the theory of modular forms [20] or the reduction theory of quadratic forms [19].

The Gaussian algorithm precisely has the property that its execution trace is invariant under lattice similarity. Let $(u_0, v_0), \ldots, (u_k, v_k)$ be the sequence of bases constructed by the Gaussian algorithm. We associate to it the sequence $(1, z_0), \ldots, (1, z_k)$ where $z_j = v_j/u_j$. The geometric transformation effected by the each step of the algorithm consists of an exchange $(u, v) \mapsto (v, u)$, a translation $v \mapsto v - m\,u$, and a possible sign change $v \mapsto \varepsilon v$ with $\varepsilon = \pm 1$. In the complex framework, this corresponds to an inversion $S : z \mapsto 1/z$, followed by a translation $z \mapsto T^{-m}z$ with $T(z) = z + 1$, and by a possible sign change $z \mapsto J_\varepsilon z$ where $J_\varepsilon(z) = \varepsilon z$. In this context, the Gaussian algorithm aims at realizing directly the conditions by bringing $z = v/u$ in the strip $\widetilde{\mathcal{B}} = \{0 \leq \Re(z) \leq \frac{1}{2}\}$.

The Gaussian reduction schema that we have just described involves a sign-changing operation $(v \mapsto -v)$. We introduce below a variant of the algorithm —the "standard" algorithm— that has the advantage of avoiding this operation whose presence complicates the analysis. We propose to return to the original Gaussian algorithm in Section 7.

The standard algorithm. The next sections are devoted to the analysis of a variant of the Gaussian algorithm that is directed towards bringing z inside the strip $\mathcal{B} = \{0 \leq \Re(z) \leq 1\}$. In order to do so, it suffices to consider a transformation U formed with an inversion S and a translation T^{-m} aimed at bringing z into $\mathcal{B}$. It is readily realized that this is achieved by the transformation

$$U(z) = \frac{1}{z} - \lfloor \Re(\frac{1}{z}) \rfloor,$$

with $\lfloor u \rfloor$ the integer part of u. This transformation U is an extension to the complex domain of the operation defining standard continued fraction expansions.

In the rest of the paper, we assume that the Gaussian algorithm is applied to complex numbers z such that $\Im(z) \neq 0$, which corresponds to nondegenerate lattices. One also operates with bases that are acute, so that z belongs to the half-plane $\Re(z) \geq 0$. For reasons explained below, see Eq. (2), it suffices to consider the situation where the reduction algorithm takes as input complex numbers from the disk $\mathcal{D}$ of diameter $[0, 1]$. The transformation U is then iterated till exit from that disk. This defines an algorithm called the *standard Gaussian algorithm* (*SGA*) because of its close connection with standard continued fractions and the standard Euclidean algorithm.

> **Algorithm** $SGA(z :$ complex$)$ [Standard Gaussian Algorithm]
> **Input:** $z \in \mathcal{D}$ (the disk of diameter $[0, 1]$)
>   while $(z \in \mathcal{D})$ do $z := U(z)$;
> **Output:** $z \in \mathcal{B} \setminus \mathcal{D}$. (the strip $\mathcal{B}$ is defined by $0 \leq \Re(z) \leq 1$)

For this algorithm, upon exit from the main iteration loop, it is no longer true that z belongs to the fundamental domain $\mathcal{F}$. However, z then lies in the union of six simple transforms of $\mathcal{F}$, namely

$$\mathcal{B} \setminus \mathcal{D} = \mathcal{F} \cup S\mathcal{F} \cup SJ\mathcal{F} \cup ST\mathcal{F} \cup TJ\mathcal{F} \cup STJ\mathcal{F}. \tag{2}$$

Thus simply adding a 6-way test produces an algorithm whose output is an element of $\mathcal{F}$. In addition, the analysis of the full reduction algorithm obtained in this way is then only a trivial variant of the analysis of the core algorithm SGA.

Probabilistic models. The question addressed here is the estimation of the number L of iterations performed by the standard algorithm. The model considered is in essence equivalent to applying the reduction algorithm to random bases, where similar bases are identified.

The *continuous model* is defined by the fact that the inputs are taken uniformly over the definition domain $\mathcal{D}$. The eventual goal is to analyse the behaviour of the algorithm under a *discrete model* where inputs are members of $\mathbb{Q}(i)$ of the form $\mathbb{Q}^{\langle N \rangle}(i) = \{\frac{a}{N} + i\frac{b}{N} \mid b \neq 0\}$, suitably restricted to $\mathcal{D}$. The random variable $L^{\langle N \rangle}$ then depends on N. However, as N gets large, it converges, both in moments and distribution, to its continuous counterparts, a fact to be proved in Section 5.

Thus, the results to be enounced later for the continuous model —that the average number of iterations is constant and that the probability distribution admits exponential tails — carries over to the more accurate discrete model. In other words, the behaviour of lattice reduction in dimension 2 is essentially insensitive to the size of the input vectors. This is a notable difference with the one-dimensional case of Euclid's algorithm.

3 Continued fractions and lattice reduction

The Gaussian algorithm is closely related to the linear fractional transformations (also called homographies) that are associated to continued fractions, and thus also to the classical continuant polynomials. In this way, a first analysis of the probability distribution and of the average cost of the algorithm can be given.

The fundamental disks. In its complex formulation, the algorithm SGA produces a sequence $z_0, z_1, \ldots, z_k$ of transforms of $z_0 \in \mathcal{D}$ obtained by iterating the transformation U. As we saw, each step corresponds to a particular transformation

$$z_{j+1} = -m_j + \frac{1}{z_j} \qquad \text{or} \qquad z_j = \frac{1}{m_j + z_{j+1}}. \tag{3}$$

While z_j is in $\mathcal{D}$, $1/z_j$ lies in the exterior of $\mathcal{D}$ and it satisfies $\Re(1/z_j) > 1$, so that we have the condition $m_j \geq 1$. Thus, from (3), there results that an execution of the Gaussian algorithm on input z_0 translates into a terminating "continued fraction" expansion

$$z_0 = \cfrac{1}{m_1 + \cfrac{1}{m_2 + \cfrac{1}{\ddots \atop m_k + z_k}}}, \tag{4}$$

where the expansion is stopped as soon as z_k lies in $\mathcal{B} \setminus \mathcal{D}$. The number of iterations, L, then assumes the value k. All the m_j are at least 1.

This leads to introducing the set $\mathcal{H}_k$ of linear fractional transformations of *depth* k (for $k \geq 1$) defined as the collection of all $h(z)$ of the form

$$h_{\mathbf{n}}(z) = h_{n_1, n_2, \ldots, n_k}(z) = \cfrac{1}{n_1 + \cfrac{1}{n_2 + \cfrac{1}{\ddots \atop n_k + z}}}, \tag{5}$$

where the $n_j \in \mathbb{N} = \{1, 2, \ldots\}$.

From the preceding discussion, we thus have the equivalence

$$z_k = U^k(z_0) \qquad \Longleftrightarrow \qquad \exists \mathbf{n} \in \mathbb{N}^k \ (z_0 = h_{\mathbf{n}}(z_k)).$$

The event $\{L \geq k+1\}$ coincides with the set of complex z such that all the $U^j(z)$, for $j = 0, \ldots, k$, lie in $\mathcal{D}$. Thus, defining $\mathcal{D}_k = U^{(-k)}(\mathcal{D})$ with $\mathcal{D}_0 = \mathcal{D}$, we have $\{L \geq k+1\} \equiv \mathcal{D}_k$. By definition, these domains form an infinite descending chain, $\mathcal{D}_0 \supset \mathcal{D}_1 \supset \mathcal{D}_2 \supset \cdots$. We also have that each $\mathcal{D}_k$ is the disjoint union of transforms of $\mathcal{D}$ by the transformations of $\mathcal{H}_k$ of (5),

$$\mathcal{D}_k = U^{(-k)}(\mathcal{D}) = \bigcup_{\mathbf{n} \in \mathbb{N}^k} h_{\mathbf{n}}(\mathcal{D}).$$

From elementary properties of geometrical inversion, $h_{\mathbf{n}}(\mathcal{D})$ is the disk of diameter $[h_{\mathbf{n}}(0), h_{\mathbf{n}}(1)]$. Within the theory of continued fractions, the interval $[h_{\mathbf{n}}(0), h_{\mathbf{n}}(1)]$ is known as a fundamental interval. A rendering of the domains, also called the fundamental disks, is given in Figure 2.

These considerations imply that, under the uniform probabilistic model of use, the probability ϖ_k that the algorithm performs at least $k+1$ iterations is

$$\varpi_k = \frac{\|\mathcal{D}_k\|}{\|\mathcal{D}\|} = \frac{4}{\pi} \sum_{\mathbf{n} \in \mathbb{N}^k} \|h_{\mathbf{n}}(\mathcal{D})\|, \tag{6}$$

where $\|\mathcal{A}\|$ denotes the area of a domain $\mathcal{A}$ of the plane.

Fig. 2. The domains $\mathcal{D}_0 \backslash \mathcal{D}_1$, $\mathcal{D}_1 \backslash \mathcal{D}_2$, $\mathcal{D}_2 \backslash \mathcal{D}_3$, $\mathcal{D}_3 \backslash \mathcal{D}_4$, $\mathcal{D}_4 \backslash \mathcal{D}_5$ represented alternatively in black and white. (The largest disk is $\mathcal{D}_0 \equiv \mathcal{D}$ which is the disk of diameter $[0,1]$.)

Continuants. Homographies of $\mathcal{H}_k$ are naturally associated with continued fractions of depth k themselves expressible in terms of *continuants*, see for instance the books by Knuth [10, p. 340] or by Rockett and Szüsz [18]. The continuant polynomials are defined by

$$Q_n(x_1, x_2, \ldots, x_n) = x_n Q_{n-1}(x_1, \ldots, x_{n-1}) + Q_{n-2}(x_1, \ldots, x_n),$$

with $Q_0 = 1$, $Q_1(x_1) = x_1$. Classically, a function $h_{\mathbf{n}} \in \mathcal{H}_k$ with $\mathbf{n} = (n_1, \ldots, n_k)$ admits the expression

$$h_{\mathbf{n}}(z) = \frac{P_k + z P_{k-1}}{Q_k + z Q_{k-1}}, \tag{7}$$

where

$$\begin{aligned} Q_k &= Q_k(n_1, \ldots, n_k), \quad Q_{k-1} = Q_{k-1}(n_1, \ldots, n_{k-1}), \\ P_k &= Q_{k-1}(n_2, \cdots, n_k), \; P_{k-1} = Q_{k-2}(n_2, \ldots, n_{k-1}). \end{aligned} \tag{8}$$

As is well–known the continuant polynomial $Q_n(x_1, \ldots, x_n)$ is also the sum of all monomials that obtain by crossing out pairs $x_i x_{i+1}$ of consecutive variables in the product $x_1 x_2 \cdots x_n$. Continuants thus satisfy the symmetry property $Q_k(x_1, \ldots, x_k) = Q_k(x_k, \ldots, x_1)$ and the determinant identity $Q_k P_{k-1} - Q_{k-1} P_k = (-1)^{k-1}$.

Probabilistic analysis. The previous considerations permit to express the probability distribution of the Gaussian algorithm in terms of continuants.

Theorem 1. *The probability ϖ_k that algorithm SGA performs more than k iterations on a random input $z \in \mathcal{D}$ is expressible as*

$$\varpi_k \equiv \Pr\{L \geq k+1\} = \sum_{n_1, \ldots, n_k} \frac{1}{Q_k^2 (Q_k + Q_{k-1})^2},$$

where $Q_k = Q_k(n_1, \ldots, n_k)$, $Q_{k-1} = Q_{k-1}(n_1, \ldots, n_{k-1})$, and the sum is over all integers $n_j \geq 1$.

The following table displays the probability distribution of SGA computed by Theorem 1 and the numerical methods of Section 5 against the result of 10^8 simulations of the algorithm.

k	$\Pr\{L \geq k+1\}$	Simulations
1	0.28986	0.28984361
2	0.04848	0.04847104
3	0.01027	0.01027170
4	0.00200	0.00200478
5	0.00040	0.00040299
6	0.00008	0.00008031
7	0.00002	0.00001569
Expectation:	1.35113	1.351094

Average-case analysis. Elementary number–theoretic considerations permit to express the expected cost of the Gaussian algorithm under a form no longer involving continuants. We have:

Theorem 2. *The mean number of iterations of algorithm SGA applied to a random $z \in \mathcal{D}$ is*

$$E\{L\} = \frac{5}{4} + \frac{180}{\pi^4} \sum_{d \geq 1} \frac{1}{d^2} \sum_{d < c < 2d} \frac{1}{c^2}.$$

Dynamic analysis. We have already mentioned the importance of the invariant measure of Gauss that has density $\frac{1}{\log 2} \frac{1}{1+x}$, in the 1–dimensional case. No such invariant measure can exist here as the reduction algorithm terminates. However, a rôle quite similar to the invariant measure of Gauss is played by a function that describes the distribution of successive transforms of the input as the reduction algorithms proceeds.

Initially, the input distribution is uniform in the disk $\mathcal{D}$, so that to z_0 is associated the constant density function over $\mathcal{D}$. Assume now that the algorithm performs at least $k+1$ iterations. Then the kth iterate is well defined and is an element of $\mathcal{D}$. A natural question is to determine its distribution inside $\mathcal{D}$. The corresponding *conditional* density function $F_k(z)$ must be proportional to $\lim_{\rho \to 0} \frac{1}{\pi \rho^2} \Pr\{z_k \in D(z, \rho)\}$, where $D(z, \rho)$ is the disk of center z and radius ρ. The proportionality factor must be taken so as to ensure that the integral of the density over $\mathcal{D}$ equals 1, so that the legitimate definition of the density function is

$$F_k(z) = \lim_{\rho \to 0} \frac{1}{\pi \rho^2} \frac{\Pr\{z_k \in D(z, \rho)\}}{\Pr\{z_k \in \mathcal{D}\}}.$$

We shall call F_k the *dynamic density* (of order k) of the algorithm.

Theorem 3. *The dynamic density F_k is given by*

$$F_k(z) = \frac{1}{\varpi_k} \sum_{\mathbf{n} \in \mathbb{N}^k} \frac{1}{|Q_{k-1}z + Q_k|^4}, \qquad where \qquad \varpi_k = \Pr\{L \geq k+1\}.$$

Furthermore, a functional relation holds for *real* x,

$$\frac{\varpi_{k+1}}{\varpi_k} F_{k+1}(x) = \sum_{m \geq 1} \frac{1}{(m+x)^4} F_k(\frac{1}{m+x}).$$

Thus, assuming that F_k admits a limit F_∞ and ϖ_{k+1}/ϖ_k converges to some constant λ, the quantities λ and F_∞ must be an eigenvalue and a corresponding eigenvector of the operator defined by the right hand side. This sharply motivates the introduction of the operator $\mathcal{G}$ in the next section, where we shall also establish the assumptions regarding F_k and ϖ_k.

4 The $\mathcal{G}$ operator

The complete analysis of the probability distribution and of the dynamics of the Gaussian algorithm depends on the introduction of an operator $\mathcal{G}_s$ formally defined by

$$\mathcal{G}_s[f](t) = \sum_{m \geq 1} \frac{1}{(m+t)^s} f(\frac{1}{m+t}), \tag{9}$$

and more specifically on the instance $s = 4$ that we simply denote by $\mathcal{G} \equiv \mathcal{G}_4$. (Continued fractions and the Euclidean algorithm correspond to the case $s = 2$.) Let V denote the open disk of center 1 and radius $\frac{3}{2}$. For all s with $\Re(s) > 1$, the operator $\mathcal{G}_s$ acts on the space $A_\infty(V)$ of functions f that are holomorphic in V and continuous on the closure $\overline{V}$ of V. The set $A_\infty(V)$ endowed with the sup-norm $\|f\| = \sup_{t \in \overline{V}} |f(t)|$ is a Banach space.

Such operators permit to "invert" the continued fraction operator U, and at the same time their functional analysis properties (related to the Perron-Frobenius theory) have useful consequences for the Gaussian algorithm. There is a close relationship between the iterates of $\mathcal{G}_s$ and continuants.

Lemma 4. *The iterates of $\mathcal{G}_s$ generate the continuants of depth k in the following sense:*

$$\mathcal{G}_s^k[f](t) = \sum_{n_1 \ldots n_k} \frac{1}{(Q_{k-1}t + Q_k)^s} f(\frac{P_{k-1}t + P_k}{Q_{k-1}t + Q_k}); \quad \mathcal{G}_s^k[f](0) = \sum_{n_1 \ldots n_k} \frac{1}{Q_k^s} f(\frac{Q_{k-1}}{Q_k}). \tag{10}$$

The quantities involved in Theorems 1, 3 precisely admit such expressions:

$$\varpi_k = \Pr\{L \geq k+1\} = \mathcal{G}^k[u](0) \qquad \text{where} \qquad u(t) = \frac{1}{(1+t)^2}$$

$$F_k(z) = \frac{1}{\varpi_k} \mathcal{G}^k[v_z(t)](0) \qquad \text{where} \qquad v_z(t) = \frac{1}{(1+tz)^2(1+t\bar{z})^2}.$$

Spectral properties of the $\mathcal{G}_s$ operators have been investigated in detail by Mayer and we globally refer to [17] and references therein. For s such that

$\Re(s) > 1$, the operators $\mathcal{G}_s$ are nuclear of order 0. In other words, they have a discrete spectrum and admit a spectral decomposition:

$$\mathcal{G}_s[f](t) = \sum_{i=1}^{\infty} \lambda_i e_i^*[f] e_i(t), \tag{11}$$

where the λ_i are the eigenvalues, $\{e_i\}$ is a basis of eigenfunctions, and the coefficients $\{e_i^*\}$ are the dual basis of $\{e_i\}$; in addition the λ_i are ρ-summable for all real $\rho > 0$: $\sum_i |\lambda_i|^\rho < +\infty$. (These quantities implicitly depend on s.)

For real $s > 1$ (we need the case $s = 4$), the operator $\mathcal{G}_s$ is in addition an operator satisfying the Perron–Frobenius property [17]: it has a unique positive dominant eigenvalue λ_1, the corresponding eigenfunction $e_1(t)$ is strictly positive on $\overline{V} \cap \mathbb{R}$, and $e_1^*[f]$ is strictly positive if f is itself positive on $\overline{V} \cap \mathbb{R}$. In particular, if λ_2 denotes the second eigenvalue (in order of absolute values), and if f is positive on $\overline{V} \cup \mathbb{R}$, one has

$$\left\| \frac{1}{\lambda_1^k} \mathcal{G}_s^k[f] - P[f] \right\| \leq \|f\| \cdot \left| \frac{\lambda_2}{\lambda_1} \right|^k, \tag{12}$$

where $P[f]$ denotes projection on the dominant eigensubspace: $P[f](t) = \lambda_1 e_1^*[f] e_1(t)$.

These considerations (with $s = 4$) apply to the continuant form of the probability distribution and to the (conditional) invariant measure of the Gaussian algorithm.

Theorem 5. *There exist real numbers c_j and λ_j with $\lambda_1 > |\lambda_2| > |\lambda_3| > \cdots$, such that*

$$\Pr\{L \geq k + 1\} = \sum_{j=1}^{\infty} c_j \lambda_j^k.$$

In particular, with $c_1 \approx 1.3$, $\lambda_1 \approx 0.1993$, and $\lambda_2 \in [-\frac{1}{20}, -\frac{1}{10}]$, one has asymptotically:

$$\Pr\{L \geq k + 1\} = c_1 \lambda_1^k \left[1 + \mathcal{O}((\frac{\lambda_2}{\lambda_1})^k) \right].$$

This theorem is in accordance with observation of the numerical data following Theorem 1, as the probabilities decay roughly like $(\frac{1}{5})^n$.

Theorem 6. *The dynamic density $F_k(z)$ converges geometrically to a (conditional) invariant density F_∞:*

$$F_\infty(x + iy) = \alpha \int_0^1 (1 - w^2) [e_1(x + iyw) + e_1(x - iyw)] \, dw.$$

There, e_1 is the eigenfunction of $\mathcal{G}$ corresponding to the dominant eigenvalue λ_1, and α the normalization constant determined by $\iint F_\infty \, dx \, dy = 1$. In particular, on the real axis, the invariant density F_∞ is proportional to the eigenfunction e_1.

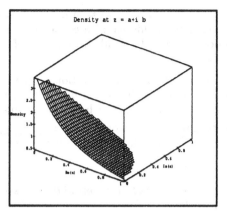

Density at z = a+i b

Fig. 3. The conditional invariant density F_∞.

5 The discrete model

The analysis of the standard algorithm under the discrete model where inputs are taken from the discrete set

$$\mathbb{Q}^{(N)}(i) = \{\frac{a}{N} + i\frac{b}{N} \mid b \neq 0\}, \tag{13}$$

is solved by a combination of two arguments: (i) by "Gauss's principle" a circle of radius x contains $\pi N^2 x^2 + \mathcal{O}(xN + 1)$ lattice points of $\mathbb{Q}^{(N)}(i)$; (ii) from worst–case bounds, the reduction algorithm performs number of iterations at most $\mathcal{O}(\log N)$.

Theorem 7. *Let $L^{(N)}$ be the number of iterations of the standard Gaussian algorithm applied to random inputs from $\mathbb{Q}^{(N)} \cap \mathcal{D}$. The random variable $L^{(N)}$ converges in moments and in distribution to the random variable L associated with the continuous model. In particular, the mean value satisfies*

$$\mu^{(N)} \equiv \mathrm{E}\{L^{(N)}\} = \mu + \mathcal{O}(\frac{\log N}{N}).$$

6 Numerical estimates

We have already cited some numerical estimates for the mean and the probability distribution of the Gaussian algorithm, as well as the approximate value $\lambda_1 \approx 0.1993$. Most of the expressions involve slowly converging sums. The purpose of this section is to give indications on series transformations that permit to evaluate some of these quantities to great accuracy, as well as on ways in which precise bounds can be proved on λ_1 using trace formulae.

A real number α is said to be *polynomial time computable* if there exists an integer r such that an approximation of α to accuracy 10^{-d} can be computed in time $\mathcal{O}(d^r)$. We let **P** denote the class of such numbers. A major problem is to find which of the constants of this paper are polynomial time computable. Effective numerical procedures usually result from proofs of membership in **P**.

The expected cost. The expected number of iterations of the Gaussian algorithm admits an expression as a sum which, once truncated till terms of order m, results in an error of $\mathcal{O}(\frac{1}{m^2})$. This sum can be expressed instead as a definite integral involving the dilogarithm function, $\mathrm{Li}_2(z) = \sum_{n=1}^{\infty} \frac{z^n}{n^2}$, itself amenable to series representations (involving Bernoulli numbers) that exhibit geometric convergence.

Theorem 8. (*i*). *The mean number of iterations of the Gaussian algorithm SGA admits the integral representation*

$$\mu = -\frac{3}{4} + \frac{180}{\pi^4} \int_0^{\infty} \frac{\mathrm{Li}_2(e^{-t}) - \mathrm{Li}_2(e^{-2t})}{1 - e^{-t}} \, t \, dt. \tag{14}$$

(*ii*). *The number μ lies in the class* **P** *of polynomial time computable numbers:*
$\mu = 1.35113\,15744\,91659\,00179\,38680\,05256\,46466\,84404\,78970\,85087 \pm 10^{-50}$.

Probability distribution. The probability distribution of the Gaussian algorithm can be expressed in terms of complicated series involving the zeta function, the resulting expressions being useful for small values of k.

Theorem 9. *The probability distribution of the number of iterations of the Gaussian algorithm has initial values:* $\Pr\{L \geq 1\} = 1$, $\Pr\{L \geq 2\} = \frac{\pi^2}{3} - 3$,

$$\Pr\{L \geq 3\} = -5 + \frac{2\pi^2}{3} - 2\zeta(3) + 2\sum_{n=0}^{\infty}(-1)^n(n+1)\zeta(n+4)(\zeta(n+2) - 1).$$

In general, each $\varpi_k = \Pr\{L \geq k+1\}$ is in the class **P**.

The proof is based on the fact that summations of analytic functions of several complex variables at integer points,

$$\varphi = \sum_{n_1,\ldots,n_k=1}^{\infty} F(\frac{1}{n_1}, \frac{1}{n_2}, \ldots, \frac{1}{n_k}),$$

can be represented as multiple sums of zeta functions at the integers. (See for instance [23] for the univariate case.) The following values have been determined in this way to great accuracy:

$$\varpi_1 = 0.28986\,81336\,96452\,87294$$
$$\varpi_2 = 0.04848\,08014\,49463\,63270$$
$$\varpi_3 = 0.01027\,81647\,79066\,59643.$$

Eigenvalues. The last numerical task is to estimate the dominant and subdominant eigenvalues, λ_1 and λ_2 that determine the rate of geometric decay of the probabilities ϖ_k. A first class of bounds is obtained by a nonlinear optimization problem based on Wirsing's approach [24] and the specific "test functions" pairs:

$$\begin{cases} \psi_a(t) = \dfrac{1}{(1+at)(1+(a+1)t)(1+(a+2)t)(1+(a+3)t)} \\ \phi_a(t) = \mathcal{G}[\psi_a](t) = \dfrac{1}{3} \dfrac{1}{(z+a+1)(z+a+2)(z+a+3)}. \end{cases} \tag{15}$$

With $a = 0.487$, the ration $\phi_a(t)/\psi_a(t)$ lies in $[0.170, 0.205]$ for $t \in [0,1]$. By iterating $\mathcal{G}$, a first bound

$$0.170 < \lambda_1 < 0.205$$

is then obtained. The more refined estimates mentioned in Theorem 5 are derived from adapting trace formulae originally due to Babenko and Mayer.

Theorem 10. (*i*). *The trace of the operator* $\mathcal{G}^k$ *satisfies*

$$\operatorname{Tr} \mathcal{G}^k \equiv \sum_i \lambda_i^k = \sum_n \frac{\tau(\mathbf{n})^4}{1 - (-1)^k \tau(\mathbf{n})^2}, \quad \tau(\mathbf{n}) = \frac{(Q_k + P_{k-1}) - \sqrt{(Q_k + P_{k-1})^2 - 4}}{2}.$$

$$\tag{16}$$

(*ii*). *Each* $\operatorname{Tr} \mathcal{G}^k$ *is computable in polynomial time. In particular* $\operatorname{Tr} \mathcal{G}$ *admits the explicit form*

$$\operatorname{Tr} \mathcal{G} = \frac{7}{2} - \frac{2}{\sqrt{5}} - \frac{7}{\sqrt{2}} + \frac{1}{2} \sum_{n=2}^{\infty} (-1)^n \frac{n-1}{n+1} \binom{2n}{n} [\zeta(2n) - 1 - \frac{1}{2^{2n}}].$$

7 The centered Gaussian algorithm

The centered Gaussian algorithm constitutes the classical implementation of the general reduction schema described in Section 2. The aim of the algorithm is to bring z in the strip $\widetilde{B} = \{z \mid 0 \le \Re(z) \le \frac{1}{2}\}$ by means of the transformation $z := \widetilde{U}(z)$ where

$$\widetilde{U}(z) = \varepsilon(\frac{1}{z})(\frac{1}{z} - \lfloor \Re(\frac{1}{z}) \rceil),$$

with $\lfloor u \rceil$ the integer nearest to u, and $\varepsilon(u)$ the sign of $\Re(u) - \lfloor \Re(u) \rceil$.

As was done with the standard algorithm, it is sufficient to restrict consideration to the *core algorithm*

Algorithm $CGA(z : \text{complex})$ [Centered Gaussian Algorithm]
Input: $z \in \widetilde{D}$ (the disk of diameter $[0, \frac{1}{2}]$)
        while $(z \in \widetilde{D})$ do $z := \widetilde{U}(z)$;
Output: $z \in \widetilde{B} \setminus \widetilde{D}$ ($\widetilde{B}$ is the strip $0 \le \Re(z) \le \frac{1}{2}$)

An analysis of the average case has been given in [22]. Methods of this paper lead to a complete analysis. The theory develops with linear fractional transformations related to centered continued fraction expansions,

$$h_{\mathbf{m},\varepsilon}(z) = \cfrac{1}{m_1 + \cfrac{\varepsilon_1}{m_2 + \cfrac{\varepsilon_2}{\ddots \cfrac{}{m_k + \varepsilon_k z}}}}, \qquad (17)$$

and the corresponding continuants $\widetilde{Q}_k$. In (17), the pairs (m_j, ε_j) satisfy the basic condition:

$$\text{if } \varepsilon > 0 \text{ then } m \geq 2; \quad \text{if } \varepsilon < 0 \text{ then } m \geq 3. \qquad (18)$$

The corresponding functional operator is then

$$\widetilde{\mathcal{G}}_s[f](t) = \sum_{m,\varepsilon} (\frac{\varepsilon}{m+t})^s f(\frac{\varepsilon}{m+t}),$$

where the pair (m, ε) satisfies the conditions (18). The operator can be proved to enjoy properties similar to those of $\mathcal{G}_s$, namely having a discrete spectrum, being nuclear, satisfying Perron-Frobenius properties for even integral s. A consequence is the following theorem:

Theorem 11. (*i*). *The mean number of iterations of the centered Gaussian algorithm CGA satisfies*

$$\widetilde{\mu} \equiv \mathrm{E}\{\widetilde{L}\} = \frac{360}{\pi^4} \sum_{d=1}^{\infty} \frac{1}{d^2} \sum_{c=\lceil d\phi^{-2} \rceil}^{\lfloor d\phi^{-1} \rfloor} \frac{1}{c^2}, \qquad \text{with } \phi = (1 + \sqrt{5})/2.$$

(*ii*). *The probability distribution decays exponentially:*

$$\widetilde{\omega}_k = \mathrm{E}\{\widetilde{L} \geq k + 1\} \sim \widetilde{c} \cdot \widetilde{\lambda}^k \qquad \text{with} \qquad \widetilde{\lambda} \approx 0.077.$$

(*iii*). *The algorithm admits a limit conditional invariant density.*

It is a perhaps surprising fact that, despite its "nonanalytical" character, the average–case constant $\widetilde{\mu}$ can be computed in polynomial time. For $\theta \in (0,1)$ with convergent sequence $\{\frac{p_n}{q_n}\}$, one has:

$$\sum_{d=1}^{\infty} \sum_{c=1}^{\lfloor d\theta \rfloor} \frac{1}{c^2 d^2} = \sum_{n=0}^{\infty} \frac{(-1)^n}{(p_n + p_{n+1})^2(q_n + q_{n+1})^2}$$
$$\int_0^1 \int_0^1 \frac{\log x \log y}{\left(1 - x^{\frac{p_n}{p_n+p_{n+1}}} y^{\frac{q_n}{q_n+q_{n+1}}}\right)\left(1 - x^{\frac{p_{n+1}}{p_n+p_{n+1}}} y^{\frac{q_{n+1}}{q_n+q_{n+1}}}\right)} \, dx \, dy,$$

a formula that results from identities of Mahler and Borwein–Borwein [13, 3]. From this, we find

$$\widetilde{\mu} = 1.08922\,14740\,95380\ldots$$

Acknowledgements. The work of Philippe Flajolet was supported by the Esprit Basic Research Action No. 7141 (Alcom II).

References

1. BABENKO, K. I. On a problem of Gauss. *Soviet Mathematical Doklady 19*, 1 (1978), 136–140.
2. BERNDT, B. C. *Ramanujan's Notebooks, Part I*. Springer Verlag, 1985.
3. BORWEIN, J. M., AND BORWEIN, P. B. Strange series and high precision fraud. *American Mathematical Monthly 99*, 7 (Aug. 1992), 622–640.
4. COHEN, H. *A Course in Computational Algebraic Number Theory*. No. 138 in Graduate Texts in Mathematics. Springer–Verlag, 1993.
5. DAUDÉ, H. *Des fractions continues à la réduction des réseaux: analyse en moyenne*. PhD thesis, Université de Caen, 1993.
6. DAUDÉ, H., AND VALLÉE, B. An upper bound on the average number of iterations of the LLL algorithm. *Theoretical Computer Science 123*, 1 (1994), 95–115.
7. EDWARDS, H. M. *Riemann's Zeta Function*. Academic Press, 1974.
8. HENSLEY, D. The number of steps in the Euclidean algorithm. Preprint, 1993.
9. KAIB, M., AND SCHNORR, C. P. A sharp worst–case analysis of the Gaussian lattice basis reduction algorithm for any norm. Preprint, 1992. To appear in *J. of Algorithms*.
10. KNUTH, D. E. *The Art of Computer Programming*, 2nd ed., vol. 2: Seminumerical Algorithms. Addison-Wesley, 1981.
11. LAGARIAS, J. C. Worst–case complexity bounds for algorithms in the theory of integral quadratic forms. *Journal of Algorithms 1*, 2 (1980), 142–186.
12. LENSTRA, A. K., LENSTRA, H. W., AND LOVÁSZ, L. Factoring polynomials with rational coefficients. *Mathematische Annalen 261* (1982), 513–534.
13. MAHLER, K. Arithmetische Eigenschaften der Lösungen einer Klasse von Funktionalgleichungen. *Mathematische Annalen 101* (1929), 342–366.
14. MAYER, D., AND ROEPSTORFF, G. On the relaxation time of Gauss's continued fraction map. I. The Hilbert space approach. *Journal of Statistical Physics 47*, 1/2 (Apr. 1987), 149–171.
15. MAYER, D., AND ROEPSTORFF, G. On the relaxation time of Gauss's continued fraction map. II. The Banach space approach (transfer operator approach). *Journal of Statistical Physics 50*, 1/2 (Jan. 1988), 331–344.
16. MAYER, D. H. On a ζ function related to the continued fraction transformation. *Bulletin de la Société Mathématique de France 104* (1976), 195–203.
17. MAYER, D. H. Continued fractions and related transformations. In *Ergodic Theory, Symbolic Dynamics and Hyperbolic Spaces*, M. K. Tim Bedford and C. Series, Eds. Oxford University Press, 1991, pp. 175–222.
18. ROCKETT, A., AND SZÜSZ, P. *Continued Fractions*. World Scientific, Singapore, 1992.
19. SCHARLAU, W., AND OPOLKA, H. *From Fermat to Minkowski, Lectures on the Theory of Numbers and its Historical Developments*. Undergraduate Texts in Mathematics. Springer-Verlag, 1984.
20. SERRE, J.-P. *A Course in Arithmetic*. Graduate Texts in Mathematics. Springer Verlag, 1973.
21. VALLÉE, B. Gauss' algorithm revisited. *Journal of Algorithms 12* (1991), 556–572.
22. VALLÉE, B., AND FLAJOLET, P. Gauss' reduction algorithm: An average case analysis. In *Proceedings of the 31st Symposium on Foundations of Computer Science* (Oct. 1990), IEEE Computer Society Press, pp. 830–839.
23. VARDI, I. *Computational Recreations in Mathematica*. Addison Wesley, 1991.
24. WIRSING, E. On the theorem of Gauss-Kusmin-Lévy and a Frobenius-type theorem for function spaces. *Acta Arithmetica 24* (1974), 507–528.
25. ZIPPEL, R. *Effective Polynomial Computations*. Kluwer Academic Publishers, Boston, 1993.

A Fast Variant of the Gaussian Reduction Algorithm

(extended abstract)

Michael Kaib*

FB Mathematik, Universität Frankfurt, 60054 Frankfurt, Germany

Abstract. We propose a fast variant of the Gaussian algorithm for the reduction of two–dimensional lattices for the l_1-, l_2- and l_∞-norm. The algorithm uses at most $O(\mathcal{M}(B)(n + \log B))$ bit operations for the l_2-norm, $O(n\,\mathcal{M}(B)\log B)$ bit operations for the l_∞-norm and in $O(n\,\log n\,\mathcal{M}(B)\log B)$ bit operations for the l_1-norm on input vectors $a, b \in \mathbb{Z}^n$ with norm at most 2^B where $\mathcal{M}(B)$ is a time bound for B-bit integer multiplication. This generalizes Schönhages fast algorithm for monotone reduction of binary quadratic forms [Proc. ISSAC 1991, ACM 1991, pp. 128-133] to the centered case and to various norms.

The basic idea is to perform most of the arithmetic on the leading bits of the integers, following the techniques of the fast gcd–algorithms due to Lehmer and Schönhage. We extend this techniques to the classical "centered" case.

The Gaussian algorithm performs reduction steps $(a, b) \mapsto (\pm(b - \mu a), a)$ where the integer μ is chosen to minimize $\| b - \mu a \|$. Our new consideration is, that the core of the Gaussian algorithm operates stable until the approximation error exceeds $\frac{1}{12}\| a \|$, what is valid for arbitrary norms. We use the characterization of the transformation matrices which Kaib and Schnorr gave in their sharp worst case analysis for the number of reduction steps for arbitrary norms [to appear in J. Algorithms]. They prove that all bases (a, b) in the interior steps satisfy $\| a \| \le \| a - b \| < \| b \|$, calling those basis *well-ordered*. We prove that the bases in all but the last three steps satisfy $\frac{5}{4}\| a \| \le \| a - b \| < \| b \| - \| a \|$ and call those bases *strictly well-ordered*. The output basis of the fast algorithm is *minimal w.r.t. some threshold* σ, i.e. its successor basis is not strictly well-ordered or its successor vector has norm less than σ. The approximation error is controlled by the *descent* of the algorithm, which is roughly the difference of the input size and $\log_2 \sigma$. The algorithm performs two recursive calls of half accuracy and half descent. We always care that the relative approximation error is at most $\frac{1}{12}$ of the norm of the actual vector. If recovering of higher accuracy yields a well–ordered but not strictly well–ordered basis, we cancel (at most 2) preceding reduction steps to obtain a strictly well–ordered basis.

A detailed description of the algorithm is given in the authors thesis and is also available as technical report at Universität Frankfurt.

---

* e-mail: kaib@cs.uni-frankfurt.de

Reducing Lattice Bases by Means of Approximations

Johannes Buchmann

Fachbereich Informatik
Universität des Saarlandes
Postfach 1150
D-66041 Saarbrücken
Germany

Abstract Let L be a k-dimensional lattice in $\mathbb{R}^m$ with basis $B = (\underline{b}_1, \ldots, \underline{b}_k)$. Let $A = (a_1, \ldots, a_k)$ be a rational approximation to B. Assume that A has rank k and a lattice basis reduction algorithm applied to the columns of A yields a transformation $T = (\underline{t}_1, \ldots, \underline{t}_k) \in \mathrm{GL}(k, \mathbb{Z})$ such that $A\underline{t}_i \leq s_i \lambda_i(L(A))$ where $L(A)$ is the lattice generated by the columns of A, $\lambda_i(L(A))$ is the i-th successive minimum of that lattice and $s_i \geq 1$, $1 \leq i \leq k$. For $c > 0$ we determine which precision of A is necessary to guarantee that $B\underline{t}_i \leq (1+c)s_i\lambda_i(L)$, $1 \leq i \leq k$. As an application it is shown that Korkine-Zolotaref-reduction and LLL-reduction of a non integer lattice basis can be effected almost as fast as such reductions of an integer lattice basis.

1 Introduction

Let $m, k \in \mathbb{N}$. Let L be a *k-dimensional lattice* in $\mathbb{R}^m$, $B = (\underline{b}_1, \ldots, \underline{b}_k)$ a *basis* of L, i.e. the elements of B are linearly independent and

$$L = L(B) = \sum_{i=1}^{k} \underline{b}_i \, \mathbb{Z} \ .$$

Such a basis is uniquely determined up to multiplication from the right by elements of the set $\mathrm{GL}(k, \mathbb{Z})$ of all invertible $k \times k$-matrices with integer coefficients. Such matrices are called *unimodular transformations*.

It is a natural question to ask whether there are bases which are particularly simple, e.g. which have elements of short euclidean length or a special form or which are even uniquely determined by certain conditions. Those bases are usually called *reduced bases*. There are various notions of reduction such as Minkowski-reduction, Korkine-Zolotarev-reduction, LLL-reduction etc. (see [GrLoSchri]). If L is an *integral lattice*, i.e. $L \subset \mathbb{Z}^m$, algorithms for computing those bases are known. If L is not integral one can apply reduction algorithms to a rational approximation of a basis of L. In addition to the reduced approximation, which is, in general, useless, the algorithms yield a *reducing unimodular transformation*. In this paper we determine which precision guarantees that the reducing unimodular transformation of the approximate basis is a reducing transformation of the original basis. Reducing approximate lattice bases is important in many applications, e.g. for computing small integral bases or small systems of fundamental units in number fields.

The problem of reducing lattice bases by means of approximations has been previously considered in [BuPo] and [BuKe]. There, only the case of LLL-reduction was treated and the method is less efficient than the method presented here.

2   The Results

To state the results precisely we introduce some notation.

For $\underline{b} = (b_1, \ldots, b_m)$, $\underline{c} = (c_1, \ldots, c_m) \in \mathbb{R}^m$ the *inner product* of $\underline{b}$ and $\underline{c}$ is

$$\langle \underline{b}, \underline{c} \rangle = \sum_{i=1}^{m} b_i c_i \ .$$

The *(euclidean) length* of $\underline{b}$ is

$$\|\underline{b}\| = \langle \underline{b}, \underline{b} \rangle^{1/2} \ .$$

For any set or sequence S of vectors in $\mathbb{R}^m$ whose elements generate a subspace of dimension k and for $1 \leq i \leq k$ the ith *successive minimum* of S is

$$\lambda_i(S) = \min\{r \in \mathbb{R}_{>0} : \text{there are } i \text{ lin. indep. vectors in } S \text{ of length } \leq r\} \ .$$

For $C \in \mathbb{R}^{m \times k}$ we define

$$\|C\| = \left(\sum_{i=1}^{m} \sum_{j=1}^{k} c_{ij}^2 \right)^{1/2} \ .$$

Let L be a lattice in $\mathbb{R}^m$ of dimension k, $B = (\underline{b}_1, \ldots, \underline{b}_k)$ a basis of L. The determinant of L is

$$\det L = |\det(\langle \underline{b}_i, \underline{b}_j \rangle)_{1 \leq i,j \leq k}|^{1/2} \ .$$

The determinant is the volume of the parallelepiped spanned by the elements of B. It is an invariant of L. The *defect* of B is

$$\text{dft}(B) = \frac{\prod_{i=1}^{k} \|\underline{b}_i\|}{\det L} \ .$$

The defect measures how far the basis vectors are from being pairwise orthogonal; in particular $\text{dft}(B) = 1$ if and only if the basis vectors are pairwise orthogonal. Bases with small defect have short elements.

Let $A = (\underline{a}_1, \ldots, \underline{a}_k) \in \mathbb{R}^{m \times k}$ with

$$\|B - A\| < 2^{-q} \tag{1}$$

for some $q \in \mathbb{N}$. In the main theorems of this paper we answer the following questions: When is A of rank k? How close are the successive minima of $L(A)$ to those of L? What is the effect if we apply a reducing transformation for the approximate basis A to the original basis B?

Theorem 1. *Let* $0 < c \leq 1$ *and*

$$q \geq q_1(B,c) = \lceil k + \log(\lambda_k(L)/\lambda_1(L)) + \log \mathrm{dft}(B) + \log(1/\lambda_1(B)) + \log(1/c) \rceil$$

then the rank of A is k and

$$\lambda_i(L(A)) \leq (1+c)\lambda_i(L), \quad 1 \leq i \leq k .$$

To formulate the next theorem we assume that A is of rank k (this will be guaranteed by the choice of q) and that $T = (\underline{t}_1, \ldots, \underline{t}_k) \in \mathbb{Z}^{k \times k}$ is non singular and such that

$$\|A\underline{t}_i\| \leq s_i \lambda_i(L(A)), \quad 1 \leq i \leq k,$$

where $(s_1, \ldots, s_k)$ is a non decreasing sequence of real numbers ≥ 1. Let $s = s_k$.

Theorem 2. *Let* $0 < c \leq 1$ *and*

$$q \geq q_2(B,c,s) = \lceil k + \log(\lambda_k(L)/\lambda_1(L)) + \log \mathrm{dft}(B) + \log(1/\lambda_1(B)) + \log(3s/c) \rceil.$$

Then

$$\|\underline{t}_i\| \leq (3/2)^{k+1}\sqrt{k}s_i \mathrm{dft}(B)\lambda_i(L)/\lambda_1(B), \quad 1 \leq i \leq k$$

and

$$\|B\underline{t}_i\| \leq (1+c)s_i\lambda_i(L), \quad 1 \leq i \leq k .$$

This result is applied in the special cases where A is a rational approximation to B and the transformation T is obtained by Korkine-Zolotaref-reduction and LLL-reduction of A. Let $p \in \mathbb{Z}_{>0}$ and let b be a real number. A rational number a is called a *rational approximation of precision p to b* if

$$|a - b| < 2^{-p}$$

and

$$a \in 2^{-(p+1)}\mathbb{Z} .$$

Such an approximation always exists.
Put

$$\beta = \log(\lambda_k(B)/\lambda_1(L))$$

and for $c > 0, s \geq 1$ define

$$q_0(B,c,s) = \lceil q_2(B,c,s) + \log(\sqrt{km}) \rceil .$$

Corollary 3. *There is an algorithm which for $0 < c \leq 1$ when given an approx-imation of precision $p \geq q_0(B, c, \sqrt{(k+3)/4})$ to B computes $T = (\underline{t}_1, \ldots, \underline{t}_k) \in$ $\mathrm{GL}(k, \mathbb{Z})$ such that*

$$B\underline{t}_i \leq (1+c)\sqrt{(i+3)/4}\,\lambda_i(L), \quad 1 \leq i \leq k$$

and

$$\lambda_i(T) \leq (3/2)^{k+1}\sqrt{k(i+3)/4}\,\mathrm{dft}(B)\lambda_i(L)/\lambda_1(B), \quad 1 \leq i \leq k \ .$$

The algorithm performs

$$mk^{O(k)}(\log m + \beta + \log(1/c))$$

arithmetic operations on numbers of binary length

$$O\left(k^2\left(k\left(\log k + \beta\right) + \log(1/c) + \log m\right)\right) \ .$$

Corollary 4. *There is an algorithm which for $0 < c \leq 1$ when given an approx-imation of precision $p \geq q_0(B, c, 2^{(k-1)/2})$ to B computes $T \in \mathrm{GL}(k, \mathbb{Z})$ such that*

$$\lambda_i(BT) \leq (1+c)\,2^{(k-1)/2}\,\lambda_i(L), \quad 1 \leq i \leq k$$

and

$$\lambda_i(T) \leq (3/2)^{k+1}\sqrt{k}\,2^{(k-1)/2}\mathrm{dft}(B)\lambda_i(L)/\lambda_1(B), \quad 1 \leq i \leq k \ .$$

The algorithm performs

$$O\left(mk^3\left(k\left(\log k + \beta\right) + \log m + \log(1/c)\right)\right)$$

arithmetic operations on numbers of binary length

$$O\left(k\left(k\left(\log k + \beta\right) + \log(1/c) + \log m\right)\right) \ .$$

3   The Proof

Let A, B be as in Section 2. We recall a few results from linear algebra.

Lemma 5 (Cauchy-Schwarz inequality). *Let $\underline{a}, \underline{b} \in \mathbb{R}^m$ then*

$$\langle \underline{a}, \underline{b} \rangle \leq \|a\|\,\|b\| \ .$$

Let V be the vector space generated by the elements of B. Let $C = (\underline{c}_1, \ldots, \underline{c}_k)$ be a sequence of vectors in V. Then there is $T \in \mathbb{R}^{k \times k}$ such that $C = BT$. For the purpose of this proof we define

$$\det C = \det L \det T \ .$$

That determinant has the following properties.

Lemma 6. *1. The vectors in C are linearly dependent if and only if $\det C = 0$.*

2. *For $1 \leq i \leq k$ the map $\mathbb{R}^m \to \mathbb{R}$, $\underline{c} \mapsto \det(\underline{c}_1, \ldots, \underline{c}_{i-1}, \underline{c}, \underline{c}_{i+1}, \ldots, \underline{c}_k)$ is a homomorphism of $\mathbb{R}$-vector spaces, i.e. $\det$ is multilinear.*

Lemma 7 (Cramer's rule). *Let $\underline{t} = (t_1, \ldots, t_k) \in \mathbb{R}^k$, $\underline{b} \in \mathbb{R}^m$ with $\underline{b} = B\underline{t}$ then*

$$t_i = \frac{\det(\underline{b}_1, \ldots, \underline{b}_{i-1}, \underline{b}, \underline{b}_{i+1}, \ldots, \underline{b}_k)}{\det L} .$$

Lemma 8 (Gram-Schmidt orthogonalization). *Let $C = (\underline{c}_1, \ldots, \underline{c}_k)$ be a basis of V. Define*

$$\underline{c}_1^* = \underline{c}_1,$$

$$\underline{c}_i^* = \underline{c}_i - \sum_{j=1}^{i-1} \frac{\langle \underline{c}_i, \underline{c}_j^* \rangle}{\langle \underline{c}_j^*, \underline{c}_j^* \rangle} \underline{c}_j^*, \quad 2 \leq i \leq k,$$

and $C^* = (\underline{c}_1^*, \ldots, \underline{c}_k^*)$. Then the elements of C^* are pairwise orthogonal and we have

$$\|\underline{c}_i^*\| \leq \|\underline{c}_i\|, \quad 1 \leq i \leq k,$$

and

$$\det C = \det C^* = \prod_{i=1}^{k} \|\underline{c}_i^*\| \leq \prod_{i=1}^{k} \|\underline{c}_i\| .$$

In the previous statement, the basis C^* is called the *Gram-Schmidt orthogonalization* of C. The last inequality is *Hadamard's inequality*. As an immediate consequence of Cramer's rule and Hadamard's inequality we obtain an upper bound for the solution of a linear system. For a vector $\underline{v} = (v_1, \ldots, v_m) \in \mathbb{R}^m$ set

$$|\underline{v}| = \max\{|v_i| : 1 \leq i \leq m\} .$$

Note that

$$\|\underline{v}\| \leq \sqrt{m}|\underline{v}| .$$

Corollary 9. *Let $\underline{b} \in \mathbb{R}^m$, $\underline{t} \in \mathbb{R}^k$ such that $B\underline{t} = \underline{b}$. Then*

$$|\underline{t}| \leq \|\underline{b}\| \, \mathrm{dft}(B)/\lambda_1(B) .$$

For any $\underline{t} \in \mathbb{R}^k$ and $C \in \mathbb{R}^{m \times k}$ we have $\|C\underline{t}\| \leq \|C\| \|\underline{t}\|$. Therefore, we obtain from (1)

$$\|B\underline{t} - A\underline{t}\| < 2^{-q}\|\underline{t}\| .$$

This implies the next lemma.

Lemma 10. *If $\underline{t} \in \mathbb{R}^k$, $c > 0$ and*

$$q \geq q_3(B, \underline{t}, c) = \lceil \log \|\underline{t}\| - \log(c\|B\underline{t}\|) \rceil$$

then

$$\|A\underline{t} - B\underline{t}\| < c \cdot \|B\underline{t}\| .$$

Next we prove an upper bound for $\det B - \det A$.

Lemma 11. *If*

$$q \geq \lceil -\log \lambda_1(B) \rceil$$

then

$$|\det A - \det B| < 2^{-q+k}\mathrm{dft}(B)\det B/\lambda_1(B) \ .$$

Proof. In Lemma 10 choose $c = 1$ and for $\underline{t}$ any unit vector. Then $q_3(B, \underline{t}, c) \leq \lceil -\log \lambda_1(B) \rceil$. Hence, if $q \geq \lceil -\log \lambda_1(B) \rceil$ then we have by Lemma 10 for $1 \leq i \leq k$

$$\|\underline{a}_i\| < 2\|\underline{b}_i\| \ .$$

This and Hadamard's inequality implies

$$\begin{aligned}
|\det B - \det A| &= |\sum_{j=1}^{k}(\det(\underline{b}_1, \ldots, \underline{b}_{j-1}, \underline{a}_j, \ldots, \underline{a}_k) \\
&\quad - \det(\underline{b}_1, \ldots, \underline{b}_j, \underline{a}_{j+1}, \ldots, \underline{a}_k))| \\
&\leq \sum_{j=1}^{k} |\det(\underline{b}_1, \ldots, \underline{b}_{j-1}, \underline{a}_j - \underline{b}_j, \underline{a}_{j+1}, \ldots, \underline{a}_k)| \\
&\leq \sum_{j=1}^{k}(\|\underline{a}_1\| \cdots \|\underline{a}_{j-1}\| \|\underline{b}_j - \underline{a}_j\| \|\underline{b}_{j+1}\| \cdots \|\underline{b}_k\|) \\
&< 2^{-q} \sum_{j=1}^{k} 2^{j-1} \prod_{\substack{i=1 \\ i \neq j}}^{k} \|\underline{b}_i\| \\
&\leq 2^{-q}(2^k - 1)\mathrm{dft}(B)\det B/\lambda_1(B) \\
&< 2^{-q+k}\mathrm{dft}(B)\det B/\lambda_1(B) \ .
\end{aligned}$$

$\square$

Corollary 12. *Let $c > 0$. If*

$$q \geq q_4(B, c) = \lceil k + \log(\mathrm{dft}(B)/\lambda_1(B)) - \log c \rceil$$

then

$$|\det A - \det B| < c|\det B| \ .$$

Now we prove Theorem 1. In order for A to be non singular it suffices by Corollary 12 to have

$$q \geq q_4(B, 1) = \lceil k + \log(\mathrm{dft}(B)/\lambda_1(B)) \rceil \ .$$

Let $U = (\underline{u}_1, \ldots, \underline{u}_k) \in \mathbf{Z}^{k \times k}$ be non singular with

$$\|B\underline{u}_i\| = \lambda_i(L), \quad 1 \leq i \leq k \ .$$

Applying Lemma 10 we find that for

$$q \geq \log \lambda_k(U) - \log(c\lambda_1(L)) \tag{2}$$

we have

$$\|A\underline{u}_i\| \le (1+c)\|B\underline{u}_i\| = (1+c)\lambda_i(L), \quad 1 \le i \le k$$

and therefore

$$\lambda_i(L(A)) \le (1+c)\lambda_i(L), \quad 1 \le i \le k \ .$$

It follows from Corollary 9 that

$$\lambda_k(U) \le \sqrt{k}\lambda_k(L)\mathrm{dft}(B)/\lambda_1(B) \ .$$

Therefore $q \ge q_1(B,c)$ implies (2). This concludes the proof of Theorem 1.

To prove Theorem 2 we need the following result.

Lemma 13. *If*

$$q \ge q_1(B, 1/3)$$

then

$$\mathrm{dft}(A)/\lambda_1(A) \le (3/2)^k \mathrm{dft}(B)/\lambda_1(B), \quad 1 \le i \le k \ .$$

Proof. By Corollary 12 we have

$$1/\det A \le (3/2)/\det B \ .$$

Finally, by Lemma 10 it follows that

$$\|\underline{a}_i\| \le (4/3)\|\underline{b}_i\|, \quad 1 \le i \le k \ .$$

This implies the assertion. □

Now we can prove Theorem 2. First, note that

$$q_2(B,c,s_k) \ge q_1(B,c/3) \ . \tag{3}$$

So A is non singular by Theorem 1.

Applying Lemma 10 we find that for

$$q \ge \log \lambda_k(T) - \log((c/3)\lambda_1(L))$$

we have

$$\|B\underline{t}_i\| < (1/(1-c/3))\|A\underline{t}_i\| \le (s_i/(1-c/3))\lambda_i(L(A)) \ .$$

Theorem 1 and (3) implies that

$$\lambda_i(L(A)) \le (1+c/3)\lambda_i(L), \quad 1 \le i \le k \ .$$

Since

$$(1+c/3)/(1-c/3) \le 1+c$$

this implies

$$\|B\underline{t}_i\| < (1+c)s_i\lambda_i(L)$$

for
$$q \geq \max\{\log \lambda_k(T) - \log((c/3)\lambda_1(L)), q_2(B, c, s_k)\} \ .$$

It remains to estimate $\|\underline{t}_i\|$. It follows from Corollary 9 that
$$\|\underline{t}_i\| \leq \sqrt{k} s_i \lambda_i(L(A)) \mathrm{dft}(A)/\lambda_1(A) \ .$$

So the assertion follows from Lemma 13 and Theorem 1.

Now can prove Corollary 3. Given a basis $C = (\underline{c}_1, \ldots, \underline{c}_k) \in \mathbf{Z}^{m \times k}$ of a k-dimensional integral lattice Λ. The algorithm of [He] determines a reduced basis $D \in \mathbf{Z}^{m \times k}$ such that
$$\|\underline{d}_i\| \leq \sqrt{(i+3)/4}\,\lambda_i(\Lambda), \quad 1 \leq i \leq k \ .$$

The algorithm performs at most $mk^{O(k)} \log |C|$ arithmetic operations on integers of binary length $O(k^2(\log k + \log |C|))$. Here, $|C|$ denotes the maximum of the absolute values of the entries of C. Choose a rational approximation A to B of precision
$$p = q_0(B, c, \sqrt{(k+3)/4}) \ .$$

Then
$$\|A - B\| < 2^{-q_2(B,c,s)} \ .$$

Hence, if we apply the reduction algorithm of [He] to
$$C = 2^{p+1} A$$

and determine the reducing transformation $T = (\underline{t}_1, \ldots, \underline{t}_k) \in \mathrm{GL}(k, \mathbf{Z})$ then by Theorem 2 we have
$$\|B\underline{t}_i\| \leq (1+c)\sqrt{(i+3)/4}\,\lambda_i(L), \quad 1 \leq i \leq k \ .$$

The transformation T can be determined by applying each reducing elementary transformation not only to the basis but also to the $k \times k$ identity matrix. We analyse the running time. We first estimate $\log |C|$. By Lemma 10 we have $|A| < 2\,\lambda_k(B)$ and therefore
$$\log |C| = O(k + \log m + \log(\lambda_k(L)/\lambda_1(L)) +$$
$$\log(\lambda_k(B)/\lambda_1(B)) + \log(\mathrm{dft}(B)) + \log(1/c)) \ .$$

Since
$$\lambda_i(L) \leq \lambda_i(B), \quad 1 \leq i \leq k$$

and
$$|\det(B)| = \det L \geq \left(\lambda_1(L)/\gamma_k^{1/2}\right)^k$$

where $\gamma_k = O(k)$ (see [LaLeSch] and [WaGr68]) it follows that
$$\log |C| = O(k(\log k + \beta) + \log m + \log(1/c))$$

with β from Corollary 3.

The number of arithmetic operations to find the transformation is bounded by the number of arithmetic operations to find the reduced basis, namely

$$mk^{O(k)}(\log m + \beta + \log(1/c)) \ .$$

Let C' be a basis which is computed in the course of the reduction. Let $T' \in \mathrm{GL}(k, \mathbf{Z})$ with $CT' = C'$. It follows from Corollary 9 that

$$|T'| \le \lambda_k(C')\mathrm{dft}(A)/(2^{p+1}\lambda_1(A)) \ .$$

Since

$$\log|C'| = O\left(k^2\left(k\left(\log k + \beta\right) + \log m + \log(1/c)\right)\right)$$

Lemma 13 implies that

$$\log|T'| = O\left(k^2\left(k\left(\log k + \beta\right) + \log m + \log(1/c)\right)\right) \ .$$

To prove Corollary 4 we apply the same method using the following result of [LeLeLo]. Given a basis $C = (\underline{c}_1, \ldots, \underline{c}_k) \in \mathbf{Z}^{m \times k}$ of a k-dimensional integral lattice Λ. The algorithm of [LeLeLo] determines a reduced basis $D \in \mathbf{Z}^{m \times k}$ such that

$$\|\underline{d}_i\| \le 2^{(k-1)/2}\lambda_i(\Lambda), \quad 1 \le i \le k \ .$$

The algorithm performs at most $O(k^3 m \log|C|)$ arithmetic operations on integers of binary length $O(k \log|C|)$.

References

[BuPo] J. Buchmann, M. Pohst: *Computing a lattice basis from a system of generating vectors*, Proceedings EUROCAL 87, Springer Lecture Notes in Computer Science **378**, 54-63, 1989.

[BuKe] J. Buchmann, V. Kessler: *Computing a reduced lattice basis from a generating system*, preprint, 1993.

[LeLeLo] A.K. Lenstra, H.W. Lenstra Jr., L. Lovász: *Factoring polynomials with rational coefficients*, Math. Ann. **261**, 515-534, 1982.

[GrLoSchri] M. Grötschel, L. Lovász, A. Schrijver: *Geometric algorithms and combinatorial optimization*, Springer-Verlag Berlin Heidelberg, 1988.

[HaMcCu] J.L. Hafner, K.S. McCurley: *Asymptotically fast triangularization of matrices over rings*, SIAM J. Computation **20**, no. 6, 1068-1083, 1991.

[He] B. Helfrich, *Algorithms to construct Minkowski reduced and Hermite reduced lattice bases*, Theoretical Computer Science **41**, 125-139, 1985.

[Po] M. Pohst: *A modification of the LLL Reduction Algorithm*, J. Symbolic Computation, **4**, 123-127, 1987.

[LaLeSch] J.C. Lagarias, H.W. Lenstra Jr., C.P. Schnorr: *Korkine-Zolotarev bases and successive minima of a lattice and its dual lattice*, Combinatorica **10**, no.4, 333-348, 1990.

[WaGr68] B.L. van der Waerden and H. Gross: *Studien zur Theorie der Quadratischen Formen*, Birkhäuser, Basel, 1968.

Analysis of a Left-Shift Binary GCD Algorithm

Jeffrey Shallit[1] and Jonathan Sorenson[2]

[1] Department of Computer Science, University of Waterloo,
Waterloo, Ontario N2L 3G1, Canada, shallit@graceland.uwaterloo.ca
[2] Department of Mathematics and Computer Science, Butler University,
4600 Sunset Avenue, Indianapolis, IN 46208, USA, sorenson@butler.edu

Abstract. We introduce a new left-shift binary algorithm, LSBGCD, for computing the greatest common divisor of two integers, and we provide an analysis of the worst-case behavior of this algorithm. The analysis depends on a theorem of Ramharter about the extremal behavior of certain continuants.

1 Introduction

There are many polynomial-time algorithms known for computing the greatest common divisor of two positive integers: for example, Euclid's algorithm, Euclid's algorithm with least remainder, Stein's "right-shift" binary algorithm [11], and Brent's "left-shift" binary algorithm [1]. The analysis of these algorithms has received much attention; for example, see [4, 5] for Euclid's algorithm; [2] for Euclid's algorithm with least remainder; and [1, 9] for the left- and right-shift binary algorithms.

In particular, it is desirable to determine the *worst-case* behavior of such algorithms: for each n, find the minimal input that requires the algorithm to perform n "steps". Of course, it is not immediately clear what is meant by "minimal input," since these algorithms typically have two inputs.[3] One reasonable definition is the *lexicographically least* input: we say (x, y) is lexicographically less than or equal to (x', y') if $x < x'$, or $x = x'$ and $y \leq y'$.

In this paper, we present and analyze a new kind of left-shift binary algorithm, LSBGCD, for computing the greatest common divisor. The LSBGCD algorithm is, in some sense, a "least-remainder" variant of the usual left-shift algorithm introduced by Brent [1]. It is also efficient in practice (see Sect. 7). Our worst-case analysis is of independent interest, since it shows that the worst-case inputs do not arise from the truncation of the continued fraction expansion of an irrational number, as is the case with the ordinary and least-remainder Euclidean algorithms.

Below is a Pascal-like pseudocode description of the LSBGCD algorithm.

---

[3] Nor is the definition of "step" immediately obvious. However, it is frequently taken to be an instance of the most expensive operation, such as a division.

LSBGCD

> INPUT: Positive integers u and v, $u \geq v$
> OUTPUT: $\gcd(u, v)$
>
> while $v \neq 0$ do
>   find $e \geq 0$ such that $2^e v \leq u < 2^{e+1} v$;
> $t := \min\{u - 2^e v, 2^{e+1} v - u\}$;
> $u := v$; $v := t$;
>   if $u < v$ then interchange(u, v);
> return(u);

Correctness of the algorithm is left to the reader.

As a consequence of our main result, the number of iterations used by this algorithm is at most $O(\log(uv))$. Since each iteration can be computed using at most $O(\log(uv))$ bit operations, LSBGCD uses at most $O(\log^2(uv))$ bit operations to compute $\gcd(u, v)$. This algorithm has a k-ary extension (see Sorenson [10]), making LSBGCD the special case of $k = 2$.

Our main result is the following.

Theorem 1. *Let $x_n > y_n \geq 0$, and let (x_n, y_n) be the lexicographically least pair of integers that requires the LSBGCD algorithm to perform n iterations. Then both of the sequences (x_n) and (y_n) satisfy linear recurrences of order 16 with characteristic polynomial $(z^4 - 1)(z^8 - 14z^4 + 1)(z^4 - 4z^2 + 1)$. Furthermore, letting $\alpha = 2 + \sqrt{3}$ and $\beta = 2 - \sqrt{3}$, we have*

$$x_{2n+1} = \frac{\sqrt{3}}{6}(\alpha^{n+1} - \beta^{n+1}) \; ;$$

$$y_{2n+1} = \frac{\sqrt{3}}{6}((1 + \sqrt{3})\alpha^n - (1 - \sqrt{3})\beta^n) \; ;$$

$$x_{4n} = \frac{2 + (4 - \sqrt{3})\alpha^{2n+1} + (4 + \sqrt{3})\beta^{2n+1}}{12} \; ;$$

$$y_{4n} = \frac{-2 + (1 + 3\sqrt{3})\alpha^{2n} + (1 - 3\sqrt{3})\beta^{2n}}{12} \; ;$$

$$x_{4n+2} = \frac{4 + (4 - \sqrt{3})\alpha^{2n+2} + (4 + \sqrt{3})\beta^{2n+2}}{12} \; ;$$

$$y_{4n+2} = \frac{-10 + (1 + 3\sqrt{3})\alpha^{2n+1} + (1 - 3\sqrt{3})\beta^{2n+1}}{12} \; .$$

The proof of this theorem has two main steps:

1. First, we show that, for worst-case inputs, LSBGCD always assigns either $t := u - v$ or $t := u - 2v$ in its main loop, and never performs an interchange. After introducing necessary notation in Sect. 2, this is proven in Sects. 3 and 4.

2. Second, we observe that, for worst-case inputs (x_n, y_n), the steps performed by the algorithm correspond to the continued fraction expansion of x_n/y_n, and therefore must have a special form. Applying a theorem of Ramharter [7] then allows us to complete the proof.

With this strategy, we complete the proof of our main result in Sects. 5 and 6.

We conclude in Sect. 7 with some remarks on the implementation of the LSBGCD algorithm.

2   Notation, Definitions, and an Example

In order to determine the smallest pair (u, v), we will model computations performed by this algorithm in terms of linear transformations on $\mathbb{N} \times \mathbb{N}$.

One loop iteration of the algorithm must perform precisely one of the following two transformations:

$$A_e(u, v) := (v, u - 2^e v) \quad \text{occurs when } u - 2^e v \leq 2^{e+1} v - u$$
$$B_e(u, v) := (v, 2^{e+1} v - u) \quad \text{occurs when } u - 2^e v > 2^{e+1} v - u$$

In addition, either of these is followed by the interchange transformation

$$I(u, v) := (v, u)$$

if the new pair (u, v) satisfies $u < v$.

Thus, a computation of the algorithm can be expressed as the composition of A_e, B_e, and I transformations. If f and g are transformations, we write $fg(u, v)$ for $f(g(u, v))$. So one iteration of the algorithm performs one of A_e, B_e, IA_e, or IB_e. We refer to one of these four options as a *step*, and we refer to any composition of steps as a *partial computation*. The *length* of a partial computation is the number of steps it contains, and corresponds to the number of loop iterations performed by the algorithm.

Example

Consider the computation of the algorithm on the inputs $(47, 40)$. The following table summarizes the steps performed by the algorithm:

u	v	t	Step
47	40	7	A_0
40	7	12	IA_2
12	7	2	B_0
7	2	1	B_1
2	1	0	A_1
1	0		

We write this computation as

$$A_1 B_1 B_0 I A_2 A_0 \ .$$

This partial computation has length 5.

Loosely speaking, our goal is to construct a long partial computation where the corresponding inputs are as small as possible. We do this by starting with the pair $(1,0)$ and applying the inverses of steps to obtain the corresponding input. The inverses of our three transformations are:

$$A_e^{-1}(x,y) := (2^e x + y, x)$$
$$B_e^{-1}(x,y) := (2^{e+1} x - y, x)$$
$$I^{-1}(x,y) := (y,x) = I(x,y)$$

Let f be the partial computation $A_1 B_1 A_0$. Observe that $f^{-1}(1,0) = (9,7)$. If we trace the algorithm on $(9,7)$, we in fact obtain the computation f, of length 3. However, it is possible to construct partial computations which would never occur as actual computations of the algorithm. The shortest such example is $A_0 A_0$. Inverting this computation on $(1,0)$, we obtain the pair $(2,1)$, and the algorithm actually performs step A_1 on this input.

Let f be a partial computation of length n. Define $(u,v) = f^{-1}(x,y)$, and let $(u_1, v_1), \ldots, (u_n, v_n)$ be the successive values taken by (u,v) after the 1st, 2nd, $\ldots$, nth iterations of the algorithm. We say f is *valid* on the pair (x,y) if $u \geq v$, $(u_n, v_n) = (x,y)$, and the n iterations performed by the algorithm correspond to the partial computation f. Thus, $A_1 B_1 A_0$ is valid on $(1,0)$, and $A_0 A_0$ is not. Also notice that $A_1 A_1$ is valid on $(10,5)$, and in fact on any pair (x,y) where $x \geq y$ except the pair $(0,0)$.

In the following section, we give necessary and sufficient conditions for constructing valid partial computations. We assume throughout that $x, y \geq 0$.

3 Valid Computations

We begin by examining the properties of the A_e and B_e transformations.

Lemma 2. *The algorithm will perform the transformation A_e resulting in the pair (x,y) if and only if $y \leq 2^{e-1} x$.*

Proof. Let $(u,v) = A_e^{-1}(x,y)$ so that $u = 2^e x + y$ and $v = x$. The algorithm performs A_e on (u,v) if and only if $u \geq 2^e v$ and $u - 2^e v \leq 2^{e+1} v - u$. But these two inequalities are true if and only if $2^e v \leq u \leq (3/2) \cdot 2^e v$, which in turn is true if and only if $2^e x \leq 2^e x + y \leq (3/2) \cdot 2^e x$. But this is equivalent to $y \geq 0$ and $y \leq 2^{e-1} x$. $\qquad \square$

Lemma 3. *The algorithm will perform the transformation B_e resulting in the pair (x,y) if and only if $y < 2^{e-1} x$ and $y > 0$.*

Proof. Let $(u,v) = B_e^{-1}(x,y)$ so that $u = 2^{e+1}x - y$ and $v = x$. The algorithm performs B_e on (u,v) if and only if $u < 2^{e+1}v$ and $u - 2^e v > 2^{e+1}v - u$. But these two inequalities are true if and only if $(3/2) \cdot 2^e v < u < 2^{e+1}v$, which in turn is true if and only if $(3/2) \cdot 2^e x < 2^{e+1}x - y < 2^{e+1}x$. But this is equivalent to $y > 0$ and $y < 2^{e-1}x$. $\qquad\square$

Theorem 4. *A partial computation consisting of a single step is valid on the pair (x,y) if and only if $x \geq y$ and the condition indicated below holds:*

Step	Condition
A_e	$y \leq 2^{e-1}x$
B_e	$y < 2^{e-1}x$ and $y > 0$
IA_e	$x \leq 2^{e-1}y$ and $e > 1$
IB_e	$x < 2^{e-1}y$, $x > 0$, and $e > 1$

Proof. The proof for steps A_e and B_e follows from a direct application of Lemmas 2 and 3.

Let $(u,v) = f^{-1}(x,y)$, where f is either IA_e or IB_e. For f to be valid on (x,y), we must have $x > y$ because if $x = y$ no interchange takes place. Thus $e > 1$. The rest of the proof is easily completed with the use of Lemmas 2 and 3. $\qquad\square$

Corollary 5. *The partial computations $A_0 A_0$, $A_0 B_0$, $B_0 A_0$, and $B_0 B_0$ are not valid for any pairs (x,y).*

Proof. We prove the corollary for $A_0 B_0$. The other three cases are similar.

Let $(w,z) = A_0^{-1}(x,y) = (x+y, x)$. For the A_0 step to be valid, we must have $y \leq x/2$. But then $z \leq w \leq (3/2)z$, so $z \geq (2/3)w$. This implies the B_0 step is not valid by Theorem 4 (since $z > w/2$), and hence the partial computation is invalid. $\qquad\square$

Lemma 6. *Let $f = s_1 s_2 \cdots s_n$ be a partial computation of length n, where each s_i is an A_e step, and no two consecutive steps are both A_0. Then f is valid on (x,y) if s_1 is valid on (x,y).*

Proof. Fix some pair (x,y). Define $(x_0, y_0) = (x,y)$ and $(x_i, y_i) = s_i^{-1}(x_{i-1}, y_{i-1})$ for all i, $1 \leq i \leq n$. Then f is valid on (x,y) if s_i is valid on (x_{i-1}, y_{i-1}) for all i.

If $s_{i-1} = A_e$ with $e > 0$, then we have $y_{i-1} \leq x_{i-1}/2$, so s_i is valid on (x_{i-1}, y_{i-1}) by Theorem 4. If $s_{i-1} = A_0$, we have $y_{i-1} \leq x_{i-1}$, but since we know $s_i = A_e$ with $e > 0$, s_i is valid on (x_{i-1}, y_{i-1}) by Theorem 4.

Thus, if s_1 is valid on (x,y), so is f. $\qquad\square$

4 Steps Performed on Worst-Case Inputs

The main result of this section implies that the steps performed by the algorithm on worst-case inputs use only steps A_0 and A_1. Before we proceed, we need some technical lemmas.

We say that (a, b) is *smaller* than (c, d) if both $a \leq c$ and $a + b \leq c + d$ are true. If this is the case, we write $(a, b) \leq (c, d)$. Note that this relation is transitive, and implies lexicographic order.

Lemma 7. *For any pair (x, y) where $x > 0$ and $x \geq y$, the following are true:*

1. *If $y \leq 2^{e-1}x$ then $A_e^{-1}(x, y) \leq B_e^{-1}(x, y)$.*
2. *If $e \leq e'$ then $A_e^{-1}(x, y) \leq A_{e'}^{-1}(x, y)$.*
3. *If $e > 1$ and $x \leq 2^{e-1}y$ then $A_1^{-1}(x, y) \leq (IA_e)^{-1}(x, y)$.*

Proof. Let (x, y) be as stated above.

1. We need show that $(2^e x + y, x) \leq (2^{e+1}x - y, x)$. But this is true since $(2^{e+1}x - y) - (2^e x + y) = 2^e x - 2y \geq 0$.
2. We need to show that $(2^e x + y, x) \leq (2^{e'}x + y, x)$, which is obvious.
3. We must show that

$$A_1^{-1}(x, y) = (2x + y, x) \leq (IA_e)^{-1}(x, y) = (2^e y + x, y) \ .$$

We have

$$2^e y + x = 2(2^{e-1}y) + x \geq 2^{e-1}y + 2x \geq 2x + y$$

and similarly

$$(2^e y + x) + (y) = 2(2^{e-1}y) + x + y \geq 3x + y = (2x + y) + (x) \ .$$

□

Lemma 8. *Let g be a partial computation consisting entirely of A_e steps, with no interchange transformations $(I$'s$)$. If $(x_1, y_1) \leq (x_2, y_2)$ then $g^{-1}(x_1, y_1) \leq g^{-1}(x_2, y_2)$.*

Proof. Let $(x_1, y_1) \leq (x_2, y_2)$ and let n be the length of g. We use induction on n. If $n = 0$ there is nothing to prove, so assume $n > 0$. Then $g = fA_e$ where f is a partial computation of length $n - 1$. Let $(w_i, z_i) = f^{-1}(x_i, y_i)$ for $i = 1, 2$. By the inductive hypothesis, we have $f^{-1}(x_1, y_1) \leq f^{-1}(x_2, y_2)$ and so $(w_1, z_1) \leq (w_2, z_2)$. It remains to show that $A_e^{-1}(w_1, z_1) \leq A_e^{-1}(w_2, z_2)$. But this is true because we have $A_e^{-1}(w_1, z_1) = (2^e w_1 + z_1, w_1)$ and $A_e^{-1}(w_2, z_2) = (2^e w_2 + z_2, w_2)$, which imply $2^e w_1 + z_1 = (2^e - 1)w_1 + (w_1 + z_1) \leq (2^e - 1)w_2 + (w_2 + z_2) = 2^e w_2 + z_2$.

□

Theorem 9. *Let g be a partial computation of length n, valid on the pair (x, y), with $x \geq y$. There is another partial computation f of length n, also valid on (x, y), where f consists entirely of A_0 and A_1 computations, and $f^{-1}(x, y) \leq g^{-1}(x, y)$.*

Proof. Let g and (x, y) be as stated above. Let m denote the number of steps in g that are not A_0 or A_1. We prove the theorem using induction on m. There is nothing to prove if $m = 0$, so assume $m > 0$. Then we can write

$$g = rst$$

where t is a partial computation containing only A_0 and A_1 steps, s is one step that is not A_0 or A_1, and r is a partial computation having at most $m - 1$ steps that are not A_0 or A_1.

If $s = B_0$, let $h = A_0 t$, and otherwise let $h = A_1 t$. If we can show rh is valid on (x, y) and that $(rh)^{-1}(x, y) \le g^{-1}(x, y)$, then by the inductive hypothesis we are done.

Let $(w, z) = r^{-1}(x, y)$. We know that s is valid on (w, z).

If $s = B_0$, by Theorem 4 we have $z \le w/2$. Similarly, A_0 is valid on (w, z) too. In addition, since g is valid on (x, y), we know the first step of t (if t has positive length) must be A_1. Thus t is valid on $A_0^{-1}(w, z)$ by Lemma 6. Therefore, rh is valid on (x, y).

If $s \ne B_0$, we can only assume $z \le w$. But by Theorem 4 we know A_1 is valid on (w, z). Again, by Lemma 6 t is valid on $A_1^{-1}(w, z)$. So rh is valid on (x, y).

All that remains to show is that $h^{-1}(w, z) \le (st)^{-1}(w, z)$, but this follows from Lemma 8 (applied to t), the fact that $A_0^{-1}(w, z) \le B_0^{-1}(w, z)$, and if $s \ne B_0$, $A_1^{-1}(w, z) \le s^{-1}(w, z)$ by Lemma 7.

That completes the proof. $\qquad\square$

5 A Theorem on Continuants

In this section we exhibit a relationship between the worst case of the LSBGCD algorithm and extremal values of certain continuants. We will assume that the reader is familiar with the elementary properties of continued fractions and continuants, as explained, for example, in [3] and [4]. In particular, we use the standard abbreviation $[a_1, a_2, \ldots, a_n]$ to denote the continued fraction

$$a_1 + \cfrac{1}{a_2 + \cfrac{1}{a_3 + \cdots + \frac{1}{a_n}}} . \tag{1}$$

As we have seen in Theorem 9, to determine the lexicographically smallest pair (x, y) that requires n while-loop iterations in the LSBGCD algorithm, it suffices to consider inputs that result in the algorithm performing only steps of the form A_0 and A_1. But it is easy to see that the transformations $A_0(u, v) = (v, u - v)$ and $A_1(u, v) = (v, u - 2v)$ are precisely the same transformations performed by the ordinary Euclidean algorithm when the partial quotients are 1 or 2, respectively. Furthermore, we have seen that the sequence of steps $A_0 A_0$ is not valid for any pair (x, y); this corresponds to a restriction that the corresponding ordinary Euclidean algorithm cannot use two consecutive steps of partial quotient 1.

It follows that to determine the lexicographically least pair (x, y) with $x \ge y$, that causes the LSBGCD algorithm to perform n iterations, it suffices to determine $a_1, a_2, \ldots, a_n \in \{1, 2\}$ such that $(a_i, a_{i+1}) \ne (1, 1)$ for $1 \le i \le n$, and (x, y) is lexicographically minimized, where

$$[a_1, a_2, \ldots, a_n] = \frac{x}{y} .$$

Now it is well-known (see, e.g., [4]) that the numerator and denominator of a continued fraction can be expressed as continuant polynomials in the partial quotients a_i. We have

$$[a_1, a_2, \ldots, a_n] = \frac{Q_n(a_1, a_2, \ldots, a_n)}{Q_{n-1}(a_2, a_3, \ldots, a_n)} \; .$$

Here Q_n represents a polynomial in n variables, defined as follows:

$$\begin{aligned}
Q_0() &= 1 \\
Q_1(x_1) &= x_1 \\
Q_n(x_1, x_2, \ldots, x_n) &= x_n Q_{n-1}(x_1, x_2, \ldots, x_{n-1}) + Q_{n-2}(x_1, x_2, \ldots, x_{n-2})
\end{aligned} \tag{2}$$

for $n \geq 2$.

An important property of continuants is the following:

$$Q_n(x_1, x_2, \ldots, x_n) = Q_n(x_n, x_{n-1}, \ldots, x_1) \; . \tag{3}$$

Let P_n denote the set of all permutations $(a_1, a_2, \ldots, a_n)$ such that $a_i \in \{1, 2\}$ for $1 \leq i \leq n$, and $(a_i, a_{i+1}) \neq (1, 1)$ for $1 \leq i \leq n-1$. Then it follows that we seek to minimize the continuant $Q_n(a_1, a_2, \ldots, a_n)$ over all elements $(a_1, a_2, \ldots, a_n)$ of P_n.[4]

To do this, we employ the following theorem of Ramharter [7], which determines all the minimal values of $Q(c_1, c_2, \ldots, c_n)$, where the c_i form a permutation of a given list of (not necessarily distinct) integers:

Theorem 10 (Ramharter, 1983). *Let* $B = (b_1, b_2, \ldots, b_n)$ *denote a list of positive integers such that* $b_1 \leq b_2 \leq b_3 \leq \ldots \leq b_n$. *Let* P *denote the set of all permutations of* B. *Then (up to reversal of the order of the arguments of* Q_n*) the continuant polynomial* Q_n *uniquely attains its minimum on* P *at* $C = (c_1, c_2, \ldots, c_n)$, *where*

$$C = \begin{cases}
(b_1, b_{2k}, b_3, b_{2k-2}, \ldots, b_{2k-3}, b_4, b_{2k-1}, b_2), \\
\quad \textit{if } n = 2k; \\
(b_1, b_{4k-1}, b_3, b_{4k-3}, \ldots, b_{2k-1}, b_{2k+1}; b_{2k}, b_{2k+2}, \ldots, b_{4k-4}, b_4, b_{4k-2}, b_2), \\
\quad \textit{if } n = 4k - 1; \\
(b_1, b_{4k+1}, b_3, b_{4k-1}, \ldots, b_{2k+3}, b_{2k+1}; b_{2k+2}, b_{2k}, \ldots, b_{4k-2}, b_4, b_{4k}, b_2), \\
\quad \textit{if } n = 4k + 1.
\end{cases}$$

(The semicolon above plays the same semantic role as the comma, namely to denote concatenation. We have used the semicolon, following Ramharter, to indicate a logical break in the ordering of the coefficients.)

---

[4] In fact, because of the identity (3), there will typically be *two* minimal n-tuples, one of which is the reversal of the other.

Unfortunately, we cannot apply Ramharter's theorem immediately to our situation, since we do not yet know how many of the a_i are equal to 1 in the continuant that is minimal over all members of P_n. We now prove the following theorem:

Theorem 11. *The continuant polynomial $Q_n(a_1, a_2, \ldots, a_n)$ attains its minimum over all elements $(a_1, a_2, \ldots, a_n) \in P_n$ precisely as described below:*

$$(a_1, a_2, \ldots, a_n) = \begin{cases} (\overbrace{1,2,1,2,\ldots,1,2,1}^{n}), & \text{if } n \text{ is odd}; \\[2mm] (\overbrace{1,2,1,2,\ldots,1,2}^{n/2}, \overbrace{2,1,2,1,\ldots,2,1}^{n/2}), & \text{if } n \equiv 0\,(\mathrm{mod}\,4); \\[2mm] (\overbrace{1,2,1,2,\ldots,1,2}^{n/2\pm1}, \overbrace{2,1,2,1,\ldots,2,1}^{n/2\mp1}), & \text{if } n \equiv 2\,(\mathrm{mod}\,4). \end{cases}$$

We remark that the minimum is attained uniquely in the case $n \not\equiv 2$ (mod 4). When $n \equiv 2\,(\mathrm{mod}\,4)$, the minimum is attained at precisely two elements of P_n, one of which is the reversal of the other.

Proof. Suppose n is odd, say $n = 2k+1$. Let $A = (a_1, a_2, \ldots, a_n)$ be the permutation chosen from P_n that minimizes $Q_n(a_1, a_2, \ldots, a_n)$. Let j denote the number of a_i in A that are equal to 1. Define $R_{n,j}$ to be the set of all permutations of j 1's and $n-j$ 2's (with no restrictions on adjacency). Ramharter's theorem then gives us the continuant that is minimal over all members of $R_{n,j}$. Using the notation of Ramharter's theorem, define $b_1, b_2, \ldots, b_j = 1$, and $b_{j+1}, b_{j+2}, \ldots, b_n = 2$. There are three cases to consider: (i) $j < n/2$; (ii) $j > (n+1)/2$; and (iii) $j = (n+1)/2$. We show that the first two cases lead to a contradiction:

(i) If $j < n/2$, then it is easy to see that, according to Ramharter's theorem, the n-tuple which minimizes Q_n over all members of $R_{n,j}$ is in fact a member of P_n.

If $n = 1, 3$ then a simple computation shows that no permutation in P_n containing fewer than $n/2$ 1's can minimize Q_n.

Now suppose $n > 3$ and $n \equiv 3\,(\mathrm{mod}\,4)$. Then by Ramharter's theorem the element $A = (a_1, a_2, \ldots, a_n)$ that minimizes Q_n, over all members of $R_{n,j}$ (up to reversal of A), satisfies

$$(a_{k+1}, a_{k+2}, a_{k+3}) = (b_{k+2}, b_{k+1}, b_{k+3}) = (2, 2, 2) \ .$$

But then

$$Q_n(a_1, a_2, \ldots, a_{k+1}, 1, a_{k+3}, \ldots, a_n) < Q_n(a_1, a_2, \ldots, a_n) \ ,$$

and $(a_1, a_2, \ldots, a_{k+1}, 1, a_{k+3}, \ldots, a_n) \in P_n$, contradicting our assumption that A minimized Q_n over the set P_n.

Similarly, suppose $n > 1$ and $n \equiv 1 \pmod 4$. Then by Ramharter's theorem the element $A = (a_1, a_2, \ldots, a_n)$ that minimizes Q_n over all members of $R_{n,j}$ (up to reversal of A), satisfies

$$(a_k, a_{k+1}, a_{k+2}) = (b_{k+3}, b_{k+1}, b_{k+2}) = (2, 2, 2) \ .$$

Then

$$Q_n(a_1, a_2, \ldots, a_k, 1, a_{k+2}, \ldots, a_n) < Q_n(a_1, a_2, \ldots, a_n) \ ,$$

and $(a_1, a_2, \ldots, a_k, 1, a_{k+2}, \ldots, a_n) \in P_n$, a contradiction again.

(ii) If $j > (n+1)/2$, then $P_n \cap R_{n,j} = \emptyset$, since any permutation of n 1's and 2's that contains more than $(n+1)/2$ 1's must contain two consecutive 1's. This contradicts our assumption that A minimized the continuant Q_n over the set P_n.

(iii) It follows that $j = (n+1)/2$. Now we can use Ramharter's theorem directly, and we conclude that the member of $R_{n,(n+1)/2}$ which minimizes Q_n is

$$(\overbrace{1, 2, 1, 2, \ldots, 1, 2, 1}^{n}) \ .$$

It remains to consider the case when n is even, say $n = 2k$. Let A, j, and $R_{n,j}$ be defined as above. Again, there are three cases to consider: (i) $j < n/2$; (ii) $j > n/2$; and (iii) $j = n/2$. We will show that the first two cases lead to a contradiction:

(i) If $j < n/2$, then by Ramharter's theorem, the n-tuple which minimizes Q_n over all members of $R_{n,j}$ is in fact a member of P_n.

If $n = 2$ then a simple computation shows that no permutation in P_n containing fewer than $n/2$ 1's can minimize Q_n.

Now suppose $n > 2$ and $n \equiv 2 \pmod 4$. Then by Ramharter's theorem the element $A = (a_1, a_2, \ldots, a_n)$ that minimizes Q_n, over all members of $R_{n,j}$ (up to reversal of A), satisfies

$$(a_k, a_{k+1}, a_{k+2}) = (b_k, b_{k+1}, b_{k+2}) = (2, 2, 2) \ .$$

But then

$$Q_n(a_1, a_2, \ldots, a_k, 1, a_{k+2}, \ldots, a_n) < Q_n(a_1, a_2, \ldots, a_n) \ ,$$

and $(a_1, a_2, \ldots, a_k, 1, a_{k+2}, \ldots, a_n) \in P_n$, contradicting our assumption that A minimized Q_n over the set P_n.

Similarly, suppose $n > 0$ and $n \equiv 0 \pmod 4$. Then by Ramharter's theorem the element $A = (a_1, a_2, \ldots, a_n)$ that minimizes Q_n over all members of $R_{n,j}$ (up to reversal of A), satisfies

$$(a_k, a_{k+1}, a_{k+2}) = (b_{k+2}, b_{k+1}, b_k) = (2, 2, 2) \ .$$

Then

$$Q_n(a_1, a_2, \ldots, a_k, 1, a_{k+2}, \ldots, a_n) < Q_n(a_1, a_2, \ldots, a_n) \ ,$$

and $(a_1, a_2, \ldots, a_k, 1, a_{k+2}, \ldots, a_n) \in P_n$, a contradiction again.

(ii) If $j > n/2$, then $P_n \cap R_{n,j} = \emptyset$, since any permutation of n 1's and 2's that contains more than $n/2$ 1's must contain two consecutive 1's. This contradicts our assumption that A minimized the continuant Q_n over the set P_n.

(iii) It follows that $j = n/2$. Now we can use Ramharter's theorem directly, and we conclude that the member(s) of $R_{n,n/2}$ which minimizes Q_n is (are)

$$\begin{cases} (\overbrace{1,2,1,2,\ldots,1,2}^{n/2}, \overbrace{2,1,2,1,\ldots,2,1}^{n/2}), & \text{if } n \equiv 0 \ (\text{mod } 4); \\ (\overbrace{1,2,1,2,\ldots,1,2}^{n/2\pm 1}, \overbrace{2,1,2,1,\ldots,2,1}^{n/2\mp 1}), & \text{if } n \equiv 2 \ (\text{mod } 4). \end{cases}$$

This completes the proof of Theorem 11. $\qquad\square$

6   Proof of the Main Result

Let (x_n, y_n) denote the lexicographically least pair that requires n iterations of the LSBGCD algorithm.

It now follows immediately from Theorem 11 that if n is odd, then

$$x_n/y_n = [\ \overbrace{1,2,1,2,\ldots,1,2,1}^{n}\ ] \ .$$

Using standard techniques, we now deduce that $(x_1, y_1) = (1,1)$; $(x_3, y_3) = (4,3)$, and $(x_n, y_n) = (4x_{n-2} - x_{n-4}, 4y_{n-2} - y_{n-4})$ for n odd, $n \geq 5$.

Also, if $n \equiv 0 \,(\text{mod } 4)$, then

$$x_n/y_n = [\ \overbrace{1,2,1,2,\ldots,1,2}^{n/2}, \overbrace{2,1,2,1,\ldots,2,1}^{n/2}\ ] \ .$$

In this case, we can obtain a recurrence relation for (x_n) and (y_n) using a trick from [8], as follows: First, we observe that

$$\begin{bmatrix} x_1 & 1 \\ 1 & 0 \end{bmatrix} \begin{bmatrix} x_2 & 1 \\ 1 & 0 \end{bmatrix} \cdots \begin{bmatrix} x_n & 1 \\ 1 & 0 \end{bmatrix} = \begin{bmatrix} Q_n(x_1,\ldots,x_n) & Q_{n-1}(x_1,\ldots,x_{n-1}) \\ Q_{n-1}(x_2,\ldots,x_n) & Q_{n-2}(x_2,\ldots,x_{n-1}) \end{bmatrix}, \quad (4)$$

a fact which is easily proved by induction.

Now define $M_1 = \begin{bmatrix} 1 & 1 \\ 1 & 0 \end{bmatrix}$, and $M_2 = \begin{bmatrix} 2 & 1 \\ 1 & 0 \end{bmatrix}$, and $M(k) = (M_1 M_2)^k (M_2 M_1)^k$.

Also define

$$M(k) = \begin{bmatrix} C_k & D_k \\ E_k & F_k \end{bmatrix} .$$

Then we find

$$M(k+1) = M_1 M_2 M(k) M_2 M_1$$
$$= \begin{bmatrix} 9C_k + 3D_k + 3E_k + F_k & 6C_k + 3D_k + 2E_k + F_k \\ 6C_k + 2D_k + 3E_k + F_k & 4C_k + 2D_k + 2E_k + F_k \end{bmatrix} ;$$
$$M(k+2) = M_1 M_2 M(k+1) M_2 M_1$$
$$= \begin{bmatrix} 121C_k + 44D_k + 44E_k + 16F_k & 88C_k + 33D_k + 32E_k + 12F_k \\ 88C_k + 32D_k + 33E_k + 12F_k & 64C_k + 24D_k + 24E_k + 9F_k \end{bmatrix} ;$$

$$M(k+3) = M_1 M_2 M(k+2) M_2 M_1$$
$$= \begin{bmatrix} 1681C_k + 615D_k + 615E_k + 225F_k & 1230C_k + 451D_k + 450E_k + 165F_k \\ 1230C_k + 450D_k + 451E_k + 165F_k & 900C_k + 330D_k + 330E_k + 121F_k \end{bmatrix}.$$

It is now easy to verify that $M(k+3) = 15M(k+2) - 15M(k+1) + M(k)$; it follows that each of the sequences $(C_k), (D_k), (E_k), (F_k)$ are linear recurrences with characteristic polynomial $z^3 - 15z^2 + 15z - 1 = (z-1)(z^2 - 14z + 1)$. It follows that $(x_4, y_4) = (10, 7)$; $(x_8, y_8) = (137, 100)$; $(x_{12}, y_{12}) = (1906, 1395)$; and $(x_n, y_n) = (15x_{n-4} - 15x_{n-8} + x_{n-12}, 15y_{n-4} - 15y_{n-8} + y_{n-12})$ for $n \equiv 0 \pmod 4$, $n \geq 16$.

It remains to handle the case $n \equiv 2 \pmod 4$. This case is slightly more difficult than the preceding cases, for now Theorem 11 tells us that there are *two* distinct minimal continuants. These two continuants correspond to two distinct candidates for the lexicographically least pair (x_n, y_n) for each n with $n \equiv 2 \pmod 4$. Define

$$x'_n/y'_n = [\; \overbrace{1,2,1,2,\ldots,1,2}^{n/2-1},\; \overbrace{2,1,2,1,\ldots,2,1}^{n/2+1}\;]$$

and

$$x''_n/y''_n = [\; \overbrace{1,2,1,2,\ldots,1,2}^{n/2+1},\; \overbrace{2,1,2,1,\ldots,2,1}^{n/2-1}\;].$$

We know that $x_n = x'_n = x''_n$; to determine which is in fact the lexicographically least pair (x_n, y_n), we must determine which of the inequalities $y'_n < y''_n$ or $y'_n > y''_n$ is true.

As in the previous case, define $M_1 = \begin{bmatrix} 1 & 1 \\ 1 & 0 \end{bmatrix}$; $M_2 = \begin{bmatrix} 2 & 1 \\ 1 & 0 \end{bmatrix}$; and

$$N(k) = (M_1 M_2)^k (M_2 M_1)^{k+1} = \begin{bmatrix} R_k & S_k \\ T_k & U_k \end{bmatrix}. \tag{5}$$

It follows that $x'_{4k+2} = x''_{4k+2} = R_k$ and $y'_{4k+2} = T_k$. By taking the transpose of equation (5), we see that

$$N(k)^T = (M_1 M_2)^{k+1} (M_2 M_1)^k = \begin{bmatrix} R_k & T_k \\ S_k & U_k \end{bmatrix},$$

and hence $y''_{4k+2} = S_k$.

Now we prove by induction that $S_k - T_k = 1$ for all $k \geq 0$. Clearly this is true for $k = 0$, for then

$$N(0) = M_2 M_1 = \begin{bmatrix} 3 & 2 \\ 1 & 1 \end{bmatrix},$$

and hence $S_0 = 2$ and $T_0 = 1$. Now assume that $S_k - T_k = 1$ for all $k' \leq k$; we prove it for $k+1$. We find

$$N(k+1) = M_1 M_2 N(k) M_2 M_1$$
$$= \begin{bmatrix} 9R_k + 3S_k + 3T_k + U_k & 6R_k + 3S_k + 2T_k + U_k \\ 6R_k + 2S_k + 3T_k + U_k & 4R_k + 2S_k + 2T_k + U_k \end{bmatrix}.$$

It follows that $S_{k+1} - T_{k+1} = (6R_k + 3S_k + 2T_k + U_k) - (6R_k + 2S_k + 3T_k + U_k) = S_k - T_k = 1$, by induction. This completes the proof that $S_k - T_k = 1$ for all $k \geq 0$.

It now follows that $y'_{4k+2} < y''_{4k+2}$. Thus we have established that, for $n \equiv 2 \pmod 4$,

$$x_n/y_n = [\overbrace{1,2,1,2,\ldots,1,2}^{n/2-1}, \overbrace{2,1,2,1,\ldots,2,1}^{n/2+1}] .$$

We can obtain a recurrence relation for (x_n, y_n) as before. Omitting the calculations, it follows that $(x_2, y_2) = (3, 1)$; $(x_6, y_6) = (37, 26)$; $(x_{10}, y_{10}) = (511, 373)$; and

$$(x_n, y_n) = (15x_{n-4} - 15x_{n-8} + x_{n-12}, 15y_{n-4} - 15y_{n-8} + y_{n-12})$$

for $n \equiv 2 \pmod 4$, $n \geq 14$.

We can now prove Theorem 1 by a simple, though tedious, induction on n.

7 Remarks on Implementation

In practice, the LSBGCD algorithm is quite efficient.

The worst-case behavior can be compared roughly as follows: Euclid's algorithm performs at most (approximately) $1.44 \log_2 u$ iterations on input (u, v) with $u > v > 0$. The LSBGCD algorithm performs at most (approximately) $1.05 \log_2 u$ iterations on the same input. Thus, LSBGCD uses about 27% fewer iterations in the worst-case.

Empirically, LSBGCD performs about 15% more iterations of its main loop than Euclid's algorithm, in the average case. However, since each loop iteration can be performed using only shifts and subtractions, on most computers the LSBGCD outperforms the Euclidean algorithm, and is competitive with the binary algorithm.

In Table 1 below we give timing results for four GCD algorithms: the binary algorithm, Brent's left-shift binary algorithm, Euclid's algorithm, and LSBGCD. Five pseudo-random numbers of each of four digit sizes were used as inputs. The times given in the table are averages. A 486/33 CompuAdd PC running MS-DOS version 5 was used, and the algorithms were implemented in Turbo C++.

In Table 2, we give the average number of main loop iterations performed by each algorithm using the same inputs as in Table 1.

For additional implementation results, see Sorenson [10].

The LSBGCD algorithm is uniquely suited for extended GCD computation; it uses only left shifts and subtractions, and has a relatively low number of iterations. Nice [6] implemented five different extended GCD algorithms, including the Euclidean and binary algorithms, and found LSBGCD to be the fastest.

Table 1. Average running times in CPU seconds

Algorithm	Input Size in Decimal Digits			
	100	250	500	1000
Binary	0.049	0.201	0.660	2.38
Brent's	0.049	0.217	0.742	2.73
Euclidean	0.073	0.357	1.28	4.85
LSBGCD	0.049	0.211	0.714	2.64

Table 2. Average number of main loop iterations

Algorithm	Input Size in Decimal Digits			
	100	250	500	1000
Binary	469	1173	2341	4682
Brent's	288	731	1455	2914
Euclidean	191	494	980	1938
LSBGCD	219	550	1102	2212

8 Acknowledgments

We are most grateful to Prof. C. P. Schnorr, who ignited our interest in the subject when he told the first author about Brent's left-shift binary GCD algorithm in 1988. Robert Black read an early version of this paper and made several useful suggestions. Heinrich Rolletschek read a draft of this paper very carefully and helped us clarify our arguments.

We wish to acknowledge support for the first author by a grant from NSERC and for the second author by a Butler University Fellowship and NSF Grant CCR-9204414.

References

1. R. P. Brent. Analysis of the binary Euclidean algorithm. In J. F. Traub, editor, *Algorithms and Complexity: New Directions and Recent Results*, pages 321–355, Academic Press, New York, 1976.
2. A. Dupré. Sur le nombre de divisions à effectuer pour obtenir le plus grand commun diviseur entre deux nombres entiers. *J. Math. Pures Appl.*, 11:41–64, 1846.
3. G. H. Hardy and E. M. Wright. *An Introduction to the Theory of Numbers.* Oxford University Press, 5th edition, 1985.
4. D. E. Knuth. *The Art of Computer Programming. Volume 2: Seminumerical Algorithms.* Addison-Wesley, 1981. 2nd edition.

5. G. Lamé. Note sur la limite du nombre des divisions dans la recherche du plus grand commun diviseur entre deux nombres entiers. *C. R. Acad. Sci. Paris*, 19:867–870, 1844.

6. B. Nice. Extended greatest common divisors. 1993. Manuscript.

7. G. Ramharter. Extremal values of continuants. *Proc. Amer. Math. Soc.*, 89:189–201, 1983.

8. J. O. Shallit. On the worst case of three algorithms for computing the Jacobi symbol. *J. Symbolic Comput.*, 10:593–610, 1990.

9. J. O. Shallit and J. P. Sorenson. A binary algorithm for the Jacobi symbol. *ACM SIGSAM Bull.*, 27(1):4–11, 1993.

10. J. P. Sorenson. Two fast GCD algorithms. *J. Algorithms*, 16:110–144, 1994.

11. J. Stein. Computational problems associated with Racah algebra. *J. Comput. Phys.*, 1:397–405, 1967.

The complexity of greatest common divisor computations

Bohdan S. Majewski* and George Havas**

Key Centre for Software Technology
Department of Computer Science
The University of Queensland
Queensland 4072, Australia

Abstract. We study the complexity of expressing the greatest common divisor of n positive numbers as a linear combination of the numbers. We prove the NP-completeness of finding an optimal set of multipliers with respect to either the L_0 metric or the L_∞ norm. We present and analyze a new method for expressing the gcd of n numbers as their linear combination and give an upper bound on the size of the largest multiplier produced by this method, which is optimal.

1   Introduction

Euclid's algorithm for computing the greatest common divisor of 2 numbers is considered to be the the the oldest proper algorithm known ([Knu73]). It has been much analyzed and various versions exist that exploit different aspects. The extent of the literature on the gcd of just two numbers indicates that this seemingly simple problem has much more to it than meets the eye.

The nature and complexity of computing the gcd of two numbers is reasonably well understood and analyzed. From a variety of traditional algorithms [Knu73, Wat77, dBZ53], through algorithms that obtain excellent performance by exploiting binary representation of integers [Man88, Nor87], to algorithms intended to speed up a computation when the input numbers are very large [Knu73, Sor94], the two number gcd problem enjoys considerable attention. This shows how valuable it is that most of the questions about it are solved positively.

The gcd problem for more than two numbers is interesting in its own right (witness the research level problem in [Knu73]). Furthermore, it has important applications, for example in computing canonical normal forms of integer matrices ([HM93, HM94, Ili89]). However, although there are a few algorithms for computing the greatest common divisor of $n > 2$ numbers ([Bla63, Bra70, Ili89, Wat77]), there was no good understanding of the nature of such computations.

There are a number of problems for which efficient solutions are not readily available. Given a multiset of n numbers, how many of them do we need to take to obtain the gcd of all of them? How can we efficiently find 'good' multipliers

* Email: bohdan@cs.uq.oz.au
** Email: havas@cs.uq.oz.au; partially supported by the Australian Research Council

in an extended gcd computation? How quickly can we obtain the result? We investigate some of these problems and provide concrete answers.

2 On the complexity of finding small multipliers

One of the more challenging issues in any gcd computation is the expression of the result as an integer linear combination of the input. This is done by an extended gcd computation. For example, when divisions are replaced by logical shift operations, the resulting algorithm for just two numbers fails to produce optimal multipliers. Here by optimal we mean a definitely least solution [Lev56], where $\gcd(a_1, a_2)$ is expressed as $x_1 a_1 + x_2 a_2$ with $|x_1| \leq a_2/2$ and $|x_2| \leq a_1/2$.

The situation is worse if we consider expressing the gcd of n numbers, for arbitrary n, as a linear combination. It is not even clear how 'a definitely least solution' should be defined in this context. A number of sensible interpretations are possible. The collection of numbers is a multiset, which corresponds to the set A and size $s(a) \in Z^+$ of [GJ79, problem SP12]. By ordering the multiset we can conveniently use linear algebra, so that $\sum_{i=1}^{n} x_i a_i$ is simply the dot product of two vectors $\mathbf{x} = [x_1, \ldots, x_n]$ and $\mathbf{a} = [a_1, \ldots, a_n]$. Thus we define an optimal vector $\mathbf{x}$ with respect to a specific metric. An optimal vector with respect to the L_0 metric may represent a very poor solution with respect to the L_2 or the L_∞ norm. For the L_0 metric we at least know that an optimal solution $\mathbf{x}$ will satisfy the upper bound $|\mathbf{x}|_0 \leq \log_2(\min\{a_1, \ldots, a_n\})$. No such bounds were known for other metrics for general n.

We give proofs about the complexity of obtaining optimum multipliers with respect to the L_0 metric and the L_∞ norm. We present a new method for expressing the gcd of n numbers as their linear combination, and prove that it achieves the best possible upper bound on coordinates of $\mathbf{x}$, that is, with respect to the L_∞ norm.

Generally, the numeric variables in this paper are integers. For simplicity we assume that all a_i's are positive, which does not diminish the generality of our considerations. We start with the following problem:

MINIMUM GCD SET

INSTANCE: Multiset A of positive integers, positive integer $K \leq |A|$.
QUESTION: Does A contain a subset R, such that $|R| \leq K$ and $\gcd(r \in R) = \gcd(a \in A)$.

Theorem 1. *The MINIMUM GCD SET problem is NP-complete.*

Proof. Clearly the problem is in NP, as given a set R we can verify in polynomial time that $|R| \leq K$ and that $\gcd(r \in R) = \gcd(a \in A)$ (cf. [Bra70]).

Now we show that there exists a transformation from MINIMUM COVER [GJ79, problem SP5] to MINIMUM GCD SET, such that the answer 'yes' or 'no' for MINIMUM GCD SET is also the right answer for MINIMUM COVER. An instance of MINIMUM COVER is defined by a collection C of m subsets of a finite set S of size n and a positive integer $K \leq |C|$. We show how a polynomial time algorithm for MINIMUM GCD SET solves MINIMUM COVER in polynomial time.

Initially we select the first n prime numbers, say $p_1 = 2$, $p_2 = 3$, $\ldots$, p_n. Let $C = \{c_1, c_2, \ldots, c_m\}$, where $c_i = \{s_{i_1}, \ldots, s_{i_h}\}$ for $h > 0$. For each $c_i \in C$ construct the number $a_i = \prod_{i=1}^{n} p_i^{\alpha(i,j)}$, $j = 1, \ldots, n$, where $\alpha(i,j) = 0$ if $s_j \in c_i$, and 1 if $s_j \notin c_i$. The size of p_n is $O(\log n \log \log n)$. Therefore the length of each a_i is bounded by $O(n^{1+\epsilon})$ for any $\epsilon > 0$. As there are m such numbers, the size of the new problem is at most $O(mn^{1+\epsilon})$.

The greatest common divisor of the a_i's is, from the definition, equal to

$$p_1^{\min(\alpha(1,1),\ldots,\alpha(1,n))} \cdots p_n^{\min(\alpha(m,1),\ldots,\alpha(m,n))} = \prod_{i=1}^{n} p_i^0 = 1.$$

Now suppose that we have found a subset R with the desired property. As $\gcd(r \in R) = 1$, for each $r \in R$ there is a corresponding $c \in C$, $c = c(r)$ and $\bigcup_{r \in R} c(r) = S$. (To have $\gcd(r \in R) = 1$ we must have the factor p_i^0 in some $r \in R$ for each $i = 1, \ldots, n$, and there is a 1-1 mapping between the p_i's and s_i's.) Thus $C' = C'(R)$ is a solution to MINIMUM COVER. If however $l > K$ numbers are required in order to obtain the correct greatest common divisor there is no subset C' of C with the specified property. $\square$

Next we look at the task of finding small multipliers. First consider the following problem (satisfaction of an L_∞ norm bound):

MINIMUM GCD MULTIPLIERS

INSTANCE: Multiset A of positive integers, positive integer K.
QUESTION: Does there exist a solution to

$$\sum_{i=1}^{n} x_i a_i = \gcd(a_1, \ldots, a_n)$$

such that each $|x_i|$ is bounded by K?

The MINIMUM GCD MULTIPLIERS problem is NP-complete. To prove this we proceed in two steps. First we prove that the restricted case, where each multiplier is either $+1$ or -1, is NP-complete. Then we use the bit-pattern engineering technique of van Emde Boas [vEB81] to build a polynomial time reduction from the restricted problem to the case where each $x_i \in \{-1, 0, +1\}$. A modification of this gives the proof of our claim.

We start by exhibiting a polynomial time transformation from PARTITION [GJ79, problem SP12] to UNEVEN PARTITION (UP). UP asks if, for a given multiset A of n positive integers and positive constant c, there exists a submultiset $A' \subset A$, such that

$$\sum_{a \in A'} a = \sum_{a \in A \setminus A'} a + c.$$

There are various ways to prove the NP-completeness of UP. We use an auxiliary lemma, which is a modification of [vEB81, Lemma 1.2], since we also need it later.

Lemma 2. *Given a multiset of positive integers* $A = \{a_1, \ldots, a_n\}$ *and a positive constant* c, *if each* a_i *can be expressed as* $a_i = r_i + Mq_i$, *with* $M > K \sum_{i=1}^{n} |r_i| + c$, *then the equation* $\sum_{i=1}^{n} x_i a_i = c$, *with each* $|x_i|$ *bounded by* K, *is equivalent to the system of two equations*

$$\begin{array}{l} \sum_{i=1}^{n} x_i r_i = c \\ \sum_{i=1}^{n} x_i q_i = 0 \end{array} \quad \text{subject to } |x_i| \leq K.$$

Proof. Consider the equation $\sum_{i=1}^{n} x_i r_i + M \sum_{i=1}^{n} x_i q_i = c$ and suppose that $\sum_{i=1}^{n} x_i q_i = \alpha$. Then the left hand side of the equation is of the form $\sum_{i=1}^{n} x_i r_i + \alpha M$. As $M > K \sum_{i=1}^{n} |r_i| + c$, the expression $\sum_{i=1}^{n} x_i r_i + \alpha M$ cannot be equal to c unless $\alpha = 0$, since each $|x_i|$ is bounded by K. (The contribution to the expression made by $\alpha K \sum_{i=1}^{n} |r_i|$, for $\alpha \neq 0$, can at best be cancelled by the $\sum_{i=1}^{n} x_i r_i$, leaving a residue exceeding αc.) Hence the only possible solution is expressed by the system of two equations presented above. □

Theorem 3. *UNEVEN PARTITION is NP-complete.*

Proof. Verifying a solution for UP can be done in polynomial time, hence UP $\in$ NP. For a multiset $A = \{a_1, \ldots, a_n\}$, solving UP for the multiset $B = \{b_1, \ldots, b_{n+1}\}$, where $b_i = 0 + Ma_i$, for $1 \leq i \leq n$, and $b_{n+1} = c + M \times 0$, with $M > 2c$ (which is of the form of the problem in Lemma 2) is equivalent to finding a partition of A. This follows from substituting the appropriate r_i and q_i in the equivalent system of equations in Lemma 2. Moreover, it is easy to see that a bounded version of UP, where each x_i must be in the range $[-K, +K]$, for some positive constant K, is also NP-complete. □

Corollary 4. *The instance of extended gcd computation where we ask for all multipliers to be units is NP-complete.*

Proof. This is UNEVEN PARTITION with $c = \gcd(a_1, \ldots, a_n)$.

Next we look at the complexity of WEAK UNEVEN PARTITION (WUP). For a given multiset of integers $A = \{a_1, \ldots, a_n\}$ and a positive constant c, WUP asks for a solution to the following 0-1 integer linear programming problem:

$$\text{find } \sum_{i=1}^{n} x_i a_i - \sum_{i=1}^{n} y_i a_i = c$$
$$\text{subject to } x_i + y_i \leq 1, \text{ for } i = 1, \ldots, n.$$

WUP is a variation of WEAK PARTITION, which was proved to be NP-complete by van Emde Boas [vEB81]. Notice however that, unlike WEAK PARTITION, the parasite solution with $x_i = y_i = 0$ is impossible. Our proof closely follows the proof in [vEB81]. Due to the relative inaccessibility of that proof, and a few minor changes specific to our problem, we include the proof here.

Theorem 5. *WEAK UNEVEN PARTITION is NP-complete.*

Proof. Again, verifying a solution for WUP can be done in polynomial time, hence WUP $\in$ NP. To transform a given instance of UP for a set A into WUP we introduce five new numbers for each a_i, as defined below. We choose $M = 2\sum_{i=1}^{n} a_i + c + 1$, and a constant $d > 4$.

$$
\begin{aligned}
b_{i1} &= a_i + M\left(d^{4i-4} + d^{4i-3} + 0 \quad\quad + d^{4i-1} + 0 \quad\right) \\
b_{i2} &= 0 + M\left(0 \quad\quad + d^{4i-3} + 0 \quad\quad + 0 \quad\quad + d^{4i}\right) \\
b_{i3} &= 0 + M\left(d^{4i-4} + 0 \quad\quad + d^{4i-2} + 0 \quad\quad + 0 \quad\right) \\
b_{i4} &= a_i + M\left(0 \quad\quad + 0 \quad\quad + d^{4i-2} + d^{4i-1} + d^{4i}\right) \\
b_{i5} &= 0 + M\left(0 \quad\quad + 0 \quad\quad + 0 \quad\quad + d^{4i-1} + 0 \quad\right)
\end{aligned}
$$

In b_{n2} and b_{n4} the term Md^{4n} is replaced by Md^0, which wraps around the $4n$-th bit in the d-ary representation of the integral part of these numbers divided by M.

The definition of M allows us to use Lemma 2, while the condition that $d > 4$ avoids overflows from one d-ary digit to another (since each equation involves at most four x_{ij}'s). Repeated application of Lemma 2 replaces a single equation $\sum_{i,j} x_{ij} b_{ij} = c$ by a system of $4n + 1$ equations

$$
\begin{aligned}
&(0) \ \sum_i (x_{i1} + x_{i4}) a_i = c \\
&(i1) \ x_{i1} + x_{i3} + x_{i-1,2} + x_{i-1,4} = 0, \quad i = 2,3,\ldots,n \\
&\quad\quad x_{11} + x_{13} + x_{n2} + x_{n4} = 0 \\
&(i2) \ x_{i1} + x_{i2} = 0, \quad\quad\quad\quad\quad\quad\quad i = 1,2,\ldots,n \\
&(i3) \ x_{i3} + x_{i4} = 0, \quad\quad\quad\quad\quad\quad\quad i = 1,2,\ldots,n \\
&(i4) \ x_{i1} + x_{i4} + x_{i5} = 0, \quad\quad\quad\quad i = 1,2,\ldots,n
\end{aligned}
$$

Combining $(i1)$ with $(i2)$ and $(i3)$ gives $x_{i1} + x_{i3} = x_{i-1,1} + x_{i-1,3}$, hence the value of $x_{i1} + x_{i3}$ does not depend on the index i. Furthermore, by discarding number b_{n2}, thus obtaining $x_{11} + x_{13} + x_{n4} = 0$, we force $x_{i1} + x_{i3}$ to assume only one of the three values $\{-1, 0, +1\}$.

We now show that the only possible values for $x_{i1} + x_{i3}$ are -1 or $+1$. Suppose that $x_{i1} + x_{i3} = 0$. This case may happen for only three pairs: $\langle +1, -1\rangle$, $\langle -1, +1\rangle$ and $\langle 0, 0\rangle$. The first two pairs are impossible, as equation $(i4)$ would necessitate $|x_{i5}|$ to be as large as 2. For the last pair we have $x_{i1} + x_{i4} = x_{i1} - x_{i3} = 0$, which cannot possibly be a solution to (0). Thus $x_{i1} + x_{i3}$ must be either -1 or $+1$. It follows that $x_{i1} + x_{i4}$ is also either -1 or $+1$, so equation (0) provides a solution to UP. Consequently, the ability to solve WUP in polynomial time would mean the ability to solve UP in polynomial time. $\quad\square$

Corollary 6. *Given a multiset of n positive integers $A = \{a_1, \ldots, a_n\}$, the task of expressing $g = \gcd(a_1, \ldots, a_n)$ in the form $\sum_{i=1}^{n} x_i a_i = g$ with $x_i \in \{-1, 0, +1\}$, is NP-complete.*

Corollary 7. *Given a multiset of n positive integers $A = \{a_1, \ldots, a_n\}$, the task of expressing $g = \gcd(a_1, \ldots, a_n)$ in the form $\sum_{i=1}^{n} x_i a_i = g$ with $|x_i| \leq K$, is NP-complete.*

Proof.([vEB81]) A variant of the previous reduction is used. The constant d is replaced by Kd, and all terms Md^{4i-j} are replaced by KMd^{4i-j}, except for the numbers b_{i5}. These changes modify equation ($i4$), which now becomes:

$$(i4') \quad K(x_{i1} + x_{i4}) + x_{i5} = 0.$$

Since x_{i5} is bounded by K this enforces $|x_{i1} + x_{i4}|$ to be bounded by 1. By again discarding number b_{n2}, we force x_{n1} to 0, and consequently x_{n4} must take one of the three values: $-1, 0, +1$. If $x_{n4} = 0$ then $x_{11} + x_{13} = 0 = x_{i1} + x_{i3}$, for $i = 1, \ldots, n$. Combining this with the inequality $|x_{i1} - x_{i3}| \leq 1$ proves that no solution is possible for this choice. If x_{n4} is equal to ± 1 so is $x_{i1} + x_{i3}$, and hence $x_{i1} + x_{i4} = x_{i1} - x_{i3} \in \{-1, +1\}$, as required. For our specific corollary, we set $c = \gcd(a_1, \ldots, a_n)$. $\qquad\square$

3   Some bounds

We prove that the size of the multipliers for an arbitrary number of numbers can be kept within the same bounds as for just two numbers. This improves significantly on the bound in [Ili89], which grows with the number of numbers. While it would be nice to have a bound which decreases as the number of numbers increases, this is not possible. The multiset $\{2, 2k+1, 2k+1, \ldots, 2k+1\}$, with $k \geq 1$, serves as a simple counterexample. Our bound is constructive, and we start by giving the lemma that forms the basis of our approach.

Lemma 8. *For a pair of positive integers, $\{a_1, a_2\}$, with $\gcd(a_1, a_2) = 1$, there exists a solution to the equation*

$$x_1 a_1 + x_2 a_2 = c \tag{1}$$

with $|x_1| \leq \max(a_2/2, c)$ and $|x_2| \leq a_1/2$.

Proof. The lemma is true if $a_1 = 1$, thus consider the case where $a_1 \geq 2$. Equation (1) may be rewritten as

$$(x_1 + a_2 i)a_1 + (x_2 - a_1 i)a_2 = c$$

By setting $i = \lfloor x_2/a_1 \rceil$, we ensure that the second multiplier, $x_2(i)$ say, is equal to $x_2 - a_1 \lfloor x_2/a_1 \rceil = x_2 \operatorname{rem} a_1$, where rem denotes the least remainder operation. Thus $|x_2(i)| \leq a_1/2$. For the first multiplier, $x_1(i)$, we have

$$x_1(i) = x_1 + a_2 \left\lfloor \frac{x_2}{a_1} \right\rceil$$

$$= x_1 + \frac{a_2}{a_1}(x_2 - x_2 \operatorname{rem} a_1)$$

$$= \frac{c}{a_1} - \frac{a_2}{a_1} x_2(i)$$

If $|x_2(i)| = \frac{1}{2}a_1$ we can select i such that $x_2(i) = +\frac{1}{2}a_1$ and then $x_1(i) = c/a_1 - a_2/2$. If $c \leq a_1 a_2$ then $|x_1(i)| \leq a_2/2$, otherwise $|x_1(i)| \leq c$. Now assume that $|x_2(i)| < a_1/2$. This time we have no choice of sign for $x_2(i)$ and thus a bound on $|x_1(i)|$ is $c/a_1 + |x_2(i)|a_2/a_1$. First we take $c \leq a_2/2$. Then we have, for some positive integer α,

$$
\begin{aligned}
|x_1(i)| &\leq \frac{c}{a_1} + \frac{a_2(a_1 - \alpha)}{2a_1} \\
&\leq \frac{a_2}{2a_1} + \frac{a_2}{2} - \frac{\alpha a_2}{2a_1} \\
&= \frac{a_2}{2} - \frac{a_2}{2a_1}(\alpha - 1).
\end{aligned}
$$

As $\alpha \geq 1$ we have $|x_1(i)| \leq a_2/2$. If $c > a_2/2$ a similar argument shows that the bound $|x_1(i)| \leq c$ holds. □

Lemma 8 can be obtained as a corollary of [Lev56, Theorem 3]. However, our proof of Lemma 8 also gives us a constant time method for computing small multipliers whenever we have any solution to (1) available. This property is used in the method we describe next.

Theorem 9. *There is an optimal time, optimal space algorithm for computing the extended gcd of n integers $\{a_1, \ldots, a_n\}$, which guarantees that no multiplier is larger than the largest of the numbers divided by 2.*

Proof. We start by describing a binary tree method for computing the gcd of n numbers and then prove that the multipliers produced by the method obey the bound specified in the theorem.

Consider the following algorithm. For n numbers (we may restrict our attention to positive numbers) $\{a_1, \ldots, a_n\}$, the algorithm builds a binary tree on $2n - 1$ nodes, called a gcd tree. Label the nodes in the top to bottom and left to right fashion, so that the children of node i have labels $2i$ and $2i + 1$ and the parent of child j has the label $\lfloor j/2 \rfloor$, as in Fig. 1. We name each node by its label and the letter g, and take the liberty of associating the value in a node with the node name. Nodes g_n to g_{2n-1} (the leaves of a complete binary tree) are set to a_1 to a_n. At the end of the algorithm, node g_1 will hold $\gcd(a_1, \ldots, a_n)$, while nodes g_n to g_{2n-1} will store multipliers x_1 to x_n such that $\sum_{i=1}^n x_i a_i = g_1$.

The computation proceeds in two phases, bottom-top and top-bottom. In the bottom-top phase, the algorithm starts at the bottom of the tree. It takes two leaves with the same parent at a time. Let the selected nodes be g_{4i} and g_{4i+1} (Fig. 1). Using the standard extended Euclidean method, the algorithm computes $g_{2i} = y_{4i}g_{4i} + y_{4i+1}g_{4i+1}$ and stores g_{2i}, y_{4i} and y_{4i+1} at the nodes $2i$, $4i$ and $4i + 1$, respectively. The process works up through the levels (and computations at each level can in fact be done in parallel) until the value of g_1 is computed.

In the top-down phase, the algorithm starts at node 4 (here we assume $n \geq 3$) and proceeds down through the levels, taking two nodes with a common parent at a time. Let the selected nodes be g_{4i} and g_{4i+1}. For these two nodes the algorithm computes two multipliers z_{4i} and z_{4i+1}, such that $z_{4i}g_{4i} + z_{4i+1}g_{4i+1} = g_{2i}z_{2i}$.

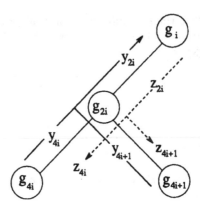

Fig. 1. Part of a gcd tree. Solid lines indicate the actions of the bottom-top phase, while dashed lines follow the top-bottom phase.

The value of z_{2i} is either directly the value of y_{2i} or the value computed in a previous step of the top to bottom phase. To compute z_{4i} and z_{4i+1}, the algorithm uses the fact that $y_{4i}g_{4i} + y_{4i+1}g_{4i+1} = g_{2i}$. The algorithm premultiplies each side by z_{2i} and then uses the technique of Lemma 8 to compute two 'small' multipliers. The phase ends when the multipliers for all leaves of the tree are computed. Finally x_i, for $i = 1$ to n, are set to z_{n+i-1}. From the description of the algorithm it is clear that both the gcd and the multipliers are computed in optimal time and space.

To complete the proof we show that each multiplier z_j computed in the top-bottom phase is no greater than $\max\{a_i\}/2$. The inductive proof uses Lemma 8. Firstly we observe that $g_1 \leq \min(a_1, \ldots, a_n) \leq \max(a_1, \ldots, a_n)$ and both g_2 and g_3, by the properties of the extended Euclidean method for two numbers, have their absolute values less than the maximum number in the input set divided by $2g_1$. Hence the theorem is true for the nodes on levels 1 and 2. Assume that this holds for levels 1 through $2i$. Now consider nodes $4i$ and $4i + 1$. The multipliers z_{4i} and z_{4i+1} must solve $z_{4i}(g_{4i}/g_{2i}) + z_{4i+1}(g_{4i+1}/g_{2i}) = z_{2i}$. By the inductive hypothesis, we have $z_{2i} \leq \max(a_1, \ldots, a_n)/2$. By Lemma 8, we can find a pair, z_{4i} and z_{4i+1}, such that either $|z_{4i+1}| \leq g_{4i}/(2g_{2i})$ and $|z_{4i}| \leq g_{4i+1}/(2g_{2i})$ or $|z_{4i+1}| \leq g_{4i}/(2g_{2i})$ and $|z_{4i}| \leq |z_{2i}| \leq \max(a_1, \ldots, a_n)/2$. Thus in either case the hypothesis holds for the new level. □

The fact that our upper bound is optimal relies on an example with repeated numbers. It is reasonable to hope that a better upper bound may apply if all numbers are required to be distinct. However the following theorem indicates problems which need to be addressed in developing an algorithm to achieve a better bound for sets of numbers (as against multisets).

Theorem 10. *The bound achieved by the method of Theorem 9 is the best possible for any method that computes the gcd of a set of numbers by computing gcds pairwise and reducing the resulting multipliers in the natural way using the associated linear equations.*

Proof. The existence an algorithm that could improve the previous bounds while manipulating at most two numbers at a time contradicts [Lev56, Theorem 3]. □

4   Conclusions

In this paper we concentrated our attention on the complexity of expressing the gcd of $n > 2$ numbers as a linear combination of them. We showed that for $n > 2$ the definition of minimum multipliers depends on the selected metric. We proved that the complexity of computing optimal multipliers with respect to the L_0 metric and the L_∞ norm is NP-complete.

On the other hand, we described and analyzed a new method for computing the gcd of $n \geq 2$ numbers, and proved that no multiplier generated by this method will exceed the largest number in the input set divided by 2. This bound is the best possible without additional restrictions on the input. It is also the tightest bound for any method that operates on two numbers at a time.

References

[Bla63] W.A. Blankinship. A new version of the Euclidean algorithm. *Amer. Math. Monthly*, 70:742–745, 1963.

[Bra70] G.H. Bradley. Algorithm and bound for the greatest common divisor of n integers. *Communications of the ACM*, 13:433–436, 1970.

[dBZ53] N.G. de Bruijn and W.M. Zaring. On invariants of g.c.d. algorithms. *Nieuw Archief voor Wiskunde*, I(3):105–112, 1953.

[GJ79] M.R. Garey and D.S. Johnson. *Computers and Intractibility: A Guide to the Theory of NP-completeness*. W.H. Freeman, San Francisco, 1979.

[HM93] G. Havas and B.S. Majewski. Integer matrix diagonalization. Technical Report TR0277, The University of Queensland, Brisbane, 1993.

[HM94] G. Havas and B.S. Majewski. Hermite normal form computation for integer matrices. Technical Report TR0295, The University of Queensland, Brisbane, 1994.

[Ili89] C.S. Iliopoulos. Worst case complexity bounds on algorithms for computing the canonical structure of finite abelian groups and the Hermite and Smith normal forms of an integer matrix. *SIAM J. Computing*, 18:658–669, 1989.

[Knu73] D.E. Knuth. *The Art of Computer Programming, Vol. 2: Seminumerical Algorithms*. Addison-Wesley, Reading, Mass., 2nd edition, 1973.

[Lev56] R.J. Levit. A minimum solution to a Diophantine equation. *Amer. Math. Monthly*, 63:647–651, 1956.

[Man88] D. M. Mandelbaum. New binary Euclidean algorithms. *Electron. Lett.*, 24:857–858, 1988.

[Nor87] G. Norton. A shift-remainder gcd algorithm. In *Proc. 5th International Conference on Applied Algebra, Algebraic Algorithms and Error-correcting Codes – AAECC'87*, Lecture Notes in Computer Science 356, pages 350–356, Menorca, Spain, 1987.

[Sor94] J. Sorenson. Two fast GCD algorithms. *Journal of Algorithms*, 16:110–144, 1994.

[vEB81] P. van Emde Boas. Another NP-complete partition problem and the complexity of computing short vectors in a lattice. Technical Report MI/UVA 81–04, The University of Amsterdam, Amsterdam, 1981.

[Wat77] M.S. Waterman. Multidimensional greatest common divisor and Lehmer algorithms. *BIT*, 17:465–478, 1977.

Explicit Formulas for Units
in certain Quadratic Number Fields

A. J. van der Poorten

ceNTRe for Number Theory Research
Macquarie University NSW 2109 Australia

Abstract. There is a class of quadratic number fields for which it is possible to find an explicit continued fraction expansion of a generator and hence an explicit formula for the fundamental unit. One therewith displays a family of quadratic fields with relatively large regulator. The formula for the fundamental unit seems far simpler than the continued fraction expansion, yet the expansion seems necessary to show the unit is fundamental. I explain what is going on and go some way towards taming the sequence of ever more complicated arguments of Yamomoto, Hendy, Bernstein, Williams, Levesque and Rhin, Levesque, Azuhata, Halter-Koch, Mollin and Williams, and Williams.

1   Introduction

It remains an interesting problem to detect infinite families of positive integers D for which one can readily describe the fundamental unit of the quadratic number field $\mathbb{Q}(\sqrt{D})$. My purpose here is to show that a sequence of known results is a little less horrible, and indeed rather simpler, than the literature might suggest. Of course everyone knows that, in the crunch, determining a fundamental unit is a matter of detailing the period of the continued fraction expansion of $\sqrt{D}$, or of $\frac{1}{2}(\sqrt{D}+1)$ if $D \equiv 1 \pmod 4$. I will show that that task is a little easier than has seemed, and that in any case, in the present examples, one may not actually need all the details.

The simplest nontrivial case of the class of cases with which I deal is the so-called S_n sequence of Shanks, where $S_n = (2^n + 1)^2 + 4 \cdot 2^n$. The most extreme generalisation that seems to be about is essentially the cases

$$M = \left(qra^n + \mu(a^k - \lambda)/q\right)^2 + 4\mu\lambda ra^n,$$

where both μ, $\lambda = \pm 1$, and $rq \mid a^k - \lambda$. In all these examples one can readily write down an element of norm ± 1 and guess that it is the fundamental unit. But showing it is an integer and proving that it is indeed fundamental appears to require producing the period explicitly. On doing that one notices curious cyclical patterns within the period.

My explanation for these phenomena runs as follows. Referring to the general example, one immediately finds ideals of respective norm a^k and $4ra^n$ in the order $\mathbb{Z}[\sqrt{M}]$. It turns out that the product of the $2n$-th power of the first and the $2k$-th power of the second has norm 1; and if $\gcd(n, k) = 1$ no 'smaller' combination, except perhaps its square root, will do [we may of course suppose that r is positive; if $r = 1$ or k is even we get the square of a fundamental unit]. The trick is to make this explicit. I do that by associating 2×2 matrices with these ideals, with determinant the respective norms, in such a way that multiplication of the matrices corresponds to multiplication of the ideals — or, equivalently, to composition of the corresponding quadratic forms. Performing the matrix multiplication strategically, automatically gives one a decomposition of the product as a constant multiple of a product of unimodular matrices. The 'strategic' multiplication of the matrices immediately explains the 'cyclical patterns' that appear. One knows, as a general principle, that our product of unimodular matrices corresponds to the period of the continued fraction expansion of $\sqrt{M}$. That leads to a second innovation. The continued fraction expansion is inadmissable in that it may well have negative and even fractional partial quotients in $\mathbb{Z}[\frac{1}{2}]$. But easy tricks, including recognising that we should rather have expanded $\frac{1}{2}\sqrt{M}$ or $\frac{1}{2}(\sqrt{M} + 1)$, according to the parity of $M \bmod 4$, readily cope with those quasi-difficulties. I get both the fundamental unit and the explicit continued fraction expansion, with its patterns nakedly exposed.

I must commence with an extensive collection of, what my French colleagues call, 'rappels'. Accordingly, let me say that I will begin by reminding the reader of the following notions and notation:

I economise on space by denoting a continued fraction

$$c_0 + \cfrac{1}{c_1 + \cfrac{1}{c_2 + \cfrac{1}{\ddots}}}$$

by $[c_0 , c_1 , c_2 , \ldots]$, and economise on effort by noting that it follows easily by induction on h that the statements

$$\begin{pmatrix} c_0 & 1 \\ 1 & 0 \end{pmatrix} \begin{pmatrix} c_1 & 1 \\ 1 & 0 \end{pmatrix} \cdots \begin{pmatrix} c_h & 1 \\ 1 & 0 \end{pmatrix} = \begin{pmatrix} x_h & x_{h-1} \\ y_h & y_{h-1} \end{pmatrix}$$

and

$$[c_0 , c_1 , \ldots , c_h] = \frac{x_h}{y_h} ,$$

for $h = 0, 1, \ldots$, are essentially the same. That transforms seemingly intricate remarks about continued fractions to simple remarks about 2×2 matrices. I insert the qualification 'essentially' inter alia because the matrices serve just to represent linear fractional transformations; thus matrices corresponding to continued fraction expansions might be constant multiples of a matrix of determinant ± 1 and those differing just by such a constant multiple correspond to the same continued fraction expansion.

Then the following lies central to the observations below. Let δ be a quadratic irrational with trace $\delta + \bar{\delta} = t$ and with norm $\delta\bar{\delta} = n$. If $\mathrm{Norm}(X - \delta Y) = \pm 1$

provides a nontrivial solution of 'Pell's equation', thus with $Y \neq 0$, and the matrix

$$\begin{pmatrix} X & -nY \\ Y & X - tY \end{pmatrix}$$

has a decomposition

$$\begin{pmatrix} X & -nY \\ Y & X - tY \end{pmatrix} = \begin{pmatrix} c_0 & 1 \\ 1 & 0 \end{pmatrix} \begin{pmatrix} c_1 & 1 \\ 1 & 0 \end{pmatrix} \cdots \begin{pmatrix} c_{r+1} & 1 \\ 1 & 0 \end{pmatrix},$$

with integers c_i, then, formally, δ has the periodic continued fraction expansion

$$\delta = [\overline{c_0, c_1, \cdots, c_{r+1}}].$$

This is easy to see. For if, say,

$$\gamma = [\overline{c_0, c_1, \cdots, c_{r+1}}] = [c_0, c_1, \cdots, c_{r+1}, \gamma],$$

the equality by virtue of the meaning of periodicity, then because

$$\gamma \longmapsto \begin{pmatrix} X & -nY \\ Y & X - tY \end{pmatrix} \begin{pmatrix} \gamma & 1 \\ 1 & 0 \end{pmatrix} = \begin{pmatrix} X\gamma - nY & X \\ Y\gamma + X - tY & Y \end{pmatrix},$$

we have

$$\gamma = \frac{X\gamma - nY}{Y\gamma + X - tY} \quad \text{so} \quad Y(\gamma^2 - t\gamma + n) = 0,$$

which is the assertion. Namely, indeed, $\gamma = \delta$.

The qualification 'formally' is necessary in that in general the decomposition may well not yield admissible, that is positive, partial quotients. Indeed, in the simplest case, when $\delta = \sqrt{D}$, with D of course some positive integer, not a square, one obtains $c_{r+1} = 0$ and the admissible expansion in positive integer partial quotients is of the shape

$$\delta = [c_0, \overline{c_1, \cdots, c_{r-1}, c_r + c_0}].$$

I will explain below that, as in this instance, there is good reason to forget one's fear of the inadmissible, because there are simple rules for taming negative and zero integral partial quotients.

Incidentally, it is easy to see that a unimodular matrix has decompositions corresponding to continued fraction expansions. That is just a matter of recalling that such a matrix is a product of elementary row transformations.

In any case, the point is this: Given an integer matrix of the shape

$$\begin{pmatrix} x & -ny \\ y & x - ty \end{pmatrix}$$

of determinant $\pm Q$ — I shall refer to such matrices as an *initial segment* of the continued fraction expansion of δ — one has a decomposition

$$\begin{pmatrix} x & -ny \\ y & x - ty \end{pmatrix} = \begin{pmatrix} x & x' \\ y & y' \end{pmatrix} \begin{pmatrix} 1 & P \\ 0 & Q \end{pmatrix}$$

where $xy' - x'y = \pm 1$ and P and Q are integers. I presume here and henceforth that δ is an algebraic integer; that is, t and n are integers. A decomposition

$$\begin{pmatrix} x & x' \\ y & y' \end{pmatrix} = \begin{pmatrix} c_0 & 1 \\ 1 & 0 \end{pmatrix} \begin{pmatrix} c_1 & 1 \\ 1 & 0 \end{pmatrix} \cdots \begin{pmatrix} c_h & 1 \\ 1 & 0 \end{pmatrix},$$

together with the preceding remarks, now is equivalent to the assertion that

$$\delta = [c_0, c_1, \cdots, c_h, (\delta + P)/Q].$$

In particular the case $Q = 1$ is about to provide the period of δ. That's clear because the corresponding convergent x_h/y_h satisfies $\mathrm{Norm}(x_h - \delta y_h) = (-1)^{h+1} Q$. It is easy to see that a product of initial segments is again an initial segment, and that is just about the main idea used below. We shall see that, loosely speaking, the larger the (positive) integers x and y, the longer the initial segment. Thus I construct the period of δ by composing initial segments. Now, even when one can see and prove that something works, one may still need to ask: What is going on here? The trick turns out to be that, whilst innocently multiplying matrices, we are in fact secretly composing quadratic forms in the principal cycle of forms. But enough of these generalities.

2 Examples

We will soon find it useful to use the definitions

$$L = \begin{pmatrix} 1 & 0 \\ 1 & 1 \end{pmatrix}, \qquad R = \begin{pmatrix} 1 & 1 \\ 0 & 1 \end{pmatrix}, \qquad \text{and} \qquad J = \begin{pmatrix} 0 & 1 \\ 1 & 0 \end{pmatrix},$$

the point being that

$$\begin{pmatrix} c & 1 \\ 1 & 0 \end{pmatrix} J = R^c, \qquad J \begin{pmatrix} c & 1 \\ 1 & 0 \end{pmatrix} = L^c, \qquad \text{whilst} \qquad J^2 = I.$$

That will allow us the alternative of viewing a product

$$\begin{pmatrix} c_0 & 1 \\ 1 & 0 \end{pmatrix} \begin{pmatrix} c_1 & 1 \\ 1 & 0 \end{pmatrix} \cdots \begin{pmatrix} c_h & 1 \\ 1 & 0 \end{pmatrix} \cdots,$$

corresponding to a continued fraction expansion, as an R–L sequence

$$R^{c_0} L^{c_1} R^{c_2} L^{c_3} \cdots .$$

Below, we will want to multiply and divide continued fraction expansions by 2, in particular to relate expansions of $\sqrt{D}$ and of $\frac{1}{2}(\sqrt{D} + 1)$. To this end I introduce the matrices

$$A = \begin{pmatrix} 2 & 0 \\ 0 & 1 \end{pmatrix} \qquad \text{and} \qquad A' = \begin{pmatrix} 1 & 0 \\ 0 & 2 \end{pmatrix},$$

noting that if $\gamma \longleftrightarrow R^{c_0} L^{c_1} R^{c_2} L^{c_3} \ldots$ then $2\gamma \longleftrightarrow A R^{c_0} L^{c_1} R^{c_2} L^{c_3} \ldots$. Now this last product is not, of course, an R–L sequence, and does not, as yet, correspond to a continued fraction expansion. However the *transition rules*

$$
\begin{aligned}
AR &= R^2 A & A'L &= L^2 A' \\
AL^2 &= LA & A'R^2 &= RA' \\
ALR &= RLA' & A'RL &= LRA
\end{aligned}
$$

allow us to transmit the multiplier A through the R–L sequence until it disappears into the $\ldots$ on the right.

Let me take the case

$$
N = \left(qa^n - \frac{a+1}{4q}\right)^2 + a^n \quad \text{with} \quad 4q \mid a+1.
$$

Plainly

$$
\sqrt{N} = [qa^n - (a+1)/4q, \, (\sqrt{N} + qa^n - (a+1)/4q)/a^n]
$$

and

$$
\sqrt{N} = [qa^n - (a+1)/4q, \, 2q, \, -(\sqrt{N} + qa^n + (a+1)/4q)/a].
$$

The latter expansion corresponds to

$$
R^{qa^n-(a+1)/4q} L^{2q} \begin{pmatrix} 1 & qa^n + (a+1)/4q \\ 0 & -a \end{pmatrix}
$$

$$
= R^{qa^n-(a+1)/4q} L^{2q} R^{-qa^{n-1}} \begin{pmatrix} 1 & 0 \\ 0 & -a \end{pmatrix} R^{(a+1)/4q}.
$$

It's easy to verify that

$$
\begin{pmatrix} 1 & 0 \\ 0 & -a \end{pmatrix} R^a = R^{-1} \begin{pmatrix} 1 & 0 \\ 0 & -a \end{pmatrix} \quad \text{and} \quad \begin{pmatrix} 1 & 0 \\ 0 & -a \end{pmatrix} L = L^{-a} \begin{pmatrix} 1 & 0 \\ 0 & -a \end{pmatrix},
$$

and thence to see that the n-th power of our initial segment is

$$
R^{qa^n-(a+1)/4q} L^{2q} R^{-2qa^{n-1}} L^{-2aq} R^{2qa^{n-2}} L^{2a^2 q} \ldots
$$

$$
\ldots R^{2(-1)^{n-1}qa} L^{2(-1)^{n-1}a^{n-1}q} R^{(-1)^n q} \begin{pmatrix} 1 & 0 \\ 0 & (-1)^n a^n \end{pmatrix} R^{(a+1)/4q}.
$$

On multiplying on the right by

$$
R^{qa^n-(a+1)/4q} J \begin{pmatrix} 1 & qa^n - (a+1)/4q \\ 0 & a^n \end{pmatrix}
$$

$$
= R^{qa^n-(a+1)/4q} L^q \begin{pmatrix} a^n & 0 \\ 0 & 1 \end{pmatrix} L^{-(a+1)/4q} J,
$$

which corresponds to the first continued fraction expansion mentioned above, we see that

$$\sqrt{N} = [qa^n - (a+1)/4q, \, 2q, \, -2qa^{n-1}, \, -2qa, \, 2qa^{n-2}, \, 2qa^2, \, \ldots,$$
$$2(-1)^{n-1}qa, \, 2(-1)^{n-1}qa^{n-1}, \, 2(-1)^n q, \, (-1)^n(qa^n - (a+1)/4q), \, 0,$$
$$(-1)^n(qa^n - (a+1)/4q), \, 0, \, (-1)^n\sqrt{N}].$$

But, one doesn't have to be a purist to take exception to the suggestion that this displays the period of $\sqrt{N}$, since the expansion is certainly polluted by nonpositive partial quotients. Therefore, let me explain how to rid an expansion of such defects. The idea is to notice that

$$-y = 0 + -y$$
$$-1/y = -1 + (y-1)/y$$
$$y/(y-1) = 1 + 1/(y-1)$$
$$y - 1 = -1 + y$$
$$1/y = 0 + 1/y$$
$$y =$$

$,$

which is

$$[\ldots, -a, b, c, \ldots] = [\ldots, 0, -1, 1, -1, 0, a, -b, -c, - \ldots].$$

In addition we need to observe that

$$\begin{pmatrix} a & 1 \\ 1 & 0 \end{pmatrix} \begin{pmatrix} 0 & 1 \\ 1 & 0 \end{pmatrix} \begin{pmatrix} b & 1 \\ 1 & 0 \end{pmatrix} = \begin{pmatrix} a+b & 1 \\ 1 & 0 \end{pmatrix},$$

so $[\ldots, a, 0, b, \ldots] = [\ldots, a+b, \ldots]$.

Hence, sequentially negating the rest of the expansion eventually yields

$$\sqrt{N} = [qa^n - (a+1)/4q, \, \overline{2q - 1, \, 1, \, 2qa^{n-1} - 1, \, 2qa - 1, \, 1,}$$
$$\overline{2qa^{n-2} - 1, \, 2qa^2 - 1, \, 1, \, \ldots, \, 2qa - 1, \, 2qa^{n-1} - 1, \, 1,}$$
$$\overline{2q - 1, \, 2(qa^n - (a+1)/4q)}].$$

Here, I actually obtain the period of $\sqrt{N}$. But the point is that it *can* be done, and not so much that I have done it. Our previous work had *already* shown that a unit of the order $\mathbb{Z}[\sqrt{N}]$ is given by

$$((qa^n - (a+1)/4q - \sqrt{N})(2q(qa^n - (a+1)/4q) + 1 - 2q\sqrt{N})^n/a^n.$$

That's plain, because all our fiddle-faddle with matrices, is really just a way to cleverly multiply their eigenvalues, namely the eigenvalues of the initial segments

$$\begin{pmatrix} qa^n - (a+1)/4q & (qa^n - (a+1)/4q)^2 + a^n \\ 1 & qa^n - (a+1)/4q \end{pmatrix} \quad \text{and of}$$

$$\begin{pmatrix} 2q(qa^n - (a+1)/4q) + 1 & ((qa^n - (a+1)/4q)^2 + a^n)2q \\ 2q & 2q(qa^n - (a+1)/4q) + 1 \end{pmatrix}^n.$$

There's nothing more to *prove*, because we have quite explicitly constructed an integer unimodular matrix providing a period of $\sqrt{N}$, and we have displayed that period, inter alia showing that it is indeed the primitive period. Mind you, our explicit composition strongly suggested that we were never about to be able to factor out the determinant until we composed with the other initial segment, so there really should have been no need to stare at the actual continued fraction expansion.

Nonetheless, it certainly remains for me to defend my methods against other more familiar techniques. I'll let that keep until I briefly discuss a generalisation of the previous example. Let me therefore consider the cases

$$M = \left(qra^n + \mu(a^k - \lambda)/q\right)^2 + 4\mu\lambda ra^n\,,$$

where both μ, $\lambda = \pm 1$, and $rq \mid a^k - \lambda$. Plainly, if $S = qra^n + \mu(a^k - \lambda)/q$ is even, we will want to deal with $\sqrt{N} = \frac{1}{2}\sqrt{M}$, whilst if it is odd then $N = M$ will do, but it will be de rigeur to attempt to cope with $\frac{1}{2}(\sqrt{N}+1)$. Since

$$\sqrt{M} = [S\,,\,(\sqrt{M}+S)/4\mu\lambda ra^n]$$
$$= [S\,,\,\tfrac{1}{2}\mu\lambda q\,,\,(\sqrt{M}+qra^n - \mu(a^k-\lambda)/q)/\lambda a^k]\,,$$

we have, dividing by 2, respectively according to the parity of S,

$$\sqrt{N} = [\tfrac{1}{2}S\,,\,(\sqrt{N}+\tfrac{1}{2}S)/\mu\lambda ra^n]$$
$$= [\tfrac{1}{2}S\,,\,\mu\lambda q\,,\,(\sqrt{N}+\tfrac{1}{2}(qra^n - \mu(a^k-\lambda)/q))/\lambda a^k]\,,$$

or

$$\tfrac{1}{2}(\sqrt{N}+1) = [\tfrac{1}{2}(S+1)\,,\,(\tfrac{1}{2}(\sqrt{N}+1)+\tfrac{1}{2}(S-1))/\mu\lambda ra^n]$$
$$= [\tfrac{1}{2}(S+1)\,,\,\mu\lambda q\,,\,(\tfrac{1}{2}(\sqrt{N}+1)+\tfrac{1}{2}(qra^n - \mu(a^k-\lambda)/q-1))/\lambda a^k]\,.$$

Nonetheless, let's see whether we can avoid splitting into these two cases for a while. Thus returning to M, we first consider powers of the initial segment corresponding to the more expanded expansion, which is, setting $S' = qra^n - \mu(a^k-\lambda)/q$,

$$R^S L^{\frac{1}{2}\mu\lambda q}\begin{pmatrix}1 & S' \\ 0 & \lambda a^k\end{pmatrix}.$$

Taking powers of this initial segment leads to

$$\cdots\begin{pmatrix}1 & S' \\ 0 & \lambda a^k\end{pmatrix} R^S L^{\frac{1}{2}\mu\lambda q}\begin{pmatrix}1 & S' \\ 0 & \lambda a^k\end{pmatrix}$$
$$= \cdots\begin{pmatrix}1 & S'+S \\ 0 & \lambda a^k\end{pmatrix} L^{\frac{1}{2}\mu\lambda q}\begin{pmatrix}1 & S' \\ 0 & \lambda a^k\end{pmatrix} = \cdots R^{2\lambda rqa^{n-k}} L^{\frac{1}{2}\mu\lambda^2 qa^k}\begin{pmatrix}1 & S' \\ 0 & \lambda^2 a^{2k}\end{pmatrix},$$

and, so if j is the least integer with $jk \geq n$, the j-th power yields

$$\cdots R^{2\lambda^{j-2}rqa^{n-(j-2)k}} L^{\frac{1}{2}\mu\lambda^{j-1}qa^{(j-2)k}} R^{2\lambda^{j-1}rqa^{n-(j-1)k}} L^{\frac{1}{2}\mu\lambda^j qa^{(j-1)k}}\begin{pmatrix}1 & S' \\ 0 & \lambda^j a^{jk}\end{pmatrix}.$$

After concatenating these j segments we'll have to turn to plan B because more than j steps 1, as above, might lead us to fiercely fractional exponents, with a's in their denominator. Those are not the very end of the world, as Shallit and I show in [13], but are to be avoided if at all practicable. In contrast, I've of course checked that the half exponents appearing above will be rendered harmless once we move to N. So it might now be a good idea to perform step 2, to multiply on the right by the first initial segment, which is

$$\begin{pmatrix} S & M \\ 1 & S \end{pmatrix} = R^S J \begin{pmatrix} 1 & S \\ 0 & 4\mu\lambda r a^n \end{pmatrix},$$

thereby obtaining

$$\cdots R^{2\lambda^{j-1} rq a^{n-(j-1)k}} L^{\frac{1}{2}\mu\lambda^j q a^{(j-1)k}} \begin{pmatrix} 1 & S' \\ 0 & \lambda^j a^{jk} \end{pmatrix} R^S J \begin{pmatrix} 1 & S \\ 0 & 4\mu\lambda r a^n \end{pmatrix}$$

$$= \cdots\cdots \begin{pmatrix} 1 & 2rq \\ 0 & \lambda^j a^{jk-n} \end{pmatrix} \begin{pmatrix} 1 & 0 \\ 0 & a^n \end{pmatrix} \begin{pmatrix} 0 & 4\mu\lambda r a^n \\ 1 & S \end{pmatrix}$$

$$= a^n \times \cdots\cdots \begin{pmatrix} 1 & 2rq \\ 0 & \lambda^j a^{jk-n} \end{pmatrix} \begin{pmatrix} 0 & 4\mu\lambda r \\ 1 & S \end{pmatrix}$$

$$= a^n \times \cdots\cdots \begin{pmatrix} 2rq & 4\mu\lambda r + 2rqS \\ \lambda^j a^{jk-n} & \lambda^j a^{jk-n} S \end{pmatrix}.$$

Let me confess that I began writing this note after using the 'quiet time' allowed me by a University Senate meeting to decide that I had the strategy to make everything work out tidily. It was a serious disturbance to me when I momentarily ground to a stop at this point when actually writing the present remarks. Luckily, I remembered that it is good style for details to be kept decently till later as a surprise for one's readers.

I may thus retreat for a moment to the special case $k \mid n$; that is, $jk = n$. Then if $r = 1$, and on reverting from M to N by division by 2, we will already have obtained a unimodular matrix, and will already have learned that a fundamental unit of $\mathbb{Q}(\sqrt{M})$ is given by

$$(\tfrac{1}{2}\mu\lambda qS + 1 - \tfrac{1}{2}\mu\lambda q\sqrt{M})^{n/k}(\tfrac{1}{2}S - \tfrac{1}{2}\sqrt{M}) / a^n.$$

A little more explicitly, and noting $\mu\lambda = \pm 1$, we may translate this claim as saying that according as S is even or odd, a fundamental unit of the order $\mathbb{Z}[\sqrt{N}]$ is given by

$$(\tfrac{1}{2}qS + \mu\lambda - q\sqrt{N})^{n/k}(\tfrac{1}{2}S - \sqrt{N}) / a^n,$$

respectively, a fundamental unit of the order $\mathbb{Z}[\tfrac{1}{2}(\sqrt{N}+1)]$ is given by

$$(\tfrac{1}{2}q(S+1) + \mu\lambda - q\cdot\tfrac{1}{2}(\sqrt{N}+1))^{n/k}(\tfrac{1}{2}(S+1) - \tfrac{1}{2}(\sqrt{N}+1)) / a^n,$$

If $r > 1$ we turn out to be halfway to the happy situation of obtaining a matrix of determinant 1 and, accordingly, will find that a fundamental unit is

$$(\tfrac{1}{2}\mu\lambda qS + 1 - \tfrac{1}{2}\mu\lambda q\sqrt{M})^{2n/k}(\tfrac{1}{2}S - \tfrac{1}{2}\sqrt{M}) / ra^{2n}.$$

To see this last claim, we return to the general case and remark jovially that any initial segment

$$\begin{pmatrix} x & My \\ y & x \end{pmatrix}$$

belonging to $\sqrt{M}$ is *false* symmetric; that is, it equals its *false* transpose[1]. Using this observation in repeating step 2 we obtain

$$\cdots \begin{pmatrix} 2rq & 4\mu\lambda r + 2rqS \\ \lambda^j & \lambda^j S \end{pmatrix} \begin{pmatrix} 4\mu\lambda ra^n & S \\ 0 & 1 \end{pmatrix} JR^S$$

$$= \cdots \begin{pmatrix} 8\mu\lambda r^2 qa^n & 4r(rq^2a^n + \mu a^k) \\ 4\mu\lambda^{j+1}ra^{jk} & 2\lambda^j a^{jk-n}S \end{pmatrix} JR^S .$$

Because $rq \mid a^k - \lambda$, the matrix is divisible by r, yielding

$$2rA \begin{pmatrix} rq^2a^n + \mu a^k & \mu\lambda rqa^n \\ \lambda^j a^{jk-n}S/r & \mu\lambda^{j+1}a^{jk} \end{pmatrix} A'R^S$$

$$= 2\mu\lambda ra^{jk-n} A \begin{pmatrix} \lambda a^{n-(j-1)k} & rqa^{2n-jk} \\ \lambda^{j+1}(a^k - \lambda)/rq & \lambda^j a^n \end{pmatrix} L^{\mu\lambda q} A'R^S$$

$$= 2\mu\lambda ra^{jk-n} A \begin{pmatrix} \lambda a^{n-(j-1)k} & \lambda^j rq \\ \lambda^{j+1}(a^k - \lambda)/rq & a^{jk-n} \end{pmatrix} L^{\mu\lambda^{j-1}qa^{2n-jk}} \begin{pmatrix} 1 & 0 \\ 0 & \lambda^j a^{2n-jk} \end{pmatrix} A'R^S.$$

Here $A = \begin{pmatrix} 2 & 0 \\ 0 & 1 \end{pmatrix}$ and $A' = \begin{pmatrix} 1 & 0 \\ 0 & 2 \end{pmatrix}$ effect multiplication, and division, by two.

One now returns to step 1, again using false symmetry. In the case $jk = n$, j further such multiplications will yield a unimodular matrix corresponding to the primitive period of $\sqrt{N} = \frac{1}{2}\sqrt{M}$, respectively to the primitive period of $\frac{1}{2}(\sqrt{N} + 1) = \frac{1}{2}(\sqrt{M} + 1)$. Moreover, in that case the unimodular matrix at the centre of the period decomposes in a friendly way to yield

$$\begin{pmatrix} \lambda a^{n-(j-1)k} & \lambda^j rq \\ \lambda^{j+1}(a^k - \lambda)/rq & a^{jk-n} \end{pmatrix} L^{\mu\lambda^{j-1}qa^{2n-jk}} = \begin{pmatrix} \lambda a^k & \lambda^j rq \\ \lambda^{j+1}(a^k - \lambda)/rq & 1 \end{pmatrix} L^{\mu\lambda^{j-1}qa^n}$$

$$= R^{\lambda^j rq} L^{\lambda^{j-1}(a^k - \lambda)/rq} L^{\mu\lambda^{j-1}qa^n} = R^{\lambda^j rq} L^{\mu\lambda^{j-1}S/r}.$$

For example, if $M = (rqa^3 + (a - 1)/q)^2 + 4ra^3$, with $S = rqa^3 + (a - 1)/q$ odd, then

$$\frac{1}{2}(\sqrt{N} + 1) = \frac{1}{2}(\sqrt{M} + 1) = [\frac{1}{2}(rqa^3 + (a - 1)/q + 1), \overline{q, rqa^2, qa, rqa,}$$

$$\overline{qa^2, rq, qa^3 + (a - 1)/rq, rq, qa^2, \dots, rqa^3 + (a - 1)/q}].$$

Here I have not detailed the almost trivial manner in which one carries out the various divisions and multiplications by 2.

In the general case, when $\gcd(k, n) = 1$ and $r \neq 1$ we have rather more work yet to do. Let me illustrate that work, and the preceding claims, by dealing with the example

$$M = (qra^7 + \mu(a^3 - \lambda)/q)^2 + 4\mu\lambda ra^7 ,$$

---

[1] These notions will be familiar to those of us dealing with recalcitrant undergraduates who refuse to acknowledge which diagonal is the main diagonal.

where both μ, $\lambda = \pm 1$, and $rq \mid a^3 - \lambda$.

As above, we set $S = qra^7 + \mu(a^3 - \lambda)/q$ and $S' = qra^7 - \mu(a^3 - \lambda)/q$. All's well to begin with, and from above we see that

$$\sqrt{M} = [\, S \,,\, \tfrac{1}{2}\mu\lambda q \,,\, 2\lambda rqa^4 \,,\, \tfrac{1}{2}\mu qa^3 \,,\, 2\lambda rqa \,,\, \tfrac{1}{2}\mu qa^6 \,,\, \ldots \,],$$

with the matrix

$$A \begin{pmatrix} \lambda a & \lambda rq \\ (a^3 - \lambda)/rq & a^2 \end{pmatrix} L^{\mu qa^5} \begin{pmatrix} 1 & 0 \\ 0 & \lambda a^5 \end{pmatrix} A' R^S$$

struggling to provide the next partial quotients. For example, if $a = 5$, $\lambda = -1$, $q = 7$, $r = 3$ then

$$\begin{pmatrix} \lambda a & \lambda rq \\ (a^3 - \lambda)/rq & a^2 \end{pmatrix} = \begin{pmatrix} -5 & -21 \\ 6 & 25 \end{pmatrix} = R^{-1} L^6 R^4 \,.$$

We next study

$$\begin{pmatrix} 1 & 0 \\ 0 & \lambda a^5 \end{pmatrix} A' R^S \begin{pmatrix} \lambda a^3 & S' \\ 0 & 1 \end{pmatrix} L^{\frac{1}{2}\mu\lambda q} R^S = \begin{pmatrix} \lambda a^3 & 2rqa^7 \\ 0 & 2\lambda a^5 \end{pmatrix} L^{\frac{1}{2}\mu\lambda q} R^S$$

$$= \lambda a^3 R^{\lambda rqa^2} \begin{pmatrix} 1 & 0 \\ 0 & 2a^2 \end{pmatrix} L^{\frac{1}{2}\mu\lambda q} R^S = \lambda a^3 R^{\lambda rqa^2} L^{\mu\lambda qa^2} \begin{pmatrix} 1 & 0 \\ 0 & 2a^2 \end{pmatrix} R^S .$$

Another step 1 provides

$$\cdots \begin{pmatrix} 1 & 0 \\ 0 & 2a^2 \end{pmatrix} R^S \begin{pmatrix} \lambda a^3 & S' \\ 0 & 1 \end{pmatrix} L^{\frac{1}{2}\mu\lambda q} R^S = \cdots \begin{pmatrix} \lambda a^3 & 2rqa^7 \\ 0 & 2a^2 \end{pmatrix} L^{\frac{1}{2}\mu\lambda q} R^S$$

$$= \cdots a^2 R^{rqa^5} \begin{pmatrix} \lambda a & 0 \\ 0 & 1 \end{pmatrix} A' L^{\frac{1}{2}\mu\lambda q} R^S .$$

Next we obtain, after some little work,

$$\cdots \begin{pmatrix} \lambda a & 0 \\ 0 & 1 \end{pmatrix} A' L^{\frac{1}{2}\mu\lambda q} R^S \cdot R^S L^{\frac{1}{2}\mu\lambda q} \begin{pmatrix} 1 & 0 \\ 0 & \lambda a^3 \end{pmatrix} R^{S'}$$

$$= \cdots \begin{pmatrix} \lambda a & \mu(a^3 - \lambda)/q \\ \mu\lambda q & a^2 \end{pmatrix} R^{\lambda rqa^6} L^{\mu qa} \begin{pmatrix} 1 & 0 \\ 0 & 2a^4 \end{pmatrix} R^{S'} .$$

With the parameters as above, and $\mu = -1$, the first matrix decomposes as $R^{-1} L^3 R L R^3$. Finally, another step 1, and then a step 2 yields

$$\cdots \begin{pmatrix} 1 & 0 \\ 0 & 2a^4 \end{pmatrix} R^{2rqa^7} L^{\frac{1}{2}\mu\lambda q} \begin{pmatrix} 1 & 0 \\ 0 & \lambda a^3 \end{pmatrix} R^{2rqa^7} \begin{pmatrix} 4\mu\lambda ra^7 & 0 \\ 0 & 1 \end{pmatrix} L^S J$$

$$= \cdots R^{rqa^3} L^{\mu\lambda qa^4} \begin{pmatrix} 1 & 0 \\ 0 & 2\lambda a^7 \end{pmatrix} R^{2rqa^7} \begin{pmatrix} 4\mu\lambda ra^7 & 0 \\ 0 & 1 \end{pmatrix} L^S J$$

$$= \cdots 2\lambda a^7 R^{rqa^3} L^{\mu\lambda qa^4} R^{\lambda rq} A L^{\mu S/r} J .$$

If $r \neq 1$ we are at halfway and append, in false transpose, the present product. If $r = 1$ we already have a unimodular product. With the suggested choice of

$a = 5$, $\lambda = -1$, $q = 7$, $r = 3$ and $\mu = -1$ for the parameters we have S is odd, so we take $M = N$ and have obtained

$$\tfrac{1}{2}(\sqrt{N}+1) \longleftrightarrow R^{\frac{1}{2}(S+1)} L^{\mu\lambda q} R^{\lambda rqa^4} L^{\mu qa^3} R^{rqa} L^{\mu\lambda qa^6} \begin{pmatrix} \lambda a & \lambda rq \\ (a^3-\lambda)/rq & a^2 \end{pmatrix} L^{\mu qa^5}$$

$$\times R^{\lambda rqa^2} L^{\mu\lambda qa^2} R^{rqa^5} \begin{pmatrix} \lambda a & \mu(a^3-\lambda)/q \\ \mu\lambda q & a^2 \end{pmatrix} R^{\lambda rqa^6} L^{\mu qa} R^{rqa^3} L^{\mu\lambda qa^4} R^{\lambda rq} L^{\frac{1}{2}\mu S/r} J \cdots$$

$$\longleftrightarrow [\,820304\,,\,7\,,\,-13125\,,\,-875\,,\,105\,,\,109375\,,\,-1\,,\,6\,,\,4\,,\,-21875\,,$$
$$-\,525\,,\,175\,,\,65625\,,\,0\,,\,-1\,,\,3\,,\,1\,,\,1\,,\,3\,,\,0\,,$$
$$-\,328125\,,\,-35\,,\,2625\,,\,4375\,,\,-21\,,\,-273434\tfrac{1}{2}\,,\,0\,,\,\ldots\,]$$

On our rendering the inadmissible partial quotients admissible, and adjoining the second half of the period, we see that

$$\tfrac{1}{2}(\sqrt{2691592265949}+1) = [\,820304\,,\,\overline{6\,,\,1\,,\,13124\,,\,874\,,\,1\,,\,104\,,\,109373\,,\,1\,,}$$
$$\overline{4\,,\,3\,,\,1\,,\,21874\,,\,524\,,\,1\,,\,174\,,\,65624\,,\,3\,,\,2\,,\,328121\,,\,34\,,\,1\,,\,2624\,,\,4374\,,\,1\,,\,20\,,}$$
$$\overline{546869\,,\,20\,,\,1\,,\,4374\,,\,2624\,,\,1\,,\,34\,,\,328121\,,\,2\,,\,3\,,\,65624\,,\,174\,,\,1\,,\,524\,,\,21874\,,}$$
$$\overline{1\,,\,3\,,\,4\,,\,1\,,\,109373\,,\,104\,,\,1\,,\,874\,,\,13124\,,\,1\,,\,6}\,,\,1640607\,]$$

It is an interesting exercise to use the present analysis to count the length $\ell(\sqrt{M})$ of the period; the discussion in [10] and [18] makes an intriguing contrast. The cyclical patterns in the expansion are well explained. They arise from multiplication (or division) by $\pm a^k$; or, if we mean the grosser pattern, by ra^n. We also see that in the general case a unit of $\mathbb{Q}(\sqrt{M})$ is given by

$$(\tfrac{1}{2}\mu\lambda qS + 1 - \tfrac{1}{2}\mu\lambda q\sqrt{M}\,)^{2n}(\tfrac{1}{2}S - \tfrac{1}{2}\sqrt{M}\,)^{2k} \,/\, a^{2nk}r^k\,.$$

This unit is fundamental unless $r = 1$ or if k is even, in which cases we did not need to double the matrix product in order to get a unimodular matrix. In those cases the just cited unit is the square of a fundamental unit.

3 'What's it all about, Alfie?'[2]

It is, I suppose, a matter of taste whether one thinks it simpler to manipulate 2×2 matrices or continued fractions. So let me not make exaggerated claims for matrices or R–L sequences. It is manifest however that the matrix viewpoint makes it natural to deal with nonpositive partial quotients, to accept certain intermediate convergents as being as important as convergents proper, and thence to see patterns that remain heavily disguised by more constipated approaches. It's interesting that one can deal with all cases more or less at once, even if doing so buries simple computations in a multitude of parameters. Certainly, the claims for the present approach might have been more convincing had I restricted myself to illustrative simple examples.

---

[2] A song of the sixties; to be precise, the song from the film *Alfie* (1966), starring Michael Caine, Shelley Winters etc.

However, we do get real insight from the talk about multiplying continued fractions by rationals, and from the fact that the matrix description makes this appropriately explicit. We *see* that the patterns that had been observed in the present expansions are in effect cases of *twisted* periodicity, with sequences of partial quotients repeating themselves, though twisted by multiplication by some power of a. Recent other work could have predicted that fact. In 'Halfway to a solution of $X^2 - DY^2 = -3$' [8], Rick Mollin, Hugh Williams, and the author find new infrastructure in the period of the continued fraction expansions of quadratic irrationals; see also my [12], and additional work still in preparation. There we find that the second half of the expansion of δ up to some complete quotient $(\delta + P)/Q$, is essentially the reverse of the first half, 'twisted' by application of some linear transformation of determinant $\pm Q$. Mollin and Williams [9] had shown that the cases M discussed above (when $r = 1$) include all those with continued fraction expansions yielding three or more consecutive complete quotients with Q's each a power of the same integer a. It is therefore no wonder that we sequentially obtain our expansion by multiplying earlier segments of the expansion by powers of a.

I am occasionally asked who 'invented' the matrix approach to continued fractions. I feel that the correct answer is that the approach was invented by those who invented matrices. It is after all obvious to anyone who has learned matrix notation that the traditional recurrence formulas for the convergents invite a matrix formulation. However, for what it is worth, I was alerted to these notions by Harold Stark's textbook [15]; and was particularly influenced by his remark (I think it was an exercise) that a matrix constructed from a solution to Pell's equation corresponds to the period of the relevant square root. It is the converse of that remark, namely that a unimodular matrix of a certain shape — called an 'initial segment' in my discussions above — corresponds to the period, that plays a primary role in the present paper. Multiplication and R–L sequences derive from a much neglected paper of Raney [14]. Mind you, we didn't need any of his detailed ideas here, but they certainly materially influence my outlook. Going back into prehistory one finds [5] which deals with matrices and continued fractions in a style not very different from mine; I saw it first only recently. Other papers have an emphasis which seems rather different from my present one but can be usefully combined with mine. For example, in [11], a paper inspired by a matrixless manuscript of Jerry Minkus, I make good use of remarks of Walters [16], to tell stories about continued fraction expansions of certain powers of e.

But, notation — no matter how felicitous — cannot be 'what it's all about'. The truth is that the matrices obtained from initial segments of the continued fraction expansion are quadratic forms in disguise, and multiplication of those matrices corresponds to composition of forms within the principal cycle. To be precise, if $(\delta + P)/Q$ is the complete quotient at end of the segment, then the corresponding quadratic form is one of

$$\pm Q X^2 \mp (t + 2P)XY \pm \big((n + tP + P^2)/Q\big)Y^2,$$

with the sign chosen so that $\pm Q$ is the determinant of the matrix. If one dislikes forms, very little is lost in thinking of the matrix as corresponding to the $\mathbb{Z}$-

module $\langle Q, \delta + P \rangle$, and noting that this is an ideal of the order $\mathbb{Z}[\delta]$. One may then consider oneself to be multiplying fractional ideals rather than composing forms. All one needs know is that one has reached the end of the period of δ if and only if $Q = \pm 1$. Given that we only want a formula for a fundamental unit, we give some details of the continued fraction expansions principally because that detail is intrinsically interesting but also because it provides evidence confirming our claim that we are dealing with a primitive period.

Thus, if one's interest is just to find a fundamental unit, as I pretend in this note's title and introduction, then a cleaner argument than the computations I present might have us 'composing' ideals.

The product of two ideals $\langle Q, P+\delta \rangle$ and $\langle Q', P'+\delta \rangle$ is generated over $\mathbb{Z}$ by the quantities QQ', $Q'(P+\delta)$, $Q(P'+\delta)$ and $(P+\delta)(P'+\delta) = PP'-n+(t+P+P')\delta$.

Set $G = \gcd(Q, Q', t + P + P')$. One may verify, by studying the classical formulaires or from first principles, that the product is a rational integer multiple, namely by G, of $\langle q, p + \delta \rangle$ where $q = QQ'/G^2$ and p satisfies the congruences

$$p \equiv P \pmod{Q/G}, p \equiv P' \pmod{Q'/G}$$

$$\text{and} \quad (P - p)(P' - p) \equiv (n + tp + p^2) \pmod{QQ'/G}.$$

The first pair of congruences determines p modulo $QQ'/G(Q, Q')$. The last congruence decides which of the remaining $(Q, Q')/G$ possibilities for $p \mod q$ is to be taken.

Correspondingly, a product of quadratic forms

$$Qx^2 - (t + 2P)xy + ((n + tP + P^2)/Q)y^2$$

$$\text{and } Q'x'^2 - (t + 2P')x'y' + ((n + tP' + P'^2)/Q')y'^2$$

together with a substitution

$$X = Axx' + Bxy' + Cx'y + Dyy' \text{ and } Y = A'xx' + B'xy' + C'x'y + D'yy',$$

with integer coefficient $A, \ldots$ and $A', \ldots$, not all sharing a common factor, yields a form $qX^2 - (t + 2p)XY + ((n + tp + p^2)/q)Y^2$ known as a *compound* of the given forms. In fact the Grassmann co-ordinates of the substitution matrix $\left(\begin{smallmatrix} A & B & C & D \\ A' & B' & C' & D' \end{smallmatrix} \right)$ are determined (they are essentially the six coefficients of the given forms), so the substitution is determined up to multiplication by a 2×2 unimodular integer matrix. Thus the compound form is defined up to *equivalence* and we see that compounding is well defined on equivalence classes of forms of the same discriminant. I think of the particular case, where the stated forms yield the compound form $qX^2 - (2p + t)XY + ((n + tp + p^2)/q)Y^2$, as *composition*.

The present work in fact began with my my studying composition of the quadratic forms corresponding to the matrices multiplied above. That study did not seem all that much easier than the work detailed in the present manuscript; and it did not readily allow me to detail the continued fraction expansion. Nor did it seem completely obvious that the units produced were fundamental. On the other hand, those considerations motivated the rather eccentric sequence in which I multiplied the matrices — after all, any sequence would have done, initial segments commute.

4 Concluding Remarks

In May 1993 I chafed Hugh Williams about those 'horrible continued fractions' of his and Mollin, and was quietened by being handed the manuscript [18]. In retaliation I claimed that I could do it all neatly, succinctly, and without distinguishing a multitude of cases. It took me only some nine months to see that this is indeed an immediate and obvious application of methods I had been using elsewhere. I suggest that my approach gives some insight into what is going on, and that it goes a considerable way towards taming the sequence of ever more complicated arguments of Yamomoto [19], Hendy [4], Bernstein [2], Williams [17], Lévesque and Rhin [7], Lévesque [6], Azuhata [1], Halter-Koch [3], Mollin and Williams [9] and of Williams [18]. Such a 'taming' seems important if ideas of the present genre are to produce families of quadratic number fields with yet relatively larger regulator.

5 Acknowledgments

I am grateful to Hugh Williams for showing me the present problems and for making available a preprint of [18].

This work was supported in part by grants from the Australian Research Council and by a research agreement with Digital Equipment Corporation.

1991 Mathematics subject classification: 11A55, 11Y65, 11J70

References

1. T. Azuhata, 'On the fundamental units and the class numbers of real quadratic fields II', *Tokyo J. Math.* **10** (1987), 259–270
2. L. Bernstein, 'Fundamental units and cycles in the period of real quadratic number fields II', *Pacific J. Math.* **63** (1976), 63–78
3. F. Halter-Koch, 'Einige periodische Kettenbruchentwicklungen und Grundeinheiten quadratischer Ordnung', *Abh. Math. Sem. Univ. Hamburg* **59** (1989), 157–169; and 'Reel-quadratischer Zahlkörper mit grosser Grundeinheit', *ibid.*, 171–181
4. M. D. Hendy, 'Applications of a continued fraction algorithm to some class number problems', *Math. Comp.* **28** (1974), 267–277
5. Kjell Kolden, 'Continued fractions and linear substitutions', *Arch. Math. Naturvid.* **50** (1949), 141–196
6. C. Lévesque, 'Continued fraction expansions and fundamental units', *J. Math. Phys. Sci.* **22** (1988), 11–14
7. C. Lévesque and G. Rhin, 'A few classes of periodic continued fractions', *Utilitas Math.* **30** (1986), 79–107
8. R. A. Mollin, A. J. van der Poorten, and H. C. Williams, 'Halfway to a solution of $X^2 - DY^2 = -3$', submitted to *J. Théorie des Nombres de Bordeaux,*, 32pp
9. R. A. Mollin and H. C. Williams, 'Consecutive powers in continued fractions', *Acta Arith.* **51** (1992), 233–264

10. R. A. Mollin and H. C. Williams, 'on the period length of some special continued fractions', *Séminaire de Théorie des Nombres de Bordeaux* **4** (1992)

11. A. J. van der Poorten, 'Continued fraction expansions of values of the exponential function and related fun with continued fractions', submitted to *Proc. Amer. Math. Soc.*, 9pp

12. A. J. van der Poorten, 'Explicit quadratic reciprocity', to appear in NTAMCS '93 (Number Theoretic and Algebraic Methods in Computer Science, Moscow 1993), 5pp

13. A. J . van der Poorten and J. Shallit, 'A specialised continued fraction', *Canad. J. Math.* **45** (1993), 1067-1079

14. G. N. Raney, 'On continued fractions and finite automata', *Math. Ann.* **206** (1973), 265-283

15. H. M. Stark, *An introduction to number theory* (MIT Press 1978)

16. R. F. C. Walters, 'Alternative derivation of some regular continued fractions', *J. Austral. Math. Soc.* **8** (1968), 205-212

17. H. C. Williams, 'A note on the period length of the continued fraction expansion of certain $\sqrt{D}$', *Utilitas Math.* **28** (1985), 201-209

18. H. C. Williams, 'Some generalizations of the S_n sequence of Shanks', to appear in *Acta Arith.*, 24pp

19. Y. Yamomoto, 'Real quadratic number fields with large fundamental units', *Osaka J. Math.* **8** (1971), 261-270

Factorization of Polynomials over Finite Fields in Subexponential Time under GRH

Sergei Evdokimov

St. Petersburg Institute for Informatics and Automation
of the Academy of Sciences of Russia,
the 14th liniya 39, 199178,
Saint-Petersburg, Russia

Abstract. We show assuming the Generalized Riemann Hypothethis that the factorization of a one-variable polynomial of degree n over an explicitly given finite field of cardinality q can be done in deterministic time $(n^{\log n} \log q)^{O(1)}$. Since we need the hypothesis only to take roots in finite fields in polynomial time, the result can also be formulated in the following way: a polynomial equation over a finite field can be solved "by radicals" in subexponential time.

1   Introduction

In this paper we are interested from the computational point of view in the problem of factoring a one-variable polynomial f over a finite field k into irreducible factors over k. For the theory of finite fields we refer to [8].

We say following [5] that a finite field k of cardinality $q = p^m$ where p is a prime, is *explicitly given* if, for some basis of k over its prime field $\mathbb{F}_p$, we know the product of any two basis elements expressed in the same basis. For example, if we know an irreducible polynomial $g \in \mathbb{F}_p[X]$ of degree m, then such *explicit data* is readily calculated, since $\mathbb{F}_p[X]/(g)$ is a field of cardinality p^m. It is supposed that the elements of $\mathbb{F}_p$ are represented in the conventional way, so that the field operations in $\mathbb{F}_p$ can be performed in time $(\log p)^{O(1)}$. So the field operations in an explicitly given finite field k of cardinality q can be performed in time $(\log q)^{O(1)}$. Here and below by the running time of an algorithm we mean the number of bit operations that it performs.

It is not known if there exists a deterministic polynomial-time algorithm that, given p and m, constructs explicit data for a finite field of cardinality $q = p^m$. If the Generalized Riemann Hypothesis (GRH in brief) is valid, then such an algorithm exists (see [1, 3]).

In spite of the fact that the problem of factoring polynomials over finite fields has been intensively studied for the last years (see, for instance, [3, 4, 5] where polynomial-time algorithms for some special classes of polynomials were presented), as far as we know so far only the following two general deterministic algorithms for factoring an arbitrary polynomial of degree n over an explicitly given finite field of cardinality $q = p^m$ have been known:

1. Berlekamp's algorithm [2] whose running time is bounded by $(nmp)^{O(1)}$ (assuming no unproved hypothethis).
2. Rónyai's algorithm [6] whose running time is bounded by $(n^n m \log p)^{O(1)}$ (assuming GRH).

It is easy to see that the first algorithm is polynomial for small p whearas the second one is polynomial for small n. However, both of them are exponential if no restriction is made. In this paper we prove (assuming GRH) a subexponential upper bound for the problem under consideration.

Theorem 1 (MAIN THEOREM). *Assuming GRH there exists a deterministic algorithm which, given a polynomial f of degree n over an explicitly given finite field k of cardinality q, decomposes f into irreducible factors over k in time $(n^{\log n} \log q)^{O(1)}$.*

Remark. As it follows from the proof (see Sect. 5), we need GRH only to take roots in finite fields in polynomial time. So the result can be formulated as follows: a polynomial equation over a finite field can be solved "by radicals" in subexponential time.

It is necessary to say some words about the proof of the MAIN THEOREM. We concentrate on the following case of the problem: the polynomial f has no multiple roots and is completely splitting over k (i.e. f has n distinct roots in k). The general problem is polynomial-time reduced to this one due to [2]. In order to find a root of f in this case we deal with polynomials not only over the ground field k but also over completely splitting semisimple algebras over k. Here is the idea of the algorithm. If $n = \deg f$ is even, then we apply Ronyai's result [6] permitting us to factor f into two factors over k in time $(n \log q)^{O(1)}$, and proceed with the factor of lesser degree. If n is odd, then we form an algebra $R_1 = k[X]/(f) = k[A]$ where $A = X \bmod f$. From $f(A) = 0$ it follows that

$$f(X) = (X - A)f^*(X), \quad f^* \in R_1[X] . \tag{1}$$

Since $\deg f^* = n - 1$ is even, we apply some generalization of Ronyai's result mentioned above (see Lemma 8) and factor f^* into two factors $f^* = f_1 g_1$, $\deg f_1 \le \deg g_1$ over R_1 (or we get a zero-divizor in R_1 giving a decomposition of f over k) in time $(d_1 n \log q)^{O(1)}$ where $d_1 = \dim_k R_1$ (equal n of course). We can assume that $\deg f_1$ is odd, since otherwise we could apply Lemma 8 once again. If $\deg f_1 = 1$, then $f_1(X) = X - A_1$ where $A_1 \in R_1$ is a root of f different from A. So f can be factored into two factors over k in time $(n \log q)^{O(1)}$ (since we have an explicitly given endomorphism of the algebra R_1 sending A to A_1, see Lemma 9) and we can proceed with the factor of lesser degree. If $\deg f_1 > 1$, then we form an algebra $R_2 = R_1[X]/(f_1) = R_1[A_1]$ where $A_1 = X \bmod f_1$. From $f_1(A_1) = 0$ it follows that

$$f_1(X) = (X - A_1)f_1^*(X), \quad f_1^* \in R_2[X] . \tag{2}$$

Since $\deg f_1^*$ is even, we apply Lemma 8 and factor f_1^* into two factors $f_1^* = f_2 g_2$, $\deg f_2 \le \deg g_2$ over R_2 (or we get a zero-divizor in R_2 giving a decomposition of

f_1 over R_1 or a zero-divizor in R_1) in time $(d_2 n_1 \log q)^{O(1)}$ where $d_2 = \dim_k R_2$, $n_1 = \deg f_1$. We can assume that $\deg f_2$ is odd, since otherwise we could apply Lemma 8 once again. If $\deg f_2 = 1$, then $f_2(X) = X - A_2$ where $A_2 \in R_2$ is a root of f_1 different from A_1. So by Lemma 9 f_1 can be factored into two factors over R_1 (or a zero-divizor in R_1 can be found) in time $(d_1 n_1 \log q)^{O(1)}$ and we can proceed with the factor of lesser degree. If $\deg f_2 > 1$, then we form an algebra $R_3 = R_2[X]/(f_2)$ and so on. At the end of this process we get a root of the polynomial f in the field k. The only obstacle here is increasing the dimensions of the algebras involved in the algorithm. However, it is easy to see from above that the number of successive algebra extensions is not more than $\log_2 n$. So the dimensions of the algebras involved are bounded by $n^{\log_2 n}$, whence the MAIN THEOREM follows.

It is worth noticing that all our algorithms under GRH run as follows: either the algorithm terminates and the validity of the the result obtained does not depend on any unproved hypothesis, or we find a number field K and a Dirichlet character χ of K for which GRH is not valid.

The contents of the paper are as follows. Sect. 2 contains some definitions and facts from the theory of commutative semisimple algebras over a finite field. It also contains a statement (Lemma 5) being of great use in Sect. 5. In Sect. 3 we generalize for the case of completely splitting semisimple algebras over a finite field k a well-known algorithm for taking lth roots (l is a prime) in the field k if an lth non-residue in k is given. Sect. 4 contains two statements (Lemma 8 and Lemma 9) underlying the MAIN ALGORITHM. In Sect. 5 we describe the MAIN ALGORITHM and prove the MAIN THEOREM.

2 Semisimple Commutative Algebras over a Finite Field

Here we remind some definitions and facts from the theory of semisimple algebras over a finite field (see [8]). Below we assume that all algebras are finite dimensional commutative algebras with unity.

Let k be a finite field. An algebra R over k is called *semisimple* if it contains no nilpotent element. A subalgebra and a factoralgebra of a semisimple algebra are also semisimple.

For each semisimple algebra R over k there exist uniquely determined elements $e_i \in R$, $i = 1, \ldots, d$ (called *the primitive idempotents* of the algebra R) such that: Re_i is isomorphic to a finite field extension k_i of the field k and

$$\sum_{i=1}^{d} e_i = 1, \quad e_i e_j = \begin{cases} e_i & \text{if } i = j \\ 0 & \text{otherwise} . \end{cases} \tag{3}$$

The algebra R is decomposed into the direct product of fields:

$$R = \prod_{i=1}^{d} Re_i \cong \prod_{i=1}^{d} k_i . \tag{4}$$

We say that R is *completely splitting* over k if $k_i = k$ for all i. It is easily seen from the definition that R is completely splitting over k iff $a^q = a$ for all $a \in R$ where q is the cardinality of k. For example, if f is a polynomial of degree n over k having n distinct roots in k, then the algebra $k[X]/(f)$ is semisimple and completely splitting over k (see also Lemma 3).

Definition 2. We say that a monic polynomial f over an algebra R is *separable* if its discriminant $D(f)$ is invertible in R. We call f *completely splitting* over R if there exist $a_i \in R$, $i = 1, \ldots, n$ such that

$$f(X) = \prod_{i=1}^{n} (X - a_i) \ . \tag{5}$$

(If R is not a field, then the decomposition is not unique.)

Lemma 3. *If R is a completely splitting semisimple algebra over a finite field k and f is a completely splitting separable monic polynomial over R, then the algebra $R[X]/(f)$ is also semisimple and completely splitting over k. If a monic polynomial $g \in R[X]$ divides f, then g is also separable and completely splitting over R.*

Proof. It follows from lemma's conditions that $f(X) = \prod_{i=1}^{n}(X - a_i)$ with $a_i \in R$ and $D(f) = \prod_{i,j, i \neq j}(a_i - a_j)$ is invertible in R. Set:

$$e_i^* = \prod_{j, j \neq i}(A - a_j) / \prod_{j, j \neq i}(a_i - a_j), \quad i = 1, \ldots, n \tag{6}$$

where $A = X \bmod f \in R[X]/(f)$. It is easy to check that

$$\sum_{i=1}^{n} e_i^* = 1, \quad e_i^* e_j^* = \begin{cases} e_i^* & \text{if } i = j \\ 0 & \text{otherwise} \ . \end{cases} \tag{7}$$

Hence $R[X]/(f) = \prod_{i=1}^{n} Re_i$ where $Re_i \cong R$, whence the first statement of lemma follows. If g divides f, then $D(g)$ divides $D(f)$ in R, so g is separable. Since R is isomorpfic to a direct product of fields each of which coincides with k, we can assume that $R=k$ in checking that g is completely splitting. However, in this case the statement is obvious. $\qquad\square$

Let k be an explicitly given finite field of cardinality q. We say that an algebra R over k is *explicitly given* if, for some basis of R over k, we know the product of any two basis elements expressed in the same basis. Since the field operations in k can be performed in time $(\log q)^{O(1)}$ (see Sect. 1), the algebra operations in R can be performed in time $(d \log q)^{O(1)}$ where $d = \dim_k R$. If f is a monic polynomial over R, then *explicit data* for the algebra $R[X]/(f)$ can be readily calculated from the explicit data for R in time $(dn \log q)^{O(1)}$ where $n = \deg f$.

Let R be a semisimple algebra over k and a be a zero-divisor in R. Set:

$$R_1 = Ra, \quad R_2 = \{x \in R; \; ax = 0\} \; . \tag{8}$$

Then it is easily seen that R_1 and R_2 are non-trivial ideals of R and $R = R_1 \times R_2$. Explicit data for R_1, R_2 can be calculated from explicit data for R in time $(d \log q)^{O(1)}$ by means of the ordinary linear algebra techniques.

Now we prove a statement being of great use in Sect. 5 throughout the MAIN ALGORITHM.

Definition 4. Given a monic polynomial f over an algebra R, we say that

$$f = gh, \quad g, h \in R[X] \tag{9}$$

is a *non-trivial decomposition* of f over R if both polynomials g, h are monic and of degree less than $\deg f$. In this case g and h are called *non-trivial factors* of f.

Lemma 5. *Let R be an explicitly given algebra of dimension d over an explicitly given finite field k of cardinality q and f be a monic polynomial of degree n over R. Then finding a zero-divisor in the algebra $S = R[X]/(f)$ is polynomial-time equivalent (in d, n, $\log q$) to finding either a zero-divisor in R, or a non-trivial decomposition of f over R.*

Proof. If $f = gh$ is a non-trivial decomposition of f over R, then

$$g(A)h(A) = f(A) = 0 \quad \text{where} \quad A = X \bmod f \in S \; . \tag{10}$$

Since $g(A) \neq 0$, $h(A) \neq 0$, they are zero-divizors in S. Conversely, if a is a zero-divisor in S, then it is easy to find $b \in S$, $b \neq 0$ for which $ab = 0$. We represent a and b in the form

$$a = g(A), \quad b = h(A) \tag{11}$$

where g, h are non-zero polynomials over R of degree less than n. It follows from $ab = 0$ that f divides gh. Then we try to find $\mathrm{GCD}(f, g)$ in $R[X]$ by means of the Euclid algorithm. If the algorithm fails at some step, then a zero-divisor in R is found. Otherwise, we find a monic polynomial $r \in R[X]$ dividing f. All the computation can be done in time $(dn \log q)^{O(1)}$. If $r = 1$, then $1 = uf + vg$ for some $u, v \in R[X]$ and so $h = ufh + vgh$ is divided by f, which is impossible, since $\deg h < n$. $\qquad\square$

3 Taking Roots in Completely Splitting Algebras

In this section we describe a polynomial-time algorithm for taking lth roots (l is a prime) in a completely splitting semisimple algebra over a finite field providing that some extra information is given. We need the following easy statement.

Lemma 6. *If $a_0, a_1, \ldots, a_n$ are $n+1$ distinct roots of a polynomial f of degree n over a commutative ring R, then at least one of the elements $a_i - a_j$, $i, j = 0, 1, \ldots, n$ is a zero-divisor in R.*

Proof. If $n = 0$, then lemma is clear. In the general case $f(a_0) = 0$ implies that $f(X) = (X - a_0)g(X)$ where $\deg g = n - 1$. Since $f(a_i) = 0$ for $i = 1, 2, \ldots, n$, there are two possibilities: either $a_i - a_0$ is a zero-divizor for some i, or $g(a_i) = 0$ for all $i = 1, 2, \ldots, n$. In the first case we are done, otherwise we conclude by induction. □

Proposition 7. *Let R be an explicitly given completely splitting semisimple algebra of dimension d over an explicitly given finite field k of cardinality q, l be a prime and $a \in R \setminus \{0\}$. Then:*

1. *If l is coprime to $q - 1$, then the lth root of a is uniquely determined and can be found in polynomial time.*
2. *If l divides $q - 1$ and $a^{(q-1)/l}$ is an idempotent of the algebra R, then, given an lth non-residue in the field k, at least l distinct lth roots of a can be found in polynomial time.*

Remark 1. In case 2 the term "polynomial time" means time $(dl \log q)^{O(1)}$.

Remark 2. If $a^{(q-1)/l}$ is not an idempotent of R, then the equation $X^l = a$ has no solution in R.

Proof. If l is coprime to $q - 1$, then raising to lth power is a bijection of R. The inverse map is given by raising to rth power where $lr = 1 (\bmod\ q - 1)$.

   Let l divide $q - 1$. We write $q - 1$ in the form $q - 1 = l^h m$ where m is coprime to l. First of all we remind the following well-known algorithm for taking lth roots in the field k (the idea of the algorithm is contained in [7]). Let G_l denote the l-component of the multiplicative group G of k. Since the latter is cyclic, G_l is also cyclic of order l^h. Let b be an lth non-residue in k. We assume b lying in G_l, since otherwise we could change b for $b^m \in G_l$. So b is a generator of G_l. Every $a \in G_l$ can uniquely be represented in the form

$$a = b^s \quad \text{where} \quad s = \sum_{i=0}^{h-1} s_i l^i, \quad 0 \le s_i < l . \tag{12}$$

The integers s_i can successively be found as follows. Assuming s_i for $0 \le i < u$ to be found set:

$$a = a' b^{s'} \quad \text{where} \quad s' = \sum_{i=0}^{u-1} s_i l^i . \tag{13}$$

(For $u = 0$ we have $s' = 0$, $a = a'$.) Then $(a')^{l^{h-u-1}} = \zeta^{s_u}$ where $\zeta = b^{l^{h-1}}$ is a primitive lth root of the unity of k. So the integer s_u can be found by means of l trials. If a is an lth power in G_l (i.e. $a^{(q-1)/l} = 1$), then $s_0 = 0$ in (12) and we have:

$$a = (c\zeta^j)^l \quad \text{where} \quad c = b^{s/l}, \quad j = 0, 1, \ldots, l - 1 . \tag{14}$$

The case of an arbitrary $a \in G$ is reduced to the considered one by changing a for $a^m \in G_l$.

Let $a \in R \setminus \{0\}$ and $a^{(q-1)/l} = e$ be an idempotent of the algebra R. Without loss of generality we can assume that $e = 1_R$ where 1_R is the unity of R, and a is of order dividing l^h. In this case we try to represent a in the form (12) where b now means $b \cdot 1_R$. If the algorithm is successful, then $a \in k \cdot 1_R$ and we are done. If it fails at some step u, then

$$(a')^{l^{h-u-1}} \neq \zeta^j \quad \text{for all} \quad j = 0, 1, \ldots, l-1 . \tag{15}$$

(We save the above notations.) So the polynomial $X^l - 1$ has at least $l+1$ distinct roots in R: ζ^j for $j = 0, 1, \ldots, l-1$ and $(a')^{l^{h-u-1}}$. So by Lemma 6 a zero-divizor in R can be found in polynomial time. It follows according to Sect. 2 (see (8)) that a non-trivial decomposition $R = R_1 \times R_2$ can be found in polynomial time and we can proceed in the algebras R_1 and R_2. $\qquad \square$

4 Two Main Lemmas

In this section we prove two statements underlying the MAIN ALGORITHM. Below k denotes an explicitly given finite field of cardinality q.

Lemma 8 (Assuming GRH). *Let R be an explicitly given completely splitting semisimple algebra of dimension d over a finite field k and f be a completely splitting separable monic polynomial over R of even degree n. Then in time $(dn \log q)^{O(1)}$ either a zero-divizor in R can be found, or a non-trivial decomposition of f over R can be constructed.*

Remark 1. For $R = k$ lemma was proved by Rónyai (see [6]).

Remark 2. It follows from lemma's proof, that we need GRH only to find a quadratic non-residue in the field k in polynomial time.

Proof. By Lemma 5 it suffices to find a zero-divizor in the algebra $S = R[X]/(f)$. To do this set $T = S[Y]/(f^*)$ and $B = Y \bmod f^*$ where $f^*(Y) = f(Y)/(Y - A)$, $A = X \bmod f$. So we have:

$$S = R[A], \quad T = S[B] = R[A, B] . \tag{16}$$

By Lemma 3 the algebras S, T are semisimple and completely splitting over k and $A - B$ is invertible in T. Let φ stand for the automorphism of the algebra T over R permuting A and B. Set:

$$T_\varphi = \{a \in T; \ \varphi(a) = a\} . \tag{17}$$

It is easy to see that $T_\varphi = R[A + B, AB]$ is a free module over R with the basis $\{(A + B)^i (AB)^j; \ i, j \geq 0, \ i + j \leq n - 2\}$ and $\dim_R T_\varphi = n(n-1)/2$. Since we assumed that GRH is valid, a quadratic non-residue in k can be constructed in time $(\log q)^{O(1)}$ due to [3] (without loss of generality q is supposed odd). The coefficients of the polynomial $h(Z) = (Z - A)(Z - B)$ belong to T_φ, so by Proposition 7 ($l=2$) the equation $h(Z) = 0$ can be solved in T_φ in time $(dn \log q)^{O(1)}$. Let a be a root of h lying in T_φ. Setting

$$e_1 = (A - a)/(A - B), \quad e_2 = (a - B)/(A - B) \tag{18}$$

we see that $e_1^2 = e_1$, $e_2^2 = e_2$, $e_1 e_2 = 0$, $e_1 + e_2 = 1$, $\varphi(e_1) = e_2$. Hence we have:

$$T = Te_1 \times Te_2, \quad \dim_k Te_1 = \dim_k Te_2 = dn(n-1)/2 \ . \tag{19}$$

Since $\dim_k S = dn$ and n is even, $\dim_k Te_1$ is not divided by $\dim_k S$, so S-module Te_1 cannot be free. Therefore, the Gauss algorithm applied to the set $\{B^i e_1\}_{i=0}^{n-2}$ of generators of Te_1 over S fails at some step and we get a zero-divisor in S. $\quad\square$

Lemma 9 (Assuming GRH). *Let R be an explicitly given completely splitting semisimple algebra of dimension d over a finite field k and f be a completely splitting separable monic polynomial of degree n over R. Then, given a root $B \in R[X]/(f)$ of f different from $A = X \bmod f$, in time $(dn \log q)^{O(1)}$ either a zero-divisor in R can be found, or a non-trivial decomposition of f over R can be constructed.*

Proof. By Lemma 5 it suffices to find a zero-divisor in the algebra $S = R[X]/(f)$ which is by Lemma 3 is semisimple and completely splitting over k. We define an endomorphism φ of the algebra S over R setting $\varphi(A) = B$ and consider the following sequence

$$\{A_i\}_{i=0}^{\infty} \quad \text{where} \quad A_0 = A, \ A_i = \varphi(A_{i-1}) \ \text{for} \ i \geq 1 \ . \tag{20}$$

It follows from the definitions that $A_1 = B$ and $f(A_i) = 0$ for all i. If $A_0, \ldots, A_n$ are pairwise distinct, then by Lemma 6 a zero-divisor in S can be found. Otherwise, there exist m_1, m_2 such that $0 \leq m_1 < m_2 \leq n$ and $A_{m_1} = A_{m_2}$. Let m_1, m_2 be the smallest ones of such a kind. We consider two cases.

(I) $m_2 - m_1 = 1$. In this case $A_0 - A_1$ cannot be invertible in S, because otherwise $A_i - A_{i+1} = \varphi^i(A_0 - A_1)$ should be invertible for all i, in particular for $i = m_1$. So $A_0 - A_1$ is a zero-divisor in S and we are done.

(II) $m_2 - m_1 \geq 2$. Set $l = m_2 - m_1$. Without loss of generality we can assume that l is a prime and $l < p$ where p is the characteristic of the field k. Let $\tilde{k}$ denote the minimal field extension of k containing a primitive lth root of 1. Explicit data for $\tilde{k}$ (together with an explicitly given inclusion map $k \to \tilde{k}$) can be constructed in time $(l \log q)^{O(1)}$ assuming GRH due to [3] (for $k = \mathbb{F}_p$ it was previously done in [4]). We consider an algebra $\tilde{S} = \tilde{k} \otimes S$ over $\tilde{k}$. It is clear that $\tilde{S}$ is semisimple and completely splitting over $\tilde{k}$. Set:

$$\alpha_j = \sum_{i=0}^{l-1} \zeta^{ij} A_{m_1+i}, \quad j = 0, 1, \ldots, l-1, \tag{21}$$

$$\tilde{S}_\varphi = \{a \in \tilde{S}; \ \varphi(a) = a\} \tag{22}$$

where ζ is a primitive root of 1 in $\tilde{k}$ and we assume that φ acts identically on $\tilde{k}$. Then $\tilde{S}_\varphi$ is a subalgebra of $\tilde{S}$ explicit data for which can be calculated from the explicit data for $\tilde{S}$ by means of the ordinary linear algebra techniques. It is easily seen from the definitions that

$$\varphi(\alpha_j) = \zeta^{-j}\alpha_j, \quad \alpha_j^l \in \tilde{S}_\varphi \text{ for all } j \ . \tag{23}$$

If $\alpha_j = 0$ for all $j \neq 0$, then $A_{m_1+i} = A_{m_1}$ for all $i \geq 0$, which is impossible, since $m_2 - m_1 \geq 2$. So there exists $j_0 > 0$ for which $\alpha_{j_0} \neq 0$. By (23) $\alpha_{j_0} \notin \tilde{S}_\varphi$. According to [3] (for $k = \mathbb{F}_p$ to [4]) an lth non-residue in the field $\tilde{k}$ can be constructed in time $(l \log q)^{O(1)}$ assuming GRH. So by Proposition 7 we can find in time $(dnl \log q)^{O(1)}$ at least l distinct lth roots of $\alpha_{j_0}^l$ lying in $\tilde{S}_\varphi$. Together with α_{j_0} it gives at least $l+1$ distinct roots of the polynomial $X^l - \alpha_{j_0}^l$ in $\tilde{S}$. So by Lemma 6 a zero-divisor a in $\tilde{S}$ can be easily found. If $\tilde{q} = q^h$ where $\tilde{q}$ is the cardinality of $\tilde{k}$, then

$$\text{Norm}_{\tilde{S}/S}(a) = \prod_{i=0}^{h-1} a^{q^i} \tag{24}$$

is a zero-divisor in S. $\qquad\qquad\qquad\qquad\qquad\qquad\qquad\qquad\qquad\qquad\qquad\qquad\qquad$ $\square$

5 Proof of the MAIN THEOREM

Here we describe the MAIN ALGORITHM and prove the MAIN THEOREM. According to Sect. 1 it suffices to find a root of a completely splitting separable monic polynomial over k.

Given on input explicit data for a completely splitting semisimple algebra R of dimension d over k and a completely splitting separable monic polynomial f of degree n over R, the MAIN ALGORITHM finds either a root of f over R, or a zero-divisor in R. If $R = k$, then the MAIN ALGORITHM finds a root of f, since there is no zero-divisor in the field k.

Denote by $T(R, f)$ the running time of the MAIN ALGORITHM with entries R, f. We shall prove below assuming GRH that

$$T(R, f) \leq (dn^{\log n} \log q)^{O(1)} \ . \tag{25}$$

This will prove the MAIN THEOREM.

To describe the MAIN ALGORITHM we first describe three procedures P1, P2, P3 according to Lemmas 5, 8, 9. Below R and f are as above.

(P1) Given on input R, f and a zero-divisor $a \in R[X]/(f)$, procedure P1(R, f, a) finds either a non-trivial decomposition of f over R, or a zero-divisor in R.

(P2) Given on input R and f where $\deg f$ is even, procedure P2(R, f) finds either a non-trivial decomposition of f over R, or a zero-divisor in R.

(P3) Given on input R, f and a root $B \in R[X]/(f)$ of the polynomial f different from $A = X \bmod f$, procedure P3(R, f, B) finds either a non-trivial decomposition of f over R, or a zero-divisor in R.

It follows from Lemmas 5, 8, 9 that all the three procedures can be done in time $(dn \log q)^{O(1)}$.

The MAIN ALGORITHM with entries R, f (denoted by MAIN(R, f) below) recursively runs as follows:

Case 1. If n is even, then apply P2(R, f). If a zero-divizor in R is found, then we are done. Otherwise a non-trivial decomposition $f = gh$, $\deg g \leq \deg h$ is found and we apply MAIN(R, g).

Case 2. If n is odd, then

1. If $n = 1$, then we are done.
2. If $n > 1$, then set $S = R[X]/(f)$, $f^*(Y) = f(Y)/(Y - A)$ where $A = X \bmod f$ and apply MAIN(S, f^*). Then
   (a) If a zero-divizor $a \in S$ is found, then apply P1(R, f, a). If a zero-divizor in R is found, then we are done, otherwise a non-trivial decomposition $f = gh$, $\deg g \leq \deg h$ is found and we apply MAIN(R, g).
   (b) If a root $B \in S$ of the polynomial f^* is found, then apply P3(R, f, B). If a zero-divizor in R is found, then we are done, otherwise a non-trivial decomposition $f = gh, \deg g \leq \deg h$ is found and we apply MAIN(R, g).

It follows from the above description that all the polynomials arising as entries of procedures P1, P2, P3 throughout MAIN(R, f) divide f and all the algebras R' arising as entries of the same procedures are of the form

$$R = R_0 \subset R_1 \subset \ldots \subset R_s = R', \quad R_{i+1} \cong R_i[X]/(f_i), \quad i = 0, 1 \ldots, s-1 \quad (26)$$

where f_i is a monic polynomial of degree n_i over R_i satisfying the following conditions:

1. n_i is odd, $n_i \geq 3$ for all i.
2. $n_{i+1} \leq (n_i - 1)/2$ for all i.
3. f_{i+1} divides f_i for all i.
4. f_0 divides f.

It follows on one hand that by Lemma 3 procedures P1, P2, P3 can really be applied and on the other hand that $s \leq \log_2 n$ and

$$\dim_k R' = d \prod_{i=0}^{s-1} n_i \leq d \prod_{i=0}^{s-1} (n/2^i) \leq dn^s \leq dn^{\log_2 n} \quad . \quad (27)$$

So by Lemmas 5, 8, 9 the running time $T(R, f)$ of MAIN(R, f) can be estimated in the following way:

$$T(R, f) \leq (M(R, f) + 1)(dn^{\log n} \log q)^{O(1)} \quad (28)$$

where $M(R, f)$ denotes the total number of calls for procedures P1, P2, P3 throughout MAIN(R, f). We shall prove in a moment that

$$M(R, f) \leq n - 1, \quad (29)$$

which will complete the proof of upper bound (25) for $T(R, f)$.

To prove (29) we use induction on n following the description of MAIN(R, f) given above. If $n = 1$, then $M(R, f) = 0$ and we are done. Assuming $n > 1$ we consider two cases.

Case 1: n is even. In this case we first apply $P2(R, f)$ and then $\text{MAIN}(R, g)$ where $\deg g \leq n/2$ (if necessary). So

$$M(R, f) \leq 1 + M(R, g) \leq 1 + n/2 - 1 \leq n - 1 \ . \tag{30}$$

Case 2: n is odd, $n \geq 3$. In this case we apply $\text{MAIN}(S, f^*)$ where S, f^* are similar to those in the description of $\text{MAIN}(R, f)$. Since $\deg f^*$ is even, we first apply $P2(S, f^*)$. Then:

1. If a zero-divizor $a \in S$ is found, then we first apply $P1(R, f, a)$ and then $\text{MAIN}(R, g)$ where $\deg g \leq (n-1)/2$ (if necessary). So in this case

$$M(R, f) \leq 2 + M(R, g) \leq 2 + (n-1)/2 - 1 \leq n - 1 \ . \tag{31}$$

2. If a decomposition $f^* = uv$ with $\deg u \leq (n-1)/2$ is found, then we apply $\text{MAIN}(S, u)$. Then:
   (a) If a zero-divizor $a \in S$ is found, then we first apply $P1(R, f, a)$ and then $\text{MAIN}(R, g)$ where $\deg g \leq (n-1)/2$ (if necessary). So in this case

   $$M(R, f) \leq 1 + M(S, u) + 1 + M(R, g) \leq 2(1 + (n-1)/2 - 1) = n - 1 \ . \tag{32}$$

   (b) If a root $B \in S$ of u is found, then we first apply $P3(R, f, B)$ and then $\text{MAIN}(R, g)$ where $\deg g \leq (n-1)/2$ (if necessary). So in this case

   $$M(R, f) \leq 1 + M(S, u) + 1 + M(R, g) \leq 2(1 + (n-1)/2 - 1) = n - 1 \ . \tag{33}$$

Inequality (29) and the MAIN THEOREM are proved. $\qquad\qquad\square$

6   Acknowledgements

The paper was done during the author's stay at University of Bonn in February-May, 1993, Report No. 8593-CS. The research was supported by the Volkswagen-Stiftung Program on Computational Complexity.

References

1. A.M. Adleman, H.W. Lenstra, Jr.: Finding irreducible polynomials over finite fields. Proc. 18th ACM Symp. on Theory of Computing (STOC) Berkeley (1986) 350-353
2. E.R. Berlekamp: Factoring polynomials over large finite fields. Math. Comp. **24** (1970) 713-735
3. S.A. Evdokimov: Factoring a solvable polynomial over a finite field and generalized Riemann hypothesis. Zapiski Nauchnych Seminarov LOMI **176** (1989) 104-117 (prepublication, 1986)
4. M.-D.A. Huang: Riemann hypothesis and finding roots over finite fields. Proc. 17th ACM Symp. on Theory of Computing (STOC) New-York (1985) 121-130
5. H.W. Lenstra, Jr.: Finding isomorphisms between finite fields. Math. Comp. **56** (1991) 329-347
6. L. Rónyai: Factoring polynomials over finite fields. Proc. **28**th IEEE Symp. on Foundations of Computer Science (FOCS) New-York (1987) 132-137
7. I.M. Vinogradov: Basic Number Theory. Moscow (1972)
8. B.L. van der Waerden: Algebra. Berlin (1966)

On Orders of Optimal Normal Basis Generators

Shuhong Gao[1] and Scott A. Vanstone[2]

[1] Department of Computer Science, University of Toronto, Toronto, Ontario, M5S
1A4, Canada E-mail: sgao@cs.toronto.edu
[2] Department of Combinatorics and Optimization, University of Waterloo, Waterloo,
Ontario, N2L 3G1, Canada E-mail: savanstone@math.uwaterloo.ca

Abstract. In several cryptographic systems (including exponential pseudorandom number generators), a fixed element of a group is repeatedly
raised to many different large powers. To make such systems secure, the
fixed element must have high order. While to implement these systems,
there should be an efficient algorithm for computing large powers of the
fixed element. In this paper, we show by computational results that the
generators of one class of optimal normal bases have exactly this desired
property.

More precisely, for any prime power q, let F_q denote the finite field of
q elements. Suppose that n is a positive integer such that $2n + 1$ is a
prime and the multiplicative group of Z_{2n+1} is generated by 2 and -1.
Let $\alpha = \gamma + \gamma^{-1}$ where γ is a primitive $(2n+1)$st root of unity in $F_{2^{2n}}$.
Then it is shown by Mullin, Onyszchuk, Vanstone and Wilson (*Discrete
Applied Math.* **22** (1988/1989), 149-161) that α generates an optimal
normal basis for F_{2^n} over F_2. [A normal basis for F_{q^n} over F_q is a basis
of the form $(\beta, \beta^q, \cdots, \beta^{q^{n-1}})$, where $\beta \in F_{q^n}$ is called a generator of
the normal basis. It is easy to show that, for any normal basis $N =
(\beta, \beta^q, \cdots, \beta^{q^{n-1}})$, there are at least $2n - 1$ nonzero entries among the n
products $\beta\beta^{q^i}$ $(0 \leq i \leq n - 1)$ expressed under N. If there are exactly
$2n-1$ nonzero entries then the normal basis is said to be optimal. Optimal
normal bases are classified by Gao and Lenstra (*Designs, Codes and
Cryptography* **2** (1992), 315-323).]

We computed the multiplicative orders of $\alpha = \gamma + \gamma^{-1}$ for $n \leq 1200$
where the above conditions are satisfied and the complete factorization
of $2^n - 1$ is available. We found that the multiplicative order of α is at
least $O((2^n - 1)/n)$, and is frequently primitive (i.e. has order exactly
$2^n - 1$).

For any positive integer e, we show that α^e can be computed in $O(n \cdot v(e))$
bit operations, where $v(e)$ is the number of 1's in the binary representation of e. In comparison, we should mention that, for an arbitrary $\beta \in F_{2^n}$, computing β^e by the square and multiply method needs
$O(n \log n \log\log n \log e)$ bit operations by using fast algorithms for multiplication. It is not known how to improve the time $O(n \cdot v(e))$ with
precomputations. The reason is that we do not have an $O(n)$ algorithm
for computing the product of two arbitrary elements in F_{2^n}.

Computing in the Jacobian of a Plane Algebraic Curve
(extended abstract)

Emil J. Volcheck

`volcheck@acm.org`

UCLA Dept. of Mathematics and
Center for Science and Art

Abstract. We describe an algorithm which extends the classical method of adjoints due to Brill and Noether for carrying out the addition operation in the Jacobian variety (represented as the divisor class group) of a plane algebraic curve defined over an algebraic number field K with arbitrary singularities. By working with conjugate sets of Puiseux expansions, we prove this method is rational in the sense that the answers it produces are defined over K. Given a curve with only ordinary multiple points and allowing precomputation of singular places, the running time of addition using this algorithm is dominated by M^7 coefficient operations in a field extension of bounded degree, where M is the larger of the degree and the genus of the curve.

1 Introduction

Every algebraic curve defines an abelian variety called the *Jacobian variety* (or Jacobian) of the curve. For example, elliptic curves are isomorphic to their Jacobian varieties. The classical chord-tangent method "adds" points on elliptic curves and hence describes the group operation. Computing in the Jacobian of curves has applications to the integration of algebraic functions [16], primality testing [1], and error-correcting codes [10, 19].

Davenport implemented an algorithm to test for torsion points on the Jacobians of a broad class of algebraic curves [7]. An efficient algorithm for computing in the Jacobian of a hyperelliptic curve was developed by Cantor [4], who also derived closed-form expressions to calculate multiples of points in the Jacobian of a hyperelliptic curve [5]. M.-D. Huang and D. Ierardi developed the first polynomial-time algorithm for computing in the Jacobian of a curve [12].

We present an algorithm for computing in the Jacobian of an algebraic curve defined over an algebraic number field K with arbitrary singularities. The theoretical basis for this algorithm was first described by Brill and Noether [2, sec. 5]. Our algorithm builds on Noether's algorithm for computing the adjoints of a curve [15], but it differs in one significant respect: we represent singularities using analytic rather than algebraic means.

Like Noether's algorithm, our algorithm is *rational* in the sense that it expresses an answer without a ground field extension. Unlike Noether's algorithm, our algorithm requires ground field extensions for intermediate calculations involving singularities. However any object or function we compute whose field of coefficients extends to K', a proper extension of K, is manipulated as part of a conjugate set, one invariant under the action of $\mathrm{Gal}(\bar{\mathbb{Q}}/K)$.

We first apply one or more of Noether's S-transformations to the curve to simplify its singularities into ordinary multiple points without extending the ground field (a technique which applies equally well to the Huang & Ierardi algorithm) [20]. Following precomputation of singular places, our algorithm adds points on the Jacobian in time dominated by M^7 coefficient operations in a field extension of bounded degree, where M is the larger of the degree n and the genus g of the curve.

In this extended abstract, we outline only a special case of the algorithm that allows arbitrary input but assumes sets of points produced as intermediate steps during the calculation satisfy certain generic conditions. These conditions do not impose any practically significant restrictions. We briefly indicate how the constructions for the general case can be made. Full details of the algorithm and its correctness may be found in [21].

2 Background

We adapt some basic definitions and results from Chevalley [6], [21], and chapter 4 of Walker [22]. See Eisenbud and Harris [9] for background on schemes and the spectrum of a ring.

For $F(x, y)$ defined over a number field K, let $C : F(x, y) = 0$ denote the corresponding curve as a K-scheme, that is

$$C = \mathrm{Spec}\, K[x, y]/F(x, y).$$

We assume C has only ordinary multiple points and that F is absolutely irreducible. Let $\bar{\mathbb{Q}}$ be the algebraic closure of K. Then

$$\bar{C} = \mathrm{Spec}\, \bar{\mathbb{Q}}[x, y]/F(x, y)$$

denotes the curve C with ground field extended to $\bar{\mathbb{Q}}$.

Definition: A *Puiseux expansion* of $\bar{C}$ centered at the point (x_0, y_0) is a fractional power series

$$y(x) = y_0 + a_1(x - x_0)^{n_1/n} + a_2(x - x_0)^{n_2/n} + \cdots,$$

for which $F(x, y(x)) = 0$. The n, n_i are integers such that $0 < n$, $0 < n_1 < n_2 < \cdots$, and no a_i vanish.

Because we work with an explicit coordinate representation, we need to distinguish between the curve as an algebraic set and as a scheme.

Definition: We refer to a (geometric) *point* of $\bar{C}$ to denote a point of $\bar{C}$ as an algebraic set. We refer to a *place* of $\bar{C}$ (resp. C) in the sense of Chevalley, to denote a closed point of $\bar{C}$ (resp. C) as a scheme. To avoid ambiguity, we may refer to a K-*place* of C and to a $\bar{\mathbb{Q}}$-*place* of $\bar{C}$.

Since C has only ordinary multiple points we can explicitly characterize a singular K-place as having the following simplified form:

Theorem 1 *Given a singular K-place at an ordinary multiple point, it has an expansion in the standard form*

$$x = x_0 + t, \ y = y_0 + \sum a_i t^i,$$

where $K \subseteq K_1 \subseteq K_2$ are fields such that

1. *$K_1 = K(x_0, y_0)$ is the coefficient field of the center point of the expansion, and*
2. *$K_2 = K_1(a_1, a_2, \dots)$ is the residue field of the expansion, which is formed by adjoining a root of the leading form of $F(x, y)$ at the center (x_0, y_0).*

Let L be the Galois closure of K_2 over K. Then the action $\mathrm{Gal}(L/K)$ fixes this set of Puiseux expansions. Any such set of expansions corresponds to one or more K-places.

See Duval [8] for related work on conjugate expansions. Note that a place at a nonsingular point will also have the above form, but with the a_i in K_1.

Definition: The center point P of a K-place Q corresponds to the set of center points of the corresponding conjugate $\bar{\mathbb{Q}}$-places. We say the *residue field* of P is the extension K_1, in the notation of the above theorem. The *degree* of P is $[K_1 : K]$. The *residue field* of a singular place Q centered at P is K_2. We say the *degree* of Q is $[K_2 : K]$.

Definition: Let a place Q have the Puiseux expansion $y(x)$. We write the *order* of a polynomial $g(x, y)$ at Q as $\mathrm{ord}_Q(g)$, i.e. the order of $g(x, y(x))$.

We give definitions that lead up to the construction of the divisor class group. We follow section 2.7 of Griffiths [11].

Definition: A *divisor* D of a curve is a formal sum of places of the curve, $D = \sum n_i Q_i$, where only finitely many n_i are non-zero. Let q_i denote the degree of the place Q_i. Then the *degree* of D is $\sum n_i q_i$.

We have defined C as an affine curve and only consider finite places. To avoid the cumbersome transition to a projective curve, let us agree that places on the line at infinity will be made finite by an appropriate coordinate transformation as needed.

Definition: We define the *intersection divisor* $I_F(g)$ of a polynomial $g(x, y)$ with the curve $C : F(x, y) = 0$ as the formal sum

$$I_F(g) = \sum_Q \mathrm{ord}_Q(g) \, Q,$$

where the sum ranges over all places Q of C.

Note that if $g(x, y)$ has degree m and $F(x, y)$ has degree n, then by Bezout's theorem, the intersection divisor will have degree mn.

Definition: The *principal divisor* of a rational function $g(x, y)/h(x, y)$ on the curve C, written $\text{div}(g/h)$ is defined to be the difference of intersection divisors

$$\text{div}(g/h) = I_F(g) - I_F(h).$$

The degree of a principal divisor is zero because both intersection divisors have the same degree. We may assume that g and h have the same degree when neither has a place at infinity.

Definition: The *divisor class group* is the quotient of the group of divisors of degree zero modulo the group of principal divisors. When two divisors D_1 and D_2 differ by a principal divisor, then they are *equivalent* in the divisor class group and we write $D_1 \equiv D_2$. An equivalence class in the divisor class group is a *divisor class*.

Definition: Given two divisors $D_1 = \sum n_P P$ and $D_2 = \sum m_P P$, we write $D_1 \geq D_2$ when $n_P \geq m_P$ for all P. When all $n_P \geq 0$, we write $D_1 \geq 0$ and say D_1 is *effective*. We add divisors by adding the multiplicities of their places, so we write $D_1 + D_2 = \sum (n_p + m_p)P$.

Definition: For a divisor D, we write $\mathcal{L}(D)$ to denote the vector space of functions

$$\mathcal{L}(D) = \{f : \text{div}(f) \geq -D\}.$$

Definition: A polynomial $G(x, y)$ satisfies the *adjoint condition* at an ordinary multiple point of order m when G has a singularity of order $m - 1$ there. When C has only ordinary multiple points, G defines an *adjoint curve* (or simply an *adjoint*) of C when G satisfies the adjoint condition at every ordinary multiple point.

We introduce the following divisor notation to describe adjoint conditions and adjoint curves on a curve C with only ordinary multiple points.

Definition: Consider all singular places of C and their center points, which are conjugate sets of points. We label them as follows: let $\{P_i\}_{i=1}^s$ denote the sets of center points, which are ordinary multiple points of order m_i. Say P_i has degree p_i as a conjugate set. Let $\{S_{ij}\}_{j=1}^{t_i}$ be the singular places centered at P_i. Let $\{\Delta_i\}_{i=1}^s$ be the divisors defined by

$$\Delta_i = \sum_{j=1}^{t_i} (m_i - 1)S_{ij}$$

and having degree δ_i. The divisor Δ_i represents the adjoint condition at P_i. Let Δ be defined as $\Delta = \sum_{i=1}^s \Delta_i$ with degree $\delta = \sum_{i=1}^s \delta_i$. We say Δ is the *adjoint divisor* because the condition $I_F(G) \geq \Delta$ implies that $G(x, y)$ satisfies the adjoint condition.

3   Algorithm

In this section, we describe addition and reduction in the Jacobian. For further properties of the Jacobian, see Griffiths [11, chap. 5].

Let C be an absolutely irreducible degree n, genus g curve defined over a number field K. We assume that C has no singular points on the line at infinity and that C has a K-rational point. We designate this point as the *zero point* of the Jacobian and write it as P_0.

If we add divisors by simply adding the multiplicities of their places, then repeated additions would produce ever larger divisors, with no limit to their size. The Riemann Inequality [11, p. 112] provides a way to represent divisors in a reduced form, so that we can follow each formal addition by a reduction operation. The Riemann Inequality also provides a way to test whether two divisors are equivalent [21].

Theorem 2 (Riemann) *Let D be a divisor of degree d on the curve C. Then*

$$\dim \mathcal{L}(D) \geq d - g + 1.$$

Consider divisors $D_1 - gP_0$ and $D_2 - gP_0$, where D_1 and D_2 are effective of degree g. Let $D_3 - 2gP_0$ be their formal sum as divisors, where D_3 is effective of degree $2g$. We perform the following reduction. Consider the divisor $D_3' = D_3 - gP_0$ of degree g, and choose a rational function $g/h \in \mathcal{L}(D_3')$. Let D_3^r be the degree g effective divisor $D_3' + \mathrm{div}(g/h)$. Then $D_3 \equiv D_3^r - gP_0$. Because D_3^r has only degree g, compared to degree $2g$ of D_3, we have a more compact representation of the sum.

3.1   Generic Assumptions

When computing $\mathcal{L}(D)$ for a fixed D and performing a reduction, we will construct a residual divisor E and a reduced divisor R, which will generically satisfy certain properties. Note that only finitely many points on C are

1. singularities,
2. points with the same x-coordinate as a singularity,
3. points with the same x-coordinate as a place of D, or
4. points on the line at infinity.

Assumption 1 *We assume that neither the residual divisor nor the reduced divisor shall contain any of these "bad" points.*

We refer to the above conditions as A1.1 through A1.4.

A divisor will generically consist of distinct points with simple multiplicity. If two points of a divisor share x-coordinates, then that represents an unusual vertical coincidence. Hence we make the following additional assumptions, which we label A2.1 and A2.2:

Assumption 2 *We assume that the residual divisor and the reduced divisor together have no vertical coincidences (A2.1). Further, each divisor shall have only simple points (A2.2).*

Any of the cases covered by these assumptions represents a subspace of strictly lower dimension (a Zariski closed set) in the space of divisors. Hence, these cases seldom occur in practice.

3.2 Data and Procedures

We describe the primary data structures and basic procedures for our addition algorithm.

We refer to a *parametrized conjugate set of points* to denote a set of points in the following parametrized form: $[r(t); \rho_1(t), \rho_2(t)]$. This represents the set of points $\{(\rho_1(\alpha), \rho_2(\alpha))\}$ for every root α of $r(t)$. For a place centered at a point of such a conjugate set, $r(t)$ generates the extension K_1 of Theorem 1.

A divisor can be represented as a list of quadruples, each quadruple containing an integer multiplicity, a parametrized conjugate set of points, a partial Puiseux expansion, and (for singular places) a factor of the leading form which generates the extension K_2 in the notation of Theorem 1.

We refer to the *coordinate polynomial* of a parametrized conjugate set of points to denote the polynomial whose roots are the x-coordinates of the individual geometric points with proper multiplicity. The coordinate polynomial of a place is the coordinate polynomial of its center point, and the coordinate polynomial of a divisor is the product (with proper multiplicity) of the coordinate polynomials for all its places.

We mention here but do not detail some basic procedures we need for the addition algorithm: calculating coordinate polynomials, calculating Puiseux expansions, evaluating a polynomial at a place, and interpolating a polynomial to vanish to a given positive multiplicity at a set of places on the curve. By modifying techniques of Sakkalis & Farouki [17] and Canny [3], we have an algorithm for representing an intersection divisor as a parametrized conjugate set of points. We refer to this operation as *parametrizing* the set of points and direct the reader to [21] for details.

3.3 Generic Algorithm

Given our generic assumptions, we now describe an algorithm **main** which will compute a reduction of a divisor down to g points.

Precomputation: [Singular places.] Compute coordinate polynomials for the singular points of C and for Δ and then parametrize the singularities. For each singular place Q with multiplicity δ_Q in the adjoint divisor Δ, compute the first δ_q terms of the Puiseux expansion at Q.

Input: the curve polynomial $F(x, y)$ of degree n; a (finite) divisor D of degree d greater than the genus g, written as a difference of two effective divisors $D^+ - D^-$ of degrees d^+ and d^-. Assume D^+ and D^- have no places in common.

Output: a basis for the functions of $\mathcal{L}(D)$ and a parametrized set of points representing an effective divisor of degree g equivalent to D.

Step 0: [Places of D.] For each place Q in D with multiplicity k, if Q is regular then compute the first k terms of the Puiseux expansion, else Q is singular and then compute an additional k terms of the expansion out to order $\delta_Q + k$. Compute the coordinate polynomials for D^+ and D^-.

Step 1: [Add the adjoint divisor to D^+, $A \leftarrow D^+ + \Delta$.] Iterate over all singular places Q, while adding the multiplicities of Q in D^+ and Δ and storing in A. Then add all regular places of D^+ to A.

Step 2: [Interpolate adjoints through D^+.] Construct a polynomial $b(x, y)$ of degree m with $I_F(b) \geq A$, where m is determined in section 4.1, by imposing linear conditions on a polynomial of undetermined coefficients and performing Gaussian elimination.

Step 3: [Subtract to find the residual divisor $E \leftarrow I_F(b) - A$.] Divide the coordinate polynomial for $I_F(b)$ by the coordinate polynomials for D^+ and Δ to get the coordinate polynomial $e(x)$ of the residual divisor E. We can parametrize E by computing the monic first subresultant $y - s(x)$ of $F(x, y)$ with $b(x, y)$, eliminating y, to obtain the parametrized conjugate set of points $[e(x); x, s(x)]$.

Step 4: [Add the adjoint divisor to D^- and E, $H \leftarrow D^- + E + \Delta$.] Iterate over all singular places Q, while adding the multiplicities of Q in D^- and Δ and storing in H. Then add the places of E and the regular places of D^- to H.

Step 5: [Interpolate adjoints through $E + D^-$.] Construct a basis of polynomials $r_i(x, y)$ of degree m such that $I_F(r_i) \geq H$, similar to Step 2. The rational functions r_i/b form a basis for $\mathcal{L}(D)$.

Step 6: [Reduce.] Choose some r from the space of the r_i. Similar to Step 3, subtraction occurs by dividing the coordinate polynomials of $I_F(r)$ by the coordinate polynomials for D^-, E, and Δ, and then parametrizing to get the reduced divisor R as a parametrized set.

3.4 Rationality

We compute in ground field extensions without explicit factorization through the technique of passive or lazy factorization [8, 13, 18]. This procedure ensures that when we introduce a Galois extension K' of the ground field K for an intermediate calculation, then we work with a set invariant under $\mathrm{Gal}(\bar{\mathbb{Q}}/K)$.

Here is one particularly useful application of lazy factorization to computing with places and parametrized conjugate set of points. Suppose we wish to explicitly compute places with their multiplicities in an intersection divisor $I_F(g)$. Let the parametrized set $P = [p(t); \pi_1(t), \pi_2(t)]$ represent the points of $I_F(g)$.

We compute terms of the Puiseux expansion at P. If $I_F(g)$ has points at which $g(x, y)$ has different orders, then calculating enough terms will produce a lazy factorization of $p(t)$. This enables us to rationally separate the places of different multiplicities in $I_F(g)$ and represent them as distinct parametrized sets.

Our algorithm is rational in the following sense:

Theorem 3 *Let $C : F(x, y) = 0$ be defined over a number field K, and let D be a divisor. Then all objects constructed during algorithm* **main** *are composed of either places or polynomials or functions defined over K on the curve.*

Proof: The theorem follows by inspection of the description of **main**. We analyze the algorithm step-by-step.

We apply lazy factorization as we calculate partial expansions at places in the precomputation step and step 0. In step 0, we calculate the coordinate polynomials by taking resultants over K.

Step 1 is a formal addition of divisors of K-places, so A is a divisor of K-places.

The interpolation of step 2 performs evaluations at K-places and solves a linear system over K. Hence, $b(x, y)$ has coefficients in K.

The parametrization in step 3 uses a subresultant chain, which is calculated over K.

Steps 4,5, and 6 are rational by analogy with steps 1,2, and 3.   □

3.5 Modular Calculations

Our algorithm adapts well for calculating modulo a large prime. Consider C to be defined over $\mathbb{F}_p$. Assuming C has only ordinary multiple points, all places are unramified. Since the ground field is perfect, all extensions of K are separable. These observations show that the basic procedures and operations of section 3.2 are valid modulo p, except for parametrization, which requires p to be large relative to n [14, 21]. Although the generic conditions of section 3.1 refer to an infinite ground field, when p is a 32-bit prime the ground field is sufficiently large for parametrization to work and for the generic conditions to almost always hold. We do not address the question of bad reduction of the Jacobian modulo p.

4   Complexity

To determine the complexity of this algorithm, we first bound the size of certain intermediate divisors calculated during the course of the algorithm, then we analyze the time required to carry out the computations. We use the notation $f \prec g$ to denote that functions f and g satisfy $f \leq Cg$ for some constant C and say f is dominated by g. We write $f \sim g$ to indicate that f and g are codominant.

4.1 The Residual Divisor

Theorem 4 *Let D be a divisor given as a difference of two effective divisors, $D = D^+ - D^-$, where D^+ is of degree d^+, D^- of degree d^-, and D of degree $d > g$, as per the input to* **main**. *Since $d^+ > d^-$, let us use $d_{\max}$ to denote d^+, which is the largest degree of the positive and negative parts. Let the N specified in* **main** *be set equal to nR, where $R = \lceil d_{\max}/n \rceil$. Then the size e of the residual divisor E computed by the algorithm* **main** *is dominated by N.*

Let M be $\max(n, d_{\max})$. Note that $M \sim N$ because $2M \geq N \geq M$. We obtain a more natural bound on e by using N rather than M. First we have two lemmas.

Lemma 5 *The number of singular places δ falls short of n^2 by an amount dominated by N, that is, $n^2 - \delta \prec N$.*

Proof: By the genus formula

$$n^2 - 3n + 2 = 2g + \delta,$$

so putting $n^2 - \delta$ together, we have

$$n^2 - \delta = 3n + 2g - 2.$$

The right-hand side is less than $5M$ since $g < d_{\max}$, hence it is also less than $5N$, which proves the lemma.

Let us define $\alpha(N) = n^2 - \delta$, where the N of $\alpha(N)$ serves to remind us that $\alpha(N) \prec N$.

Lemma 6 *The degree m of the interpolating polynomial $b(x, y)$ need be no larger than $n + B$, where B is less than R.*

Proof: We must choose m large enough to satisfy

$$\binom{m + 2}{2} - \binom{B + 2}{2} \geq d^+ + \frac{1}{2}\delta.$$

Bound d^+ by $N = Rn$. Then multiply both sides of the equation by two, substitute $m = n + B$, and simplify the left-hand side:

$$n^2 + (2B + 3)n \geq 2Rn + \delta.$$

Next, substitute $n^2 - \alpha(N)$ for δ:

$$n^2 + (2B + 3)n \geq 2Rn + n^2 - \alpha(N).$$

The key step of this lemma is to cancel the n^2 terms from each side. We do so and then move the $2Rn$ term to the left-hand side.

$$(2B - 2R + 3)n + \alpha(N) \geq 0.$$

Since $\alpha(N) > 0$, this inequality is satisfied by the lowest value of B for which $2B - 2R + 3 \geq 0$. Solving for B, we have $B \geq R - 3/2$, or $B = R - 1$ since B is an integer. This proves the lemma.

Proof of Theorem: Again from the Riemann Inequality, we have

$$e = mn - d^+ - \delta,$$

so by neglecting d^+, which is positive, $e \leq mn - \delta$. Substituting $n + B$ for m, we have

$$mn - \delta = n^2 + Bn - \delta.$$

We rewrite $n^2 - \delta$ as $\alpha(N)$ to get $Bn + \alpha(N)$ on the right-hand side. Since $Bn < N$ and $\alpha(N) < 5N$, we have $e < 6N$, so $e \prec N$. $\qquad\square$

4.2 Operation Count

Now that we have some control over the size of the residual divisor, we count the coefficient operations of the main algorithm. We count each algebraic number calculation in extensions K_1 or K_2 (using the notation of Theorem 1) as one operation. Because our operation count neglects the size of integer coefficients and the degree of the intermediate ground field extensions, our estimate is best suited for modular calculations on curves with relatively few singularities (say $\delta \prec n$).

Theorem 7 *Allowing for precomputation of singular places, the time required for* main *is dominated by N^6 operations, where $N = nR$ and $R = \lceil d_{\max}/n \rceil$.*

Proof: We focus on the most expensive operations.

The interpolation in steps 2 and 5 is a Gaussian elimination which has a number of equations (linear conditions) dominated by N^2 and a number of unknowns dominated by N^2. The number of unknowns equals the number of undetermined coefficients of the interpolating polynomial of degree $m = n + R \prec N$. This gives complexity dominated by $(N^2)^2 N^2$ or N^6.

The most costly operation in the subtractions of steps 3 and 6 is parametrizing the residual divisor of size $\prec N$ points. The subresultant chain used for parametrization runs in time dominated by N^6 operations (see Loos [14]). $\qquad\square$

4.3 Deterministic Complexity

We sketch techniques described in [21] to lift the generic assumptions.

Since singular points have an implicit representation using the partial derivatives of F, we may isolate any singular places in the residual or reduced divisors and handle them explicitly (A1.1).

In the case of any coordinate coincidences among D^+, D^-, Δ, E, or R, we may construct a coordinate transformation that eliminates these coincidences before parametrizing. This requires a resultant computation with complexity dominated by N^7. (A1.2, A1.3, A2.1).

Points at infinity can be made finite by a coordinate transformation (A1.4).

To detect multiple points in E or R, we compute sufficiently many terms of Puiseux expansions and then use lazy factorization as described in section 3.4 on rationality (A2.2).

This yields a deterministic complexity of addition of N^7, in contrast to the Huang & Ierardi algorithm which requires $n^{14}g^7$ [12].

Remark: The precomputation phase first requires parametrizing the singular points and then computing Puiseux expansions of each singular place. Let r equal the total number of (geometric) singular points. Parametrizing this set costs r^7 operations. Computing expansions at the singular places is dominated by n^8 operations [8, 21].

5 Example

In this section, we present an example of applying the generic algorithm **main** to adding divisors on a curve. The calculations were carried out in **Maple**.

The curve C specified by the polynomial,

$$
\begin{aligned}
F(x,y) = {}& -8 + 4\,y + 2\,y^2 - 3\,y^3 + 82\,y^4 \\
& -81\,y^5 + 2\,xy^3 - 2\,xy^4 + 8\,x^2 \\
& -4\,x^2 y + x^2 y^3 - 2\,x^4 + x^4 y,
\end{aligned}
$$

has two singular places, Δ_1 and Δ_2. The places are centered at the conjugate set of points $\{(-\alpha, 0), (\alpha, 0)\}$, where $\alpha^2 - 2 = 0$. The places Δ_1 and Δ_2 form two ordinary double points over a ground field extension to $\mathbb{Q}(\alpha)$. The expansions of F at Δ_1 and Δ_2 respectively begin with $y = 2\alpha\,(x-\alpha) + (4\alpha - 1)\,(x-\alpha)^2 + \cdots$ and $y = -2\alpha\,(x-\alpha) + (4\alpha + 1)\,(x-\alpha)^2 + \cdots$. We assign the zero point of this curve, P_0 to be $(1,1)$. Let P_1 and P_2 be the points $(-2,1)$ and $(3,-1)$ on the curve.

In this example, we add the two divisors $D_1 = 2 \cdot \Delta_1$ plus $D_2 = 2 \cdot P_1 + 2 \cdot P_2$. Each of D_1 and D_2 has total degree 4, so their sum has degree 8, which we reduce to a divisor of degree 4. We compute expansions up to order 4 at P_0 and up to order 2 at P_1 and P_2.

We compute the space of functions $\mathcal{L}(D)$ for $D = D_1 + D_2 - 4 \cdot P_0$. First we interpolate $b(x,y)$ of degree 4 such that $I_F(b) \geq D_1 + D_2 + \Delta_1 + \Delta_2$. This relationship imposes 10 linear conditions $(10 = 4 + 4 + 1 + 1)$ on the choice of $b(x,y)$ for which there are 15 degrees of freedom. This leaves 5 degrees of freedom to select $b(x,y)$. The b we choose has 15 terms with coefficients of up to 6 digits.

According to Bezout's theorem, b meets F in $4 \cdot 5 = 20$ places. The coordinate polynomials of $I_F(b)$ have degree 20. After dividing these coordinate polynomials by the coordinate polynomials for $D_1, D_2, \Delta_1, \Delta_2$, which have a product of degree

12 corresponding to the 12 places they represent $(12 = 4 + 4 + 2 + 2)$, we have degree 8 coordinate polynomials for the residual divisor E.

We pass b and both these polynomials to the parametrization routine which returns the parametrization $[r(x); \rho_1(x), \rho_2(x)]$, where r has degree 8 and integer coefficients of up to 28 digits, and ρ_1 and ρ_2 have degree 7 and rational coefficients with numerator and denominator of up to 136 digits.

Next we interpolate $r(x, y)$ of degree 4 such that $I_F(r) \geq E + 4 \cdot P_0 + \Delta_1 + \Delta_2$. Again we start with 15 degrees of freedom for the choice of r, but this relationship imposes 14 linear conditions $(14 = 8 + 4 + 1 + 1)$, leaving only one degree of freedom for r. That is, the dimension of $\mathcal{L}(D)$ is one. This $r(x, y)$ is degree 4 in both x and y, with 15 terms and integer coefficients of up to 25 digits.

We have again that r meets F in 20 places, so we divide the coordinate polynomials of $I_F(r)$ by the coordinate polynomials for $E, 4 \cdot P_0, \Delta_1, \Delta_2$ of degree 16 $(16 = 8 + 4 + 2 + 2)$ leaving coordinate polynomials of degree 4. We pass r and these degree 4 coordinate polynomials to the parametrization routine which returns the parametrization $[s(x); \sigma_1(x), \sigma_2(x)]$, where $s(x)$ has degree 4 with integer coefficients of up to 102 digits, and σ_1 and σ_2 have degree 3 with rational coefficients of up to 182 digits in both numerator and denominator. This parametrized set of points represents the reduced sum of $D_1 + D_2$ in the divisor class group.

Acknowledgement

I would like to acknowledge the help of my advisor Prof. David Cantor and the support I received from the National Security Agency as a Graduate Student Researcher under grant NSA MDA 904-88-H2031 (Cantor-Hales).

References

1. Leonard M. Adleman and Ming-Deh A. Huang. *Primality Testing and Abelian Varieties over Finite Fields*, volume 1512 of *Lecture Notes in Mathematics*. Springer Verlag, 1992.

2. A. Brill and M. Noether. Über die algebraischen Functionen und ihre Anwendung in der Geometrie. *Mathematische Annalen*, 7:269–310, 1874.

3. John Canny. Some algebraic and geometric computations in PSPACE. In *Proceedings of the ACM Symposium on the Theory of Computation*, pages 460–467, 1988.

4. David G. Cantor. Computing in the Jacobian of a hyperelliptic curve. *Mathematics of Computation*, 48(177):95–101, 1987.

5. David G. Cantor. On the analogue of the division polynomials for hyperelliptic curves. To appear in *Journal für die reine und angewandte Mathematik*, 1994.

6. Claude Chevalley. *Introduction to the Theory of Algebraic Functions of One Variable*, volume 6 of *Mathematical Surveys*. American Mathematical Society, 1951.

7. James Harold Davenport. *On the Integration of Algebraic Functions*, volume 102 of *Lecture Notes in Computer Science*. Springer Verlag, 1981.

8. Dominique Duval. Rational Puiseux expansions. *Compositio Mathematica*, 70:119–154, 1989.

9. David Eisenbud and Joe Harris. *Schemes: the Language of Modern Algebraic Geometry*. Wadsworth and Brooks/Cole, 1992.

10. V. D. Goppa. *Geometry and Codes*, volume 24 of *Mathematics and its Applications (Soviet Series)*. Kluwer Academic Publishers, 1988.

11. Phillip A. Griffiths. *Introduction to Algebraic Curves*. American Mathematical Society, 1989.

12. Ming-Deh Huang and Doug Ierardi. Efficient algorithms for the Riemann-Roch problem and for addition in the Jacobian of a curve. In *IEEE 32nd Annual Symposium on Foundations of Computer Science*, pages 678–687, 1991.

13. Lars Langemyr. Algorithms for a multiple algebraic extension. In Teo Mora and Carlo Traverso, editors, *Methods in Algebraic Geometry, Proceedings of MEGA 1990*, pages 235–248. Birkhäuser, 1991.

14. Rüdiger Loos. Computing in algebraic extensions. In B. Buchberger, G. E. Collins, and R. Loos, editors, *Computer Algebra: Symbolic and Algebraic Computation*. Springer Verlag, 1983.

15. Max Noether. Rationale Ausführung der Operationen in der Theorie der algebraischen Functionen. *Mathematische Annalen*, Band 23:311–358, 1884.

16. R. H. Risch. The problem of integration in finite terms. *Transactions of the AMS*, 139:167–189, 1969.

17. Takkis Sakkalis and Rida Farouki. Singular points of algebraic curves. *Journal of Symbolic Computation*, 9:405–421, 1990.

18. Jeremy Teitelbaum. The computational complexity of the resolution of plane curve singularities. *Mathematics of Computation*, 54(190):797–837, 1990.

19. M. A. Tsfasman and S. G. Vladut. *Algebraic Geometric Codes*, volume 58 of *Mathematics and its Applications (Soviet Series)*. Kluwer Academic Publishers, 1991.

20. Emil J. Volcheck. Noether's *S*-transformation simplifies curve singularities rationally: a local analysis. In G. Gonnet, editor, *Proceedings of the 1993 International Symposium on Symbolic and Algebraic Computation*. Association for Computing Machinery, 1993.

21. Emil J. Volcheck. *Resolving Singularities and Computing in the Jacobian of a Plane Algebraic Curve*. PhD thesis, UCLA, 1994. to appear in May.

22. Robert J. Walker. *Algebraic Curves*. Springer-Verlag, 1978.

Under the Assumption of the Generalized Riemann Hypothesis Verifying the Class Number Belongs to $\mathcal{NP} \cap \text{co-}\mathcal{NP}$

Christoph Thiel

Fachbereich Informatik
Universität des Saarlandes
Postfach 151150
D-66041 Saarbrücken
Germany

Abstract. We show that under the assumption of a certain Generalized Riemann Hypothesis the problem of verifying the value of the class number of an arbitrary algebraic number field $\mathcal{F}$ of arbitrary degree belongs to the complexity class $\mathcal{NP} \cap \text{co} - \mathcal{NP}$. In order to prove this result we introduce a compact representation of algebraic integers which allows us to represent a system of fundamental units by $(2 + \log_2(\Delta))^{O(1)}$ bits, where Δ is the discriminant of $\mathcal{F}$.

1 Introduction

Let $\mathcal{F}$ be an algebraic number field, i.e. a finite field extension of the field $\mathbb{Q}$ of the rational numbers. In [21] and [11] it was shown that the problem of verifying the value of the class number h of $\mathcal{F}$ belongs to the complexity class $\mathcal{NP}$ if $\mathcal{F}$ is of degree 2 and if a certain Generalized Riemann Hyptothesis (GRH) holds. That is, assuming the GRH, there exists a nondeterministic polynomial time algorithm that accepts the set of all pairs (Δ, h), where $\Delta \in \mathbb{Z}$ is squarefree and h is the class number of $\mathbb{Q}(\sqrt{\Delta})$. We will generalize this result by showing that under the assumption of the GRH the problem of verifying the value of the class number $h_{\mathcal{F}}$ of an arbitrary algebraic number field $\mathcal{F}$ of arbitrary degree belongs to $\mathcal{NP}$. To be more precise, our main results are:

Theorem 1. *The set* PRI *of all pairs* $(\mathcal{F}, A)$, *where* $\mathcal{F}$ *is an algebraic number field and* A *is a principal ideal of the ring of integers of* $\mathcal{F}$, *belongs to* $\mathcal{NP}$. *Moreover, if we assume the GRH, then the set* $\mathcal{H}$ *of all pairs* $(\mathcal{F}, h_{\mathcal{F}})$, *where* $\mathcal{F}$ *is an arbitrary algebraic number field and* $h_{\mathcal{F}}$ *is the class number of* $\mathcal{F}$, *belongs to* $\mathcal{NP}$.

From our theorem and from the results of [13] we immediately obtain

Corollary 2. *Assume the GRH. Then* $\mathcal{H}$ *and* PRI *belong to the complexity class* $\mathcal{NP} \cap co - \mathcal{NP}$.

The main tool used in our work is the theory of minima in ideals which will be reviewed in Sect. 3. In order to prove our results we will generalize the results of

[15] and introduce in Sect. 4 a short representation of algebraic integers which allows us to represent a system of fundamental units by $(2 + \log_2(\Delta))^{O(1)}$ bits where the O-constant is independent of $\mathcal{F}$. Given such a representation of a system of fundamental units we can determine in polynomial time an approximation to the regulator $R_{\mathcal{F}}$. This kind of representation solves an open problem suggested by H. W. Lenstra in [19, Problem 5.2].

After studying the binary length of short representations in Sect. 5, we will use short representations to show in Sect. 6 that there exists a short proof for the principality of a given ideal. As an application of this result we will show in Sect. 7 under the assumption of GRH: if an integer H is divisible by the class number $h_{\mathcal{F}}$, then there exists a short proof for this fact. In Sect. 8 we will use short representations of elements of a system of fundamental units to compute a rational R^* which is an approximation of $mR_{\mathcal{F}}$ for some $m \in \mathbf{Z}$. In Sect. 9 we will show how to verify that a number Θ satisfies $(6/5)h_{\mathcal{F}}R_{\mathcal{F}} \leq \Theta \leq (15/8)h_{\mathcal{F}}R_{\mathcal{F}}$. We can check that $H/h_{\mathcal{F}} = m = 1$ since $H R^* \leq \Theta$ if and only if $H/h_{\mathcal{F}} = m = 1$.

2 Preliminaries

For notions in complexity theory such as *polynomial time* or the complexity class $\mathcal{NP}$ we refer to [24]. The definitions and results in algebraic number theory used in our paper can be found for example in [22] or [18]. We use the notions of encoding data in a similar way as in [19]. For example, mathematical objects are encoded by finite sequences of rational integers (for example, a polynomial is encoded by its coefficients), which are represented in binary. The *binary length of a rational integer* $z \neq 0$ is size$(z) = \lfloor \log_2 |z| \rfloor + 2$, where the extra bit encodes the sign of z. We also set size$(0) = 2$. The *binary lenght* size(O) *of an object* O is defined to be the length of its encoding sequence. When we say that an object is input for an algorithm then this means that the appropriate encoding is its input.

An algebraic number field $\mathcal{F}$ of degree $n > 1$ is represented by a *generating polynomial* which is a monic and irreducible polynomial $f \in \mathbf{Z}[X]$ of degree n such that $\mathcal{F} \cong \mathbb{Q}[X]/f\mathbb{Q}[X]$, i.e. $\mathcal{F} = \mathbb{Q}(\rho)$ with $\rho = X + f\mathbb{Q}[X]$. The numbers $1, \rho, \rho^2, \ldots, \rho^{n-1}$ form a $\mathbb{Q}$-basis of $\mathcal{F}$ and each $\xi \in \mathcal{F}$ can be uniquely represented in the form $\xi = \frac{1}{d} \sum_{i=1}^{n} a_i \rho^{i-1}$, where $d \in \mathbf{Z}_{>0}$, $a_i \in \mathbf{Z}$ for $1 \leq i \leq n$ and $\gcd(a_1, \ldots, a_n, d) = 1$. By $\bar{\xi}$ we denote the complex conjugate of ξ.

Let $s, t, m \in \mathbf{Z}_{\geq 0}$ and let $\rho^{(1)}, \ldots, \rho^{(s)}$ be the real zeros and $\rho^{(s+1)}, \ldots, \rho^{(m)}$, $\rho^{(m+1)} = \overline{\rho^{(s+1)}}, \ldots, \rho^{(m+t)} = \overline{\rho^{(m)}}$ the non real zeros of f, then for $1 \leq j \leq n$ the mapping

$$\xi \mapsto \xi^{(j)} = \frac{1}{d} \sum_{i=1}^{n} a_i \rho^{(j)i-1}$$

is a $\mathbb{Q}$-isomorphism. Of course, $m = s + t$ and $n = s + 2t$. For $\xi \in \mathcal{F}$ we define $|\xi|_i = |\xi^{(i)}|^{e_i}$, where $e_i = 1$, if $i \in \{1, \ldots, s\}$ and $e_i = 2$, if $i \in \{s+1, \ldots, m\}$. The *trace* of ξ is $\text{Tr}(\xi) = \sum_{i=1}^{n} \xi^{(i)}$, the *norm* of ξ is $\text{N}(\xi) = \prod_{i=1}^{n} \xi^{(i)}$, and the *height* of ξ is $\text{H}(\xi) = \max\{|\xi|_i \mid 1 \leq i \leq m\}$.

The integral closure of $\mathbf{Z}$ in a number field $\mathcal{F}$ is called the *maximal order* of $\mathcal{F}$. It is encoded by a multiplication table $\mathrm{MT}(\mathcal{O}) = (w_{i,j,k}) \in \mathbf{Z}^{n \times n \times n}$ with the property that there exists a $\mathbf{Z}$-basis $\omega_1, \ldots, \omega_n$ of $\mathcal{O}$ such that

$$\omega_i \omega_j = \sum_{k=1}^{n} w_{i,j,k} \omega_k$$

for $1 \leq i, j \leq n$ (cf. [19]). We note, that $\mathrm{MT}(\mathcal{O})$ also encodes $\mathcal{F}$, because every $\mathbf{Z}$-basis of $\mathcal{O}$ is a $\mathbb{Q}$-basis of $\mathcal{F}$ as a vector space. The *discriminant* Δ of $\mathcal{O}$ is the determinant of the matrix

$$(\mathrm{Tr}(\omega_i \omega_j)) = \det(\omega_i^{(j)})^2 \in \mathbb{Q}^{n \times n} \ .$$

More generally, when A is a $\mathbf{Z}$-modul in $\mathcal{F}$ with $\mathbf{Z}$-basis $a_1, \ldots, a_n$ then we call $\Delta_A = \det(a_i^{(j)})^2$ the *discriminant* of A. Since by [19] the problem of computing the maximal order of a given number field is polynomial time equivalent to computing the largest square dividing the discriminant Δ_f of the generating polynomial f, the set of all pairs $(\mathcal{F}, \mathcal{O})$ belongs to $\mathcal{NP}$. Moreover, since by [19] and [14] we know that there is a multiplication table of $\mathcal{O}$ with size bounded by $(2 + \log_2(\Delta))^{O(1)} \leq (2 + \log_2(\Delta_f))^{O(1)}$ we may assume that on input $\mathcal{F}$ every nondeterministic polynomial time algorithm also knows $\mathcal{O}$.

We represent $\xi = \sum_{i=1}^{n} x_i \omega_i \in \mathcal{O}$ by the vector $(x_1, \ldots, x_n) \in \mathbf{Z}^n$, called the *standard representation* of ξ. Given elements in standard representation we can compute in polynomial time their sums, products and their quotients. Since there is a polynomial time algorithm that given an element of $\mathcal{O}$ computes its trace (cf. [9]), there is also a polynomial time algorithm, that given $\mathcal{O}$ computes the discriminant Δ of $\mathcal{O}$.

A *(fractional) ideal* of $\mathcal{O}$ is a subset A of $\mathcal{F}$ such that dA is a non zero ideal of the ring $\mathcal{O}$ for some $d \in \mathbf{Z}_{>0}$. The *denominator* $d(A)$ of A is the minimal such d. If for two ideals A and B of $\mathcal{O}$ there is a non zero $\alpha \in \mathcal{F}$ such that $A = \alpha B$, then we write $A \sim B$. An ideal A of $\mathcal{O}$ is called *principal* if $A \sim \mathcal{O}$. In that case any $\alpha \in \mathcal{F}$ with $A = \alpha\mathcal{O}$ is called a *generator* of A.

An ideal A of $\mathcal{O}$ is uniquely encoded (with respect to $\omega_1, \ldots, \omega_n$) by the pair $(d(A), \mathrm{HNF}(A))$, where $\mathrm{HNF}(A) = (h_{i,j}) \in \mathbf{Z}^{n \times n}$ is a matrix in Hermite normal form (cf. [27, pp. 45-51]) such that the elements $\alpha_j = \sum_{i=1}^{n} h_{i,j} \omega_i$, $1 \leq j \leq n$, form a $\mathbf{Z}$-basis of $d(A)A$. We note that there are polynomial time algorithms that given $\mathcal{O}$, $\alpha \in \mathcal{O}$ and fractional ideals A, B of $\mathcal{O}$ determine AB, A^{-1} and $\alpha\mathcal{O}$ (cf. [8]). We call $\mathrm{N}(A) = [\mathcal{O} : d(A)A]/d(A)^n$ the *norm* of the ideal A. Clearly, all entries of $\mathrm{HNF}(A)$ are bounded by $\mathrm{N}(A)d(A)^n$. We know

Lemma 3. *For all ideals A of $\mathcal{O}$ and all $\alpha \in A$, $\alpha \neq 0$, we have $|\mathrm{N}(\alpha)| \geq \mathrm{N}(A)$. For all $\beta \in \mathcal{O}$, $\beta \neq 0$, we have $|\mathrm{N}(\beta)| = \mathrm{N}(\beta\mathcal{O})$.*

The set $\mathcal{I_F}$ of (fractional) ideals of $\mathcal{O}$ is a multiplicative abelian group in which the set $\mathcal{P_F}$ of all principal ideals of $\mathcal{O}$ is a subgroup. We call the factor group $Cl_\mathcal{F} = \mathcal{I_F}/\mathcal{P_F}$ the *class group* and $h_\mathcal{F} = |\mathcal{I_F}/\mathcal{P_F}|$ the *class number* of $\mathcal{F}$.

The *unit group* $\mathcal{O}^*$ of $\mathcal{O}$ is the set of all elements of $\mathcal{O}$ whose multiplicative inverse also belongs to $\mathcal{O}$. By Dirichlet's Unit Theorem, there is a root of unity

$\delta \in \mathcal{O}^*$ and a *system of* $r = m - 1$ *fundamental units* $\epsilon_1, \ldots, \epsilon_r \in \mathcal{O}^*$ such that every unit of $\mathcal{O}$ is uniquely represented by a product of δ and of powers of the ϵ_i, $1 \leq i \leq r$. The image of $\mathcal{O}^*$ under the *logarithm map*

$$\text{Log} : \mathcal{F} \setminus \{0\} \to \mathbb{R}^r, \xi \mapsto \text{Log}\,\xi = (\ln|\xi|_1, \ldots, \ln|\xi|_r)^T,$$

where $\ln(\cdot)$ denotes the natural logarithm, is the lattice $L_{\mathcal{F}} = \sum_{i=1}^r (\text{Log}\,\epsilon_i)\mathbb{Z}$. The absolut value of the determinant of the matrix $(\text{Log}\,\epsilon_1, \ldots, \text{Log}\,\epsilon_r)$ is called the *regulator* $R_{\mathcal{F}}$ of $\mathcal{F}$.

By $|\cdot|_\infty$ we denote the maximum modulus of all entries of a given vector. We also use the euclidean norm $|\cdot|_2$. The distinction between the euclidean norm and the archimedian valuations arises from the context.

3 Minima

The main tool used in our work is the theory of minima in ideals which was described in [3]. An element μ of an ideal A is called a *minimum* in A if there exists no $\alpha \in A$, $\alpha \neq 0$, such that $|\alpha|_i < |\mu|_i$ for $1 \leq i \leq m$. If 1 is a minimum in A, then A is called *reduced*.

Proposition 4. *Let A be an ideal. The reduced ideals equivalent to A are exactly the ideals $(1/\alpha)A$ where α is a minimum in A.*

By [1, Proposition 2.2] and [2, Lemma 3.4] we know

Theorem 5. *For all minima μ in an ideal A we have*

$$|N(\mu)| \leq (2/\pi)^t N(A)\sqrt{\Delta} \ .$$

Moreover, for all $z \in \mathbb{R}^r$ and for ideals A there is at least one minimum μ in A with $\text{Log}\,\mu \in z + W$, where $W = \{x \in \mathbb{R}^r \mid |x|_\infty \leq (\log\sqrt{\Delta})/2\}$.

Lemma 6. *If A is a reduced ideal of $\mathcal{O}$, then $1/\sqrt{\Delta} \leq N(A) \leq 1$.*

Proof. Since 1 is a minimum in A we know by Lemma 3 that $1 = |N(1)| \geq N(A)$. On the other hand we get by Theorem 5 that $1 = |N(1)| \leq N(A)\sqrt{\Delta}$. $\square$

Lemma 7. *Let A be a reduced ideal of $\mathcal{O}$ and let $\text{HNF}(A) = (a_{i,j})$ be the Hermite normal form of A. Then $0 \leq |a_{i,j}| \leq d(A) \leq \sqrt{\Delta}$ for $1 \leq i, j \leq n$.*

Proof. Since A is reduced, we know that $\mathcal{O} \subseteq A$. Hence $[A : \mathcal{O}]A \subseteq \mathcal{O}$ and therefore $d(A) \leq [A : \mathcal{O}]$. By Lemma 6 and [22] we obtain

$$[A : \mathcal{O}] = \sqrt{\Delta_A}/\sqrt{\Delta} = 1/N(A) \leq \sqrt{\Delta},$$

where Δ_A is the discriminant of A. To show the left inequality we note that $d(A)\omega_i \in d(A)A$, hence, by induction we see that $a_{i,i}$ divides $d(A)$ for $1 \leq i \leq n$. The assertion follows since $\text{HNF}(A)$ is in Hermite normal form and therefore $|a_{i,j}| \leq |a_{i,i}|$ for $1 \leq i, j \leq n$. $\square$

Corollary 8. *Let A be a reduced ideal of $\mathcal{O}$. Then $\text{size}(A) \leq (n^2 + 1)\log_2(\sqrt{\Delta})$.*

4   Compact Representations

In this section we prove the existence of a short representation for any number in $\mathcal{O}$. In the next sections, this representation will be used to show our main results.

Let A be a reduced ideal of $\mathcal{O}$, let $u \in \mathbb{R}^r$, and let μ be a minimum in A with $\text{Log}\,\mu \in u + \mathcal{W}$ where

$$\mathcal{W} = \left\{ x \in \mathbb{R}^r \,|\, |x|_\infty \le (\log\sqrt{\Delta})/2 \right\} .$$

We let $\text{DOUBLE}(A, \mu, u)$ be the set of all minima β in $B = ((1/\mu))^2 A$ with $\text{Log}\,\beta \in 2u - 2\text{Log}\,\mu + \mathcal{W}$. By Theorem 5 the set $\text{DOUBLE}(A, \mu, u)$ is not empty. Clearly, for each $\beta \in \text{DOUBLE}(A, \mu, u)$ there exists a unique element $a(\beta) \in \mathcal{O}$ such that $1/\beta = a(\beta)/d(\beta)$ where $d(\beta) = d((1/\beta)B)$.

Lemma 9. *For every $\beta \in \text{DOUBLE}(A, \mu, u)$ we have $0 \le d(\beta) \le \sqrt{\Delta}$ and $H(a(\beta)) \le \Delta^{3(m+2)/4}$.*

Proof. Let $B = ((1/\mu))^2 A$. Since $(1/\beta)B$ is reduced, the first assertion follows from Lemma 7. Since $\text{Log}\,\beta \in 2u - 2\text{Log}\,\mu + \mathcal{W}$ we have $|\log|\beta|_i| \le (3/2)\log\sqrt{\Delta}$ for $1 \le i \le r$. Hence, $\Delta^{-3/4} \le |\beta|_i \le \Delta^{3/4}$ for $1 \le i \le r$. By Lemma 3 we have $|N(\beta)| = \prod_{i=1}^m |\beta|_i \ge N(B)$, thus from Lemma 6 and Theorem 5 we obtain $|\beta|_m \ge N(B)/\Delta^{3(m-1)/4} \ge 1/\Delta^{3(m+1)/4}$, implying that $H(1/\beta) \le \Delta^{3(m+1)/4}$. The second assertion follows. $\quad\square$

Lemma 10. *Let A be a reduced ideal, and let ρ be a minimum in A. If we set $u_j = (\text{Log}\,\rho)/2^{k-j+1}$ for $1 \le j \le k$ where $k = \max\{1, \lceil \log_2(|\text{Log}\,\rho|_\infty) \rceil + 2\}$, then there are $\beta_0, \ldots, \beta_k$ with $\beta_0 = 1$ and $\beta_j \in \text{DOUBLE}(A, \prod_{i=0}^{j-1}\beta_i^{2^{j-i}}, u_j)$ such that $\rho = \prod_{j=0}^k \beta_j^{2^{k-j}}$ and $\prod_{i=0}^j \beta_i^{2^{j-i}}$ is a minimum in A.*

Proof. The number 1 is a minimum in A with $\text{Log}\,1 \in \text{Log}\,u_1 + \mathcal{W}$. By Theorem 5 we have for every $\beta_1 \in \text{DOUBLE}(A, \beta_0, u_1)$ that $\rho_1 = \beta_1(\beta_0)^2$ is a minimum in A with $\text{Log}\,\rho_1 \in 2u_1 + \mathcal{W} = u_2 + \mathcal{W}$. By induction on j we obtain that for every $\beta_j \in \text{DOUBLE}(A, \prod_{i=0}^{j-1}\beta_i^{2^{j-i}}, u_j)$ the number $\rho_j = \beta_j(\prod_{i=0}^{j-1}\beta_i^{2^{j-i}})^2$ is a minimum in A with $\text{Log}\,\rho_j \in u_{j+1} + \mathcal{W}$, $1 \le j \le k$. If we fix $\beta_k = \rho/(\rho_{k-1})^2$ then $\text{Log}\,\beta_k \in 2u_k - 2\text{Log}\,\rho_{k-1} + \mathcal{W}$. Thus $\beta_k \in \text{DOUBLE}(A, \prod_{i=0}^{k-1}\beta_i^{2^{k-i}}, u_k)$ and $\rho = \prod_{j=0}^k \beta_j^{2^{k-j}}$. $\quad\square$

Theorem 11. *For $\alpha \in \mathcal{O}$ there exist $k \le \log_2(\log_2(\Delta) + (n-1)\log_2 H(\alpha)) + 2$ and $\gamma, \alpha_j \in \mathcal{O}$ and $d_j \in \mathbb{Z}$, $1 \le j \le k$, with $H(\gamma) \le |N(\alpha)|^{2/n}\Delta$, $H(\alpha_j) \le \Delta^{\frac{3}{4}(m+2)}$ and $0 < d_j \le \Delta^{1/2}$ such that*

$$\alpha = \gamma \prod_{j=1}^k \left(\frac{\alpha_j}{d_j} \right)^{2^{k-j}} .$$

Moreover, for $1 \le j \le k$ the ideal $\prod_{i=1}^{j-1}(\alpha_i/d_i)^{2^{j-i}}\mathcal{O}$ is reduced.

Proof. By Minkowsky's Convex body theorem (cf. [23, pp.33-34]) and Lemma 3 there is a minimum γ in the principal ideal $\alpha\mathcal{O}$ with $H(\gamma) \leq |N(\alpha)|^{2/n}\Delta$. If we set $\rho = \gamma/\alpha$ then by Proposition 4 we have that ρ is a minimum in $\mathcal{O}$. Since $\mathcal{O}$ is an ideal with minimum 1 we can apply Lemma 10. Using the notations of that Lemma we obtain $\rho = \prod_{j=0}^{k} \beta_j^{2^{k-j}}$.

We may assume that $k \leq \log_2 |\text{Log } \rho|_\infty + 2 \leq \log_2(\log_2 H(\rho)) + 2$. But $\rho = \gamma/\alpha$, thus we have

$$H(\rho) \leq \frac{H(\gamma)(H(\alpha))^{n-1}}{|N(\alpha)|} \leq \Delta(H(\alpha))^{n-1}.$$

Hence, we have shown that $k \leq \log_2(\log_2(\Delta) + (n-1)\log H(\alpha)) + 2$.

If we set $\alpha_j = a(\beta_j)$ and $d_j = d(\beta_j)$ then $\alpha_j \in \mathcal{O}$ and $d_j \in \mathbf{Z}$ and by Lemma 9 it follows that $H(\alpha_j) \leq \Delta^{\frac{3}{4}(m+2)}$ and $0 < d_j \leq \Delta^{1/2}$.

The last assertion is a direct consequence of Lemma 10. $\square$

A representation of $\alpha \in \mathcal{O}$ as described in Theorem 11 is called a *compact representation* of α.

Proposition 12. *There is a polynomial time algorithm that given $\mathcal{O}$ and the compact representation $\gamma \prod_{j=1}^{k}(\alpha_j/d_j)^{2^{k-j}}$ of $\alpha \in \mathcal{O}$ computes the ideal $\alpha\mathcal{O}$.*

Proof. We describe the algorithm. It starts by computing $A_0 = (\alpha_0/d_0)\mathcal{O}$. Then it determines recursively $A_j = (\alpha_j/d_j)(A_{j-1})^2$ for $1 \leq j \leq k$. Since each A_{j-1} is reduced, from Lemma 7 follows that all these steps can be done in polynomial time. Finally, the algorithm calculates $\alpha\mathcal{O} = \gamma A_k$. It is clear that the algorithm has polynomially bounded running time. $\square$

Corollary 13. *There is a polynomial time algorithm that given the compact representation of $\alpha \in \mathcal{O}$ decides whether α is a unit of $\mathcal{O}$ or not.*

Remark. Using methods similar to those in [10], [1], [2] and [3] it can be shown that there is a deterministic algorithm that given an ideal $\alpha\mathcal{O}$ and an appropriate approximation of $\text{Log }\alpha$ computes a compact representation of α in running time $(n \log \Delta \lceil |\text{Log }\alpha|_\infty \rceil)^{O(n)}$. It can also be shown that if the dimension n of the considered field $\mathcal{F}$ is fixed we can add, multiply and compare algebraic numbers given in compact representation in polynomial time. More details will be given in a subsequent paper.

5 Sizes of Algebraic Integers

Let $\mathcal{O}$ be given by the table $MT(\mathcal{O}) = (w_{i,j,k}) \in \mathbf{Z}^{n \times n \times n}$ such that there is a $\mathbf{Z}$-basis $\omega_1, \ldots, \omega_n$ of $\mathcal{O}$ with $\omega_i\omega_j = \sum_{k=1}^{n} w_{i,j,k}\omega_k$ for $1 \leq i, j \leq n$. Then the binary length of the standard representation of $\alpha = \sum_{i=1}^{n} x_i\omega_i \in \mathcal{O}$ with respect to that $\mathbf{Z}$-basis is

$$\text{size}_s(\alpha) = \sum_{i=1}^{n} \text{size}(x_i) .$$

The binary length of the compact representation $\gamma \prod_{i=1}^{k} (\alpha_i/d_i)^{2^{k-i}}$ of α with respect to that $\mathbf{Z}$-basis is

$$\text{size}_s(\gamma) + \sum_{i=1}^{k} \left(\text{size}_s(\alpha_i) + \text{size}(d_i) \right),$$

since the α_i are given in standard representation. We define $\text{size}_c(\alpha)$ to be the maximum of that quantity taken over all compact representations with respect to $\mathcal{O}$.

Proposition 14. *For $\alpha = \sum_{i=1}^{n} x_i \omega_i \in \mathcal{O} \setminus \{0\}$ we have*

$$\left| (x_1, \ldots, x_n)^T \right|_{\infty} \leq n \, 2^{4n^2} 2^{\text{size}(\mathcal{O})} \, \text{H}(\alpha) .$$

Proof. Let $\ell, h \in \{1, \ldots, n\}$ be such that $|\omega_\ell|_h = \max\{|\omega_j|_i \mid 1 \leq i, j \leq n\}$. Then

$$|\omega_\ell|_h^2 \leq \sum_{k=1}^{n} |w_{h,h,k}| \, |\omega_k|_h \leq n \, |\omega_\ell|_h \, 2^{\text{size}(\mathcal{O})}$$

and $|\omega_j|_i \leq n \, 2^{\text{size}(\mathcal{O})}$ for $1 \leq i, j \leq n$. For $\xi \in \mathcal{F}$ we write

$$\underline{\xi} = (\xi^{(1)}, \ldots \xi^{(s)}, Re(\xi^{(s+1)}), \ldots, Re(\xi^{(m)}), Im(\xi^{(s+1)}), \ldots, Im(\xi^{(m)}))^T.$$

If $\underline{\Omega}$ is the matrix $(\underline{\omega_1}, \ldots, \underline{\omega_n}) \in \mathbb{R}^{n \times n}$, then we obtain by Cramer's rule and Hadamard's inequality that $\|\underline{\Omega}^{-1}\|_{\infty} \leq 2^{4n^2} 2^{\text{size}(\mathcal{O})}$, where $\|\cdot\|_{\infty}$ denotes the maximum absolute value of all entries in a matrix. Thus, we have

$$\left| (x_1, \ldots, x_n)^T \right|_{\infty} = |\underline{\Omega}^{-1}\underline{\alpha}|_{\infty} \leq n\|\underline{\Omega}^{-1}\|_{\infty} \text{H}(\alpha) \leq n \, 2^{4n^2} 2^{\text{size}(\mathcal{O})} \, \text{H}(\alpha) .$$

$\square$

Corollary 15. *For $\alpha \in \mathcal{O} \setminus \{0\}$ we have*

$$\text{size}_s(\alpha) \leq n \left(\log_2(n) + 4n^2 + \text{size}(\mathcal{O}) + \log_2(\text{H}(\alpha)) \right),$$

$$\text{size}_c(\alpha) \leq 5n^2 \left(n^2 + n \log_2(\Delta \, | \, \text{N}(\alpha) \, |) + \text{size}(\mathcal{O}) \right) \left(3 + \log_2(\log_2(\Delta \, \text{H}(\alpha))) \right) .$$

Lemma 16. *There exists a polynomial $p \in \mathbf{Z}[x, y, z]$ such that for every algebraic number field $\mathcal{F}$ and every multiplication table of its maximal order $\mathcal{O}$ there exists a system of fundamental units $\epsilon_1, \ldots, \epsilon_r$ of $\mathcal{O}$ with*

$$\text{size}_c(\epsilon_i) < p(n, \log_2(\Delta), \text{size}(\mathcal{O})) .$$

Proof. By [28], there is a system of fundamental units $\epsilon_1, \ldots, \epsilon_r$ of $\mathcal{O}$ such that for $1 \leq i, j \leq r$ we have $|\ln |\epsilon_i|_j| < 2(5 \log_2(\Delta))^{n-1}\sqrt{\Delta}$. Since $\text{N}(\epsilon_i) = 1$, the assertion is an immediate consequence of Corollary 15. $\square$

Lemma 17. *For every principal (integral) ideal A in $\mathcal{O}$ there exists a generator $\alpha \in \mathcal{O}$ of A such that $|\log_2 \text{H}(\alpha)| \leq \log_2(| \text{N}(A) |) + 2n^2(5 \log_2(\Delta))^{n-1}\sqrt{\Delta}$.*

Proof. Let β be a generator of A. Let $\epsilon_1, \ldots, \epsilon_r$ be a system of fundamental units of $\mathcal{O}$. We set $T = \{\sum_{i=1}^{r} x_i \mathrm{Log}\,\epsilon_i \mid x_i \in \mathbb{R}, 0 \le x_i < 1 \text{ for } 1 \le i \le r\}$. Then there is a vector $\ell = \sum_{i=1}^{r} y_i \mathrm{Log}\,\epsilon_i \in L_\mathcal{F}$, where $y_i \in \mathbb{Z}$ for $1 \le i \le r$, such that $\ell' = \mathrm{Log}\,\beta - \ell \in T$. Thus we have $|\ell'|_2 \le r \max\{|\mathrm{Log}\,\epsilon_i|_2 \mid 1 \le i \le r\}$. If we set $\alpha = \beta / \prod_{i=1}^{r} \epsilon_i^{y_i}$, then α is a generator of A with $\mathrm{Log}\,\alpha = \ell'$. Since by [28] we may assume that $|\ln |\epsilon_i|_j| < 2(5 \log_2(\Delta))^{n-1}\sqrt{\Delta}$ for $1 \le i, j \le r$, we have

$$|\ln |\alpha|_i| \le 2r(5 \log_2(\Delta))^{n-1}\sqrt{\Delta} \; .$$

From Lemma 3 we obtain

$$\left|\ln |\alpha|_m\right| = \left|\ln |\mathrm{N}(A)| - \sum_{i=1}^{r} \ln |\alpha|_i\right| \le \left|\log_2 |\mathrm{N}(A)|\right| + 2r^2(5 \log_2(\Delta))^{n-1}\sqrt{\Delta} \; .$$

$\square$

Corollary 18. *There exists a fixed polynomial $q \in \mathbb{Z}[x, y, z]$ such that for every algebraic number field $\mathcal{F}$ and every multiplication table of its maximal order $\mathcal{O}$ and every principal ideal A of $\mathcal{O}$ there is a generator α of A with*

$$\mathrm{size}_c(\alpha) \le q(n, \log_2(\Delta), \mathrm{size}(\mathcal{O}), \log_2(\mathrm{N}(A))) \; .$$

6 Principal Ideal Testing

In this section we consider the following problem: Given an algebraic number field $\mathcal{F}$ and an ideal A of the maximal order $\mathcal{O}$ is there a short proof of the principality of A? We first look at the set PRI of all pairs $(\mathcal{F}, A)$, where $\mathcal{F}$ is an algebraic number field, and A is a principal ideal in the maximal order $\mathcal{O}$. Remember that as mentioned in Sect. 2 we may assume that every nondeterministic polynomial time algorithm knows $\mathcal{O}$ on input $\mathcal{F}$.

Theorem 19. *The set PRI belongs to $\mathcal{NP}$.*

Proof. Let $(\mathcal{F}, A) \in \mathrm{PRI}$, where $\mathcal{F}$ is a number field of degree n. By Lemma 18 there exists a generator α of the ideal A such that $\mathrm{size}_c(\alpha)$ is bounded by a polynomial in $\log_2(\Delta)$ and the input size, i.e. the sizes of $\mathcal{F}$, $\mathcal{O}$ and A. By Proposition 12 there is a polynomial time algorithm, which given the compact representation $\gamma \prod_{j=1}^{k}(\alpha_j/d_j)^{2^{k-j}}$ of α (cf. Sect. 4) computes $\alpha\mathcal{O}$. Thus a non-deterministic polynomial time algorithm, that accepts PRI, only has to guess the compact representation of α, to compute $\alpha\mathcal{O}$ and to compare this ideal with A. $\square$

To prove our second main result we need

Proposition 20. *Given $\mathcal{O}$, $\beta_1, \ldots, \beta_i \in \mathcal{O}$ and ideals $B_1, \ldots, B_i$ of $\mathcal{O}$, $i \in \mathbb{Z}_{>0}$, there is a polynomial time algorithm that decides whether $B_j = (1/\beta_j)(B_{j-1})^2$ for $1 \le j \le i$.*

Lemma 21. *Let $\mathcal{O}$ be the maximal order of $\mathcal{F}$, let A be an ideal of $\mathcal{O}$ and let $x = 2^i$, $i \in \mathbb{Z}_{>0}$. Then there exist $\beta_0, \beta_1, \ldots, \beta_i \in \mathcal{O}$ with*

$$\text{size}_s(\beta_j) = (2 + \log_2(\Delta))^{O(1)} + \log_2(\Delta)\text{size}(\mathcal{O})$$

and ideals $B_1, \ldots, B_i$ of $\mathcal{O}$ with

$$\text{size}(B_j) \leq (n^2 + 1)\log_2(\sqrt{\Delta})$$

for $1 \leq j \leq i$ such that $B_j = (1/\beta_j)(B_{j-1})^2$ and $A^x \sim B_i$.

Proof. Let $B_0 = (1/\beta_0)A$ and $B_j = (1/\beta_j)(B_{j-1})^2$, $j \geq 1$, where β_0 is a minimum of A and β_j is a minimum of $(B_{j-1})^2$ with

$$\text{Log }\beta_0, \text{Log }\beta_j \in \left\{ x \in \mathbb{R}^r \mid |x|_\infty \leq (\log \sqrt{\Delta})/2 \right\} .$$

Thus, $B = B_i$ is reduced and $A^x \sim B$. As in the proof of Lemma 9 we can show that $\mathbf{H}(\beta_j) \leq \Delta^{3(m+1)/4}$. Thus, by Corollary 15 we obtain

$$\text{size}_s(\beta_j) = (2 + \log_2(\Delta))^{O(1)} + \log_2(\Delta)\text{size}(\mathcal{O}) .$$

Since each B_j is reduced, from Corollary 8 it follows that

$$\text{size}(B_j) \leq (n^2 + 1)\log_2(\sqrt{\Delta}) .$$

$\square$

Theorem 22. *The set of all tuples $(\mathcal{F}, A_0, \ldots, A_\ell, x_0, \ldots, x_\ell)$ where $\mathcal{F}$ is an algebraic number field and $A_0, \ldots, A_\ell$ are ideals of the maximal order $\mathcal{O}$ and $x_0, \ldots, x_\ell \in \mathbb{Z}$, $\ell \in \mathbb{Z}_{>0}$, such that $\prod_{i=0}^{\ell} A_i^{x_i} \sim \mathcal{O}$, belongs to $\mathcal{NP}$.*

Proof. W.l.o.g. we may assume that $x_i \geq 0$ for $1 \leq i \leq \ell$. Otherwise, we replace A_i by A_i^{-1}, which can be computed in polynomial time. Let $y_i = \lceil \log_2(x_i) \rceil + 1$ and $x_{i,j} \in \{0, 1\}$, $1 \leq j \leq y_i$, such that $x_i = \sum_{j=0}^{y_i} x_{i,j} 2^j$. Then

$$\prod_{i=0}^{\ell} A_i^{x_i} = \prod_{i=0}^{\ell} \prod_{j=0}^{y_i} (A_i^{x_{i,j}})^{2^j} .$$

Thus, from Proposition 20 and Lemma 21 it follows that there is a nondeterministic polynomial time algorithm that guesses an ideal B with size bounded by a polynomial in the size of the input tuple and tests whether $\prod_{i=0}^{\ell} A_i^{x_i} \sim B$. Hence, the assertion follows from Theorem 19. $\square$

7 Multiples of the Class Number

Theorem 23. *Assume the GRH. Then the set of all pairs $(\mathcal{F}, H)$, where $\mathcal{F}$ is an algebraic number field and H is a multiple of $h_{\mathcal{F}}$, belongs to $\mathcal{NP}$.*

Proof. Let $\mathcal{F}$ be a number field of degree n and let $F = \{P_1, \ldots, P_{|F|}\}$ be the set of all prime ideals of the maximal order $\mathcal{O}$ with norm not exceeding $12(\log \Delta)^2 + 1$. By means of the methods presented in [9] and [29] the set F can be computed in time polynomially bounded by size($\mathcal{F}$), size($\mathcal{O}$) and $\log_2(\Delta)$.

We also set

$$A = \left\{ \alpha \in \mathcal{F} \setminus \{0\} \,\Big|\, \alpha\mathcal{O} = \prod_{i=1}^{|F|} P_i^{e_i}, e_i \in \mathbf{Z} \right\},$$

and for $\alpha \in A$ with $\alpha\mathcal{O} = \prod_{i=1}^{|F|} P_i^{e_i}$ we define $\varphi'(\alpha) = (e_1, \ldots, e_{|F|}) \in \mathbf{Z}^{|F|}$. If we assume the GRH then by [4] the image $L' = \varphi'(A)$ is a $|F|$-dimensional lattice of determinant $h_{\mathcal{F}}$. Thus a number H is a multiple of $h_{\mathcal{F}}$ if and only if there is $E = (e_{i,j}) \in \mathbf{Z}^{|F| \times |F|}$ with $\det(E) = H$ satisfying

$$\prod_{j=1}^{|F|} P_j^{e_{i,j}} \sim \mathcal{O} \tag{1}$$

for each $1 \leq i \leq |F|$. This is true, because the vectors $e_i = (e_{i,1}, \ldots, e_{i,|F|})$ form a sublattice Λ of L' with $|L'/\Lambda| = |\det(E)|/h_{\mathcal{F}}$.

W.l.o.g. we may assume that $\|E\|_\infty \leq \det(E)$. Then size($E$) is polynomially bounded by H and a nondeterministic polynomial time algorithm which accepts the above set only has to guess E and to verify condition (1) for $1 \leq i \leq |F|$. By Theorem 22 this can be done in polynomial time. Since the algorithm knows $\mathcal{O}$ (see Sect. 2) this proves the theorem. $\qquad \square$

8 Approximations

Since we can not compute with real values we sometimes have to work with approximations. For this purpose we give in this section some notations and results concerning the computation of approximations to algebraic numbers given in standard or compact representation.

Let $q \in \mathbf{Z}_{>0}$, and let z be a complex number. A number $z\{q\} \in 2^{-(q+1)}\mathbf{Z}[i]$ is called an *approximation of precision q* to z if $|z - z\{q\}| < 2^{-q}$. An approximation of precision q to a vector $\underline{v} \in \mathbf{C}^r$ or a matrix $A \in \mathbf{C}^{r \times r}$ is a vector $\underline{v}\{q\}$ or a matrix $A\{q\}$ whose entries are approximations of precision q to the corresponding entries of the original vector or matrix. We use the following technical result that can be deduced from the estimates in [5]:

Proposition 24. *Let $A \in \mathbf{C}^{\ell \times \ell}, \ell \in \mathbf{Z}_{>0}$, and let $A\{p\}$ be an approximation of precision $p \in \mathbf{Z}_{>0}$ to A. Then we have*

$$|\det(A) - \det(A\{p\})| \leq 2^{-p}\sqrt{\ell}\, 2^{\ell-1}\ell^{\frac{\ell+1}{2}}\|A\|_\infty^{\ell-1}.$$

Proposition 25. *Let $x \in \mathbb{R}_{>0}$, and let $x\{q\} \in \mathbb{R}_{>0}$ be an approximation of precision q to x, where $x\{q\} > x$, $q \in \mathbb{Z}_{>0}$. Let $p \in \mathbb{Z}_{>0}$ and let $\kappa \in \mathbb{R}_{>0}$ with $\kappa > \ln(2)$. If $q > -\ln(x) + \kappa p$ then we have $|\ln(x\{q\}) - \ln(x)| < 2^{-p}$.*

Lemma 26. *There is a polynomial time algorithm that given a number field $\mathcal{F}$, the maximal order $\mathcal{O}$ of $\mathcal{F}$, the compact representation $\gamma \prod_{j=1}^{k}(\alpha_j/d_j)^{2^{k-j}}$ of an element $\alpha \in \mathcal{O}$ and an integer 2^p, where $p \in \mathbb{Z}_{>0}$, computes an approximation of precision p to $\operatorname{Log} \alpha$.*

Proof. If we set $q' = p + 2k + 2$, then

$$(\operatorname{Log} \gamma)\{q'\} + \sum_{j=1}^{k} 2^{k-j} \big((\operatorname{Log} \alpha_j)\{q'\} - (\operatorname{Log} d_j)\{q'\}\big)$$

is an approximation of precision p to $\operatorname{Log} \alpha$. Thus, the algorithm first computes upper bounds λ_j of $\mathbf{H}(\alpha_j)$ and an upper bound δ of $\mathbf{H}(\gamma)$. Then it determines for $1 \leq i \leq n$ approximations of precision $2(\lceil \log_2(\lambda_j) \rceil + \ln(2)q')$ to $\alpha_j^{(i)}$ respectively of precision $2(\lceil \log_2(\delta) \rceil + \ln(2)q')$ to $\gamma^{(i)}$. These steps can be performed in polynomial time (cf. [7] or [17]). Therefore, from Proposition 25 and [26] the assertion follows. □

Theorem 27. *There is a polynomial time algorithm that given $\mathcal{O}$, the compact representations of r independent units $\epsilon_1, \ldots, \epsilon_r$ of $\mathcal{O}$ and a number 2^q, $q \in \mathbb{Z}_{>0}$ computes an approximation of precision q to a multiple of $R_{\mathcal{F}}$.*

Proof. The lattice $\sum_{i=1}^{r} \operatorname{Log} \epsilon_i \mathbb{Z}$ is a sublattice of the lattice $L_{\mathcal{F}}$. Therefore the determinant of the matrix $\mathbf{REG} = (\operatorname{Log} \epsilon_1, \ldots, \operatorname{Log} \epsilon_r)$ is a multiple of $R_{\mathcal{F}}$. We describe an algorithm that approximates that determinant. First, the algorithm determines an approximation of precision 1 to the matrix $\mathbf{REG}$ and computes the bound $\lambda = \lceil \|\mathbf{REG}\{1\}\|_\infty \rceil + 1 \geq \|\mathbf{REG}\|_\infty$. Then it computes an approximation $\mathbf{REG}\{p\}$ of precision $p = (n(\log_2(\lambda) + n + 2) + q + 1$ to $\mathbf{REG}$. Since $n > r$ we obtain from Proposition 24 that $|\det(\mathbf{REG}) - \det(\mathbf{REG}\{p\})| < 2^{-q}$. Finally, the algorithm computes the determinant of $\mathbf{REG}\{q\}$. According to Lemma 26, the algorithm only needs polynomial time. □

9 An Approximation of $h_{\mathcal{F}} R_{\mathcal{F}}$

Theorem 28. *Assume the GRH. Then there is a polynomial time algorithm that given an algebraic number field $\mathcal{F}$ and the number w of roots of unity in $\mathcal{F}$ computes a number $\Theta \in \mathbb{R}_{>0}$ such that $(6/5)h_{\mathcal{F}} R_{\mathcal{F}} \leq \Theta \leq (15/8)h_{\mathcal{F}} R_{\mathcal{F}}$.*

Proof. By [12] the product $h_{\mathcal{F}} R_{\mathcal{F}}$ can be expressed by means of the analytic class number formula $h_{\mathcal{F}} R_{\mathcal{F}} = C_{\mathcal{F}} \prod_{p \in P} E(p)$, where $C_{\mathcal{F}} = \frac{w\sqrt{\Delta}}{2^s(2\pi)^t}$, P is the set of rational primes and $E(p)$ is the Euler factor belonging to p. Since s and t can be determined in polynomial time from the generating polynomial f of $\mathcal{F}$, we only have to describe a method for computing an approximation of $\prod_{p\in P} E(p)$. For this purpose let L be the normal closure of $\mathcal{F}$. We choose $Q \in \mathbb{Z}_{>0}$ and split $\prod_{p\in P} E(p) = F(Q)T(Q)$, where

$$F(Q) = \prod_{\substack{p\in P, p\le Q}} E(p) \prod_{\substack{p\in P, p>Q \\ p \text{ ramified in } L}} E(p)$$

and

$$T(Q) = \prod_{\substack{p\in P, p>Q \\ p \text{ unramified in } L}} E(p).$$

Then we have $h_{\mathcal{F}} R_{\mathcal{F}} = C_F F(Q)T(Q)$. From [12] (cf. inequality (3.4) and Theorem 3.1) we obtain that $|\log(T(Q))| \le (c_3 \log(\Delta))/\sqrt{Q}$, where $c_3 = n^{O(1)}$. Therefore, if we set

$$Q = (c_3 \log(\Delta)/\log(5/4))^2 \quad \text{and} \quad \Theta = (3/2)C_{\mathcal{F}}F(Q),$$

then $(6/5)h_{\mathcal{F}}R_{\mathcal{F}} \le \Theta \le (15/8)h_{\mathcal{F}}R_{\mathcal{F}}$. By the methods of [12] the value $E(p)$ can be computed in time polynomially bounded by size$(\mathcal{F})$, size$(\mathcal{O})$ and p for every $p \le Q$. Hence, there is a polynomial time algorithm that computes Θ. $\square$

10 Proof of the Main Results

Theorem 29. *Assume the GRH. Then the set $\mathcal{H}$ of all pairs $(\mathcal{F}, h_{\mathcal{F}})$ belongs to $\mathcal{NP}$.*

Proof. We describe a nondeterministic algorithm that accepts the set $\mathcal{H}$. We know that there is a nondeterministic polynomial time algorithm which guesses on input of $\mathcal{F}$ the representation of the maximal order $\mathcal{O}$ of $\mathcal{F}$ and verifies it. There also exists a nondeterministic polynomial time algorithm that guesses the compact representations of r elements $\epsilon_1, \ldots, \epsilon_r$ such that size$_c(\epsilon_i)$ satisfies the bounds of Lemma 16 for $1 \le i \le r$. By Corollary 13, the algorithm can test in polynomial time, if the ϵ_i are units of $\mathcal{O}$. Finally, we note that there is a nondeterministic algorithm which guesses a number $w \le n(n + 1)/2$ and w elements of $\mathcal{F}$ and tests if they are roots of unity. Since the height of that elements is 1 this can be done in polynomial time, too.

The algorithm that accepts the set $\mathcal{H}$ starts the above algorithms. After that initialization, it computes by the method described in the proof of Theorem 27 an approximation R^* of precision 10 to the absolut value of the determinant of the matrix $\mathbf{REG} = (\mathrm{Log}\,\epsilon_1, \ldots, \mathrm{Log}\,\epsilon_r)$. From [30] we know that $R_{\mathcal{F}} > 0.05$. Thus, the units ϵ_i are independent if and only if $R^* 2^{-10}$. If the units are not independent our algorithm begins an endless loop. Otherwise, R^* is an approximation of precision 10 to a multiple of $R_{\mathcal{F}}$. Using the algorithm of Theorem 28 it computes an approximation Θ of $h_{\mathcal{F}} R_{\mathcal{F}}$ such that $(6/5)h_{\mathcal{F}}R_{\mathcal{F}} < \Theta < (15/8)h_{\mathcal{F}}R_{\mathcal{F}}$. Then the algorithm verifies according to Theorem 23 that H is a multiple of the class number $h_{\mathcal{F}}$. Obviously, $H = h_{\mathcal{F}}$ if and only if $HR^* \le \Theta$. $\square$

11 Acknowledgement

This research was supported by the Deutsche Forschungsgemeinschaft.

References

1. J. Buchmann, *On the Computation of Units and Class Numbers by a Generalization of Lagrange's Algorithm*, Journal of Number Theory, Vol. 26, No. 1 (1987) 8-30

2. J. Buchmann, *On the Period Length of the Generalized Lagrange Algorithm*, Journal of Number Theory, Vol. 26, No. 1 (1987) 31-37

3. J. Buchmann, *Zur Komplexität der Berechnung von Einheiten und Klassenzahlen algebraischer Zahlkörper*, Habiliationsschrift, Düsseldorf (1987)

4. J. Buchmann, *A subexponential algorithm for the determination of class group and regulator of algebraic number fields*, Séminaire de Théorie des Nombres, Paris 1988-1989, Birkhäuser Verlag (1990)

5. J. Buchmann, *Reducing lattice bases by means of approximations*, in preparation.

6. J. Buchmann, *Number Theoretic Algorithms, Algebraic Number Theory*, Lecture Notes, Saarbrücken, WS 88/89.

7. J. Buchmann, *Algorithms in algebraic number theory*, Manuscript (1992)

8. J. Buchmann, H. W. Lenstra, Jr., *Approximating rings of integers in number fields*, in preparation.

9. J. Buchmann, H. W. Lenstra, Jr., *Computing maximal orders and decomposing primes in number fields*, in preparation.

10. J. Buchmann, H. C. Williams, *On Principal Ideal testing in Algebraic Number Fields*, Journal of Symbolic Computation, Vol. 4, No. 1 (1987) 11-19

11. J. Buchmann, H. C. Williams, *On the existence of a short proof for the value of the class number and regulator of a real quadratic field*, NATO Advanced Science Institutes Series C, Vol. 256, Kluwer, Dordrecht (1989) 327-345

12. J. Buchmann, H. C. Williams, *On the Computation of the Class Number of an Algebraic Number Field*, Math. Comp., v. 53 (1989) 679-688

13. J. Buchmann, H. C. Williams, *Some Remarks Concerning the Complexity of Computing Class Groups of Quadratic Fields*, Journal of Complexity 7, (1991) 311-315

14. J. Buchmann, Oliver van Sprang, *On short representations of orders and number fields*, Manuscript (1992)

15. J. Buchmann, H. C. Williams, C. Thiel, *Short representation of quadratic integers*, to appear in Proceedings of CANT 1992.

16. H. Cohen, *A Course in Computational algebraic number theory*, Springer Verlag (1993)

17. G. Ge, *Algorithms Related to Multiplicative Representations of Algebraic Numbers*, PhD. Thesis, University of California at Berkeley (1993)

18. S. Lang, Algebraic number theory, Springer-Verlag,New York (1986)

19. H. W. Lenstra, Jr., *Algorithms in algebraic number theory*, Bulletin (New Series) of The American Mathematical Society **26** no. 4 (1992) 211-244

20. A. K. Lenstra, H. W. Lenstra Jr., L. Lovasz, *Factoring polynomials with rational coefficients*, Math. Ann. 261 (1982) 515-534

21. K. S. McCurley, *Cryptographic key distribution and computation in class groups*, NATO Advanced Science Institutes Series C, Vol. 256, Kluwer, Dordrecht (1989) 459-479

22. W. Narkiewiecz, *Elementary and Analytic Theory of Algebraic Numbers*, Polish Scientific Publishers, Warszawa (1974) 70

23. J. Neukirch, *Algebraische Zahlentheorie*, Springer Verlag, Berlin (1992)

24. H. R. Lewis, C. H. Papadimitriou, *Elements of the theory of computation*, Prentice-Hall (1981)

25. A. Schönhage, *Factorization of Univariate Integer Polynomials by Diophantine Approximation and an Improved Basis Reduction Algorithm*, ICALP (1984)

26. A. Schönhage, *Numerik analytischer Funktionen und Komplexität*, Jber. d. Dt. Math. Verein **92** (1990)

27. A. Schrijver, *Theory of linear and integer programming*, Wiley&Sons Ltd., Chichester (1987) 45-51

28. C. L. Siegel, *Gesammelte Abhandlungen IV*, Berlin, New York, Heidelberg (1979) 66-81

29. D. Weber, *Ein Algorithmus zur Zerlegung von Primzahlen in Primideale*, Universität des Saarlandes, Masterthesis (1993)

30. R. Zimmert, *Ideale kleiner Norm in Idealklassen und eine Regulatorabschätzung*, Invent. math. **62** (1981) 367-380

Calculating the Class Number of Certain Hilbert Class Fields

Farshid Hajir

California Institute of Technology, Pasadena CA 91125

Let k be a complex quadratic field with discriminant $-p$ where p is a prime congruent to 3 modulo 4. The class number h of k is odd. Assume that the ideal class group of k is cyclic. There is a unique cylic unramified extension H of k of degree $h = [H : k]$, called the Hilbert class field of k; H is in fact a dihedral extension of degree $2h$ over the field of rational numbers. Let h_H be the class number of H, and E_H the group of units of its ring of integers. Elliptic units can be used to calculate h_H and (simultaneously) a minimal set of generators for E_H. An account of an implementation of this procedure [1] for calculating the parity of h_H for prime $h \leq 19$ and $p \leq 15000$ will be given.

The method is based on an explicit formula for an elliptic unit ε (defined via special values of certain elliptic modular functions) in E_H whose h conjugates over k span a subgroup of E_H with index precisely h_H. Previous algorithms proceeded by first computing ε^{24} and its conjugates, then used lengthy trial and error calculations to determine the approximate numerical value of ε and its conjugates. Thanks to the explicit formula mentioned above (which is proved via the reciprocity law of complex multiplication), we are able to avoid this step and therefore compute the minimal polynomial of ε more efficiently.

Once this is done, given a small prime l, it is not too difficult to compute the power of l in h_H, especially if one takes advantage of properties of the Galois module structure of E_H. Therefore, the most prominent obstacle in rendering the complete calculation of h_H (via this method) practical for large h is the lack of a reasonably small upper bound for h_H, or, equivalently, the lack of a large lower bound for the regulator of H.

Our computer calculations turned up only four Hilbert class fields with even class number. The first three occur for $p = 283, 331, 643$: the corresponding h is 3 and the ideal class group of H is isomorphic to the Klein four group. These examples were known to Berwick in 1927 (!) who used calculating machines to factor the "singular moduli" (special values of the modular invariant j). The fourth example is $p = 14947$ for which $h = 17$; here H is a field of degree 34 whose ideal class group has a subgroup isomorphic to 8 copies of the cyclic group of order 2.

References

1. Hajir, F.: Unramified Elliptic Units. Thesis, MIT 1993

Efficient Checking of Computations in Number Theory

Leonard M. Adleman, Ming-Deh Huang and Kireeti Kompella

Department of Computer Science
University of Southern California
Los Angeles, CA 90089-0782

Abstract. We present *efficient* program checkers for two important number theory problems, integer GCD and modular exponentiation. The notion of *program checkers* was introduced by Manuel Blum as a new approach to the problem of program correctness. Our result regarding checking integer GCD answers an open problem posed by Blum; furthermore, the checker we give is a *constant query* checker. The other result paves the way for more robust cryptographic programs, as modular exponentiation is the basis for several cryptosystems, and introduces a new technique for writing checkers, the *tester-checker*. Taken together, these results lay the foundation for more reliable number-theoretic programs. Further details can found in [AHK].

References

[AHK] Adleman, L., Huang, M.-D., and Kompella, K., Efficient Checkers for Number-Theoretic Computations, to appear in *Information and Computation*.

Constructing Elliptic Curves
with Given Group Order
over Large Finite Fields

Georg-Johann Lay and Horst G. Zimmer

Fachbereich 9 Mathematik
Universität des Saarlandes
Postfach 151150
D–66041 Saarbrücken, Germany

Abstract. A procedure is developed for constructing elliptic curves with given group order over large finite fields. The generality of the construction allows an arbitrary choice of the parameters involved. For instance, it is possible to specify the finite field, the group order or the class number of the endomorphism ring of the elliptic curve. This is important for various applications in computational number theory and cryptography. Moreover, we give a method that yields all representations of a given integer as a norm in an imaginary quadratic field.

1 Introduction

A fundamental task in computational number theory is the construction of elliptic curves with given group order. A solution to this task, accomplished by the first author [11], has various important applications. Here, we only mention

1. the construction of large primes (of magnitude up to 10^{1000})
2. the primality proving/testing via an algorithm of Golwasser, Kilian and Atkin (cf. [6], [16], [17])
3. the determination of elliptic curves over finite fields to be used in cryptosystems which are based on discrete logarithms

In the present paper[1], based on [11], the following tasks are solved:

1. Given an integer $m > 3$, find a prime p and an elliptic curve E over the finite field $\mathbb{F}_p$ of order $\#E(\mathbb{F}_p) = m$.
2. Given two integers n and c_{max}, find an elliptic curve E over the finite field $\mathbb{F}_{2^n}$ with $\#E(\mathbb{F}_{2^n}) = c \cdot q$ with q a prime and $c \leqslant c_{max}$.
3. Given an integer $n > 1$, decide whether or not there is a prime $p > 3$ and an elliptic curve E over $\mathbb{F}_p$ with structure $E(\mathbb{F}_p) \simeq (\mathbb{Z}/n\mathbb{Z}) \times (\mathbb{Z}/n\mathbb{Z})$.
4. Given a prime $p > 3$ and an integer m with $|p + 1 - m| < 2\sqrt{p}$, build an elliptic curve over $\mathbb{F}_p$ whose group order is m and whose endomorphism ring has small class number.

---

[1] As Atkin pointed out to us, he has developed related algorithms called Schoof-Atkin1 and Schoof-Atkin1-Elkies-Atkin2.

Of course, if we are given an elliptic curve over a finite field, a combined Schoof–Shanks algorithm[2] (see [19]) yields its group order. However, this order and hence the elliptic curve may not be sufficient for the applications mentioned above. On the other hand, there is an algorithm of Spallek [21] that searches for elliptic curves suitable for cryptographic purposes. But with this algorithm it is neither possible to prescribe the group order nor to specify the cardinality of the field of definition. Furthermore, Spallek's algorithm excludes curves of j–invariants 0 and 12^3, and there is no version of it in characteristic 2 or 3.

For cryptographic applications, supersingular elliptic curves are to be avoided (cf. [12]). Therefore, we concentrate on the construction of ordinary elliptic curves. Yet, in [11], supersingular curves are also considered.

2 Elliptic Curves over Finite Fields

In this section we state some basic properties of elliptic curves over finite fields. Some of the theorems are adjusted to our application.

Theorem 1. *Let E be an elliptic curve over a finite field. Then the endomorphism ring (always considered over the algebraic closure) $\operatorname{End} E$ of E is either an order in an imaginary quadratic field K or an order in a quaternion algebra.*

In the first case E is called *ordinary* and in the second case *supersingular*. Note that, in any case, E has complex multiplication since $\operatorname{End} E$ is strictly larger than $\mathbb{Z}$. The following theorem enables us to distinguish between ordinary and supersingular curves. For a complete classification of supersingular curves we refer to [9, chap. 13, §3–§7].

Theorem 2. *1. An elliptic curve over $\mathbb{F}_p$, $p > 3$, is supersingular, iff its group has cardinality $p + 1$.*
2. An elliptic curve over $\mathbb{F}_{2^n}$ is supersingular, iff its j–invariant is zero.

Of course, in order to construct a curve with given order $m = \#E(\mathbb{F}_q)$, $q = p^n$, we make use of the fact that the Riemann hypothesis for the ζ–function of E over $\mathbb{F}_q$ is true (see [7] and [8]). In fact, we have

Theorem 3 (Hasse's theorem). *The order $m = \#E(\mathbb{F}_q)$ of an elliptic curve E over $\mathbb{F}_q$ satisfies the inequality*

$$|q + 1 - m| \leqslant 2\sqrt{q}. \tag{1}$$

Elliptic curves E over $\mathbb{F}_q$ can be obtained by reducing suitable elliptic curves over algebraic number fields.

Theorem 4. *Let K be an imaginary quadratic field and $H_{\mathcal{O}}$ be the ring class field associated to an order $\mathcal{O}$ in K. Denote by p a rational prime which splits completely in K and by $\mathfrak{P}$ a prime of $H_{\mathcal{O}}$ above p with residue degree $f = f_{\mathfrak{P}|p}$*

---

[2] implemented, e.g. in SIMATH (cf. [20])

and such that $[\mathcal{O}_K : \mathcal{O}] \notin \mathfrak{P}$. Let $\mathcal{E}$ be an elliptic curve over $H_{\mathcal{O}}$ which has complex multiplication by $\mathcal{O}$ and good, ordinary reduction at $\mathfrak{P}$. Then there is an element $\pi \in \mathcal{O} \backslash p\mathcal{O}$ satisfying the system of norm equations

$$
\begin{aligned}
q &= N_K(\pi) \\
\#E(\mathbb{F}_q) &= N_K(1 - \pi)
\end{aligned} \tag{2}
$$

for the $\mathfrak{P}$-reduced curve E of $\mathcal{E}$, where $q = p^f$. The endomorphism ring of $\mathcal{E}$ is stable under the reduction map $\mathcal{E} \xrightarrow{\mathfrak{P}} E$, i.e. $\operatorname{End} \mathcal{E} = \operatorname{End} E = \mathcal{O}$. Moreover, every elliptic curve over $\mathbb{F}_q$ with endomorphism ring $\mathcal{O}$ arises in this way.

Let $\pi_q \in \operatorname{End} E$ be the Frobenius endomorphism acting on $E(\mathbb{F}_q)$ by raising the coordinates of a rational point to the qth power:

$$
\pi_q : (x, y) \mapsto (x^q, y^q). \tag{3}
$$

In fact, π_q is the image of $\pi \in \operatorname{End} \mathcal{E}$ under the projection $\mathcal{E} \xrightarrow{\mathfrak{P}} E$ and the endomorphism $1 - \pi \in \operatorname{End} \mathcal{E}$ maps to $1 - \pi_q \in \operatorname{End} E$, where 1 is the identity endomorphism of both $\mathcal{E}$ and E. The crucial point with respect to the construction is that the group of rational points of E is given as the kernel of $1 - \pi_q$. Thus, we want to find an elliptic curve $\mathcal{E}$ over $H_{\mathcal{O}}$ such that the reduced curve E over $\mathbb{F}_q$ has the preassigned order

$$
m = \#E(\mathbb{F}_q) = \#\ker(1 - \pi_q). \tag{4}
$$

First of all, we must find the imaginary quadratic field K of Theorem 4 when m and q are given.

Theorem 5. *The imaginary quadratic field K of Theorem 4 is given by*

$$
K = \mathbb{Q}\left(\sqrt{(q + 1 - m)^2 - 4q}\right). \tag{5}
$$

Concerning the group structure, we have

Theorem 6 (Hasse). *Let E be an elliptic curve over the finite field $\mathbb{F}_p$. Then the structure of E as an abelian group is given by*

$$
E(\mathbb{F}_p) \simeq \mathbb{Z}/n_1\mathbb{Z} \times \mathbb{Z}/n_2\mathbb{Z},
$$

where n_1 and n_2 are positive integers such that $n_1 \mid n_2$ and $n_1 \mid \gcd(\#E(\mathbb{F}_p), p - 1)$.

As a matter of fact, elliptic curves over prime fields are almost always cyclic. A hint showing how to find non-cyclic curves is given in [11, 16]:

Theorem 7. *Let E be an elliptic curve over $\mathbb{F}_p$ with $n := n_1 = n_2$, where n_1 and n_2 are as in the preceding theorem. Then one of the statements*

1. *$\operatorname{End} E$ is an order of $\mathbb{Q}(\sqrt{-1})$ and $p = n^2 + 1$*
2. *$\operatorname{End} E$ is an order of $\mathbb{Q}(\sqrt{-3})$ and $p = n^2 \pm n + 1$*

is true.

3 Construction of Ring Class Fields

This section deals with the effective construction of the ring class field $H_{\mathcal{O}}$ associated to an order $\mathcal{O}$ in an imaginary quadratic field K by means of a minimal polynomial of a generating element.

3.1 Ideal Classes and Quadratic Forms

The class group of an order $\mathcal{O}$ in K (i.e. the group of invertible fractional $\mathcal{O}$-ideals), its class number and discriminant will be denoted by $\mathcal{C}\ell(\mathcal{O})$, $h(\mathcal{O})$ and $\delta(\mathcal{O})$, respectively. In the special case in which $\mathcal{O} = \mathcal{O}_K$ is the maximal order of K we shall use the abbreviating notations $\mathcal{C}\ell_K = \mathcal{C}\ell(\mathcal{O}_K)$, $h_K = h(\mathcal{O}_K)$ and $\delta_K = \delta(\mathcal{O}_K)$. Ideal classes of $\mathcal{O}$ will be represented by $\mathrm{SL}_2(\mathbb{Z})$–equivalence classes $[Q]$ of (in this paper always positive definite binary) quadratic forms $Q = (a, b, c)$ of discriminant $\delta(\mathcal{O}) = b^2 - 4ac$. To each quadratic form we associate the number $\tau_Q = (-b + \sqrt{\delta})/2a$ which is the unique root of $Q(\tau, 1) = a\tau^2 + b\tau + c$ lying in the upper half complex plane $\mathfrak{H}$.

Theorem 8. *Let $\mathcal{O}$ be the order of discriminant δ in the imaginary quadratic field $K = \mathbb{Q}(\sqrt{\delta})$.*

1. *If $Q = (a, b, c) : (x, y) \mapsto ax^2 + bxy + cy^2$ with $a, b, c \in \mathbb{Z}$ is a quadratic form of discriminant δ, then $[1, \tau_Q]$ is an invertible fractional $\mathcal{O}$–ideal.*
2. *The map sending Q to $[1, \tau_Q]$ induces an isomorphism between the form class group $C(\delta)$ of all quadratic forms of discriminant δ and the ideal class group $\mathcal{C}\ell(\mathcal{O})$.*

3.2 Dedekind's η–Function and Weber's Functions $\mathfrak{f}$, $\mathfrak{f}_1$ and $\mathfrak{f}_2$

For $\tau \in \mathfrak{H}$ and $a \in \mathbb{Q}$, we put $q = e^{2\pi i \tau}$ and $q^a = e^{2\pi i \tau a}$. Let $\zeta_n = e^{2\pi i/n}$. The Dedekind η–function is defined by (cf. Weber [22, §34])

$$\eta(\tau) = q^{1/24} \prod_{n=1}^{\infty}(1 - q^n) = q^{1/24} \sum_{n=-\infty}^{\infty} (-1)^n q^{(3n^2+n)/2} . \tag{6}$$

Note that (6) converges for $\tau \in \mathfrak{H}$. The classical Weber functions $\mathfrak{f}$, $\mathfrak{f}_1$ and $\mathfrak{f}_2$ are defined in terms of η as follows:

$$\mathfrak{f}(\tau) = \zeta_{48}^{-1} \frac{\eta((\tau+1)/2)}{\eta(\tau)} \tag{7}$$

$$\mathfrak{f}_1(\tau) = \frac{\eta(\tau/2)}{\eta(\tau)} \tag{8}$$

$$\mathfrak{f}_2(\tau) = \sqrt{2}\frac{\eta(2\tau)}{\eta(\tau)} \tag{9}$$

The elliptic modular function j, the cube root γ_2 of j and the function γ_3 are connected to the Weber functions and to j via (cf. [22, §54]):

$$\gamma_2 = \frac{\mathfrak{f}^{24} - 16}{\mathfrak{f}^8} = \frac{\mathfrak{f}_1^{24} + 16}{\mathfrak{f}_1^8} = \frac{\mathfrak{f}_2^{24} + 16}{\mathfrak{f}_2^8} \tag{10}$$

$$\gamma_3 = \frac{(\mathfrak{f}^{24} + 8)(\mathfrak{f}_1^8 - \mathfrak{f}_2^8)}{\mathfrak{f}^8} \tag{11}$$

$$j = \gamma_2^3 = \gamma_3^2 + 1728 \tag{12}$$

The action of the generators of $SL_2(\mathbb{Z})$ on these functions is given by

$$
\begin{aligned}
\eta(\tau+1) &= \zeta_{24}\, \eta(\tau) & \eta(-1/\tau) &= \sqrt{-i\tau}\, \eta(\tau) \\
\mathfrak{f}(\tau+1) &= \zeta_{48}^{-1}\, \mathfrak{f}_1(\tau) & \mathfrak{f}(-1/\tau) &= \mathfrak{f}(\tau) \\
\mathfrak{f}_1(\tau+1) &= \zeta_{48}^{-1}\, \mathfrak{f}(\tau) & \mathfrak{f}_1(-1/\tau) &= \mathfrak{f}_2(\tau) \\
\mathfrak{f}_2(\tau+1) &= \zeta_{24}\, \mathfrak{f}_2(\tau) & \mathfrak{f}_2(-1/\tau) &= \mathfrak{f}_1(\tau) \\
\gamma_3(\tau+1) &= -\gamma_3(\tau) & \gamma_3(-1/\tau) &= -\gamma_3(\tau) \\
\gamma_2(\tau+1) &= \zeta_3^{-1}\, \gamma_2(\tau) & \gamma_2(-1/\tau) &= -\gamma_2(\tau) \\
j(\tau+1) &= j(\tau) & j(-1/\tau) &= j(\tau)
\end{aligned}
\tag{13}
$$

We also recall the useful relations

$$\mathfrak{f} \cdot \mathfrak{f}_1 \cdot \mathfrak{f}_2 = \sqrt{2} \tag{14}$$

$$\mathfrak{f}_1(2\tau) = \mathfrak{f} \cdot \mathfrak{f}_1(\tau) \tag{15}$$

and

$$\mathfrak{f}^8 = \mathfrak{f}_1^8 + \mathfrak{f}_2^8. \tag{16}$$

3.3 Reduced Class Equations

Let $u(z)$ denote a modular function and $w = w(\mathcal{O}) = \tau_{Q_0}$ be the generator of $\mathcal{O}$ such that $\mathcal{O} = \mathbb{Z} + w\mathbb{Z} = \mathbb{Z} + [\mathcal{O}_K : \mathcal{O}]w_K\mathbb{Z}$, where $w_K = w(\mathcal{O}_K)$. Weber calls $u(w)$ a *class invariant*, if $u(w) \in H_{\mathcal{O}} = K(j(w))$. We then write $W_\delta[u]$ for the minimal polynomial of $u(w)$ over $\mathbb{Q}$. Note that $W_\delta[j]$ is just the usual class equation corresponding to the discriminant δ. We define (cf. [22], [23])

$$\gamma_2^*(Q) = \zeta_3^{(a-c+a^2c)b}\gamma_2(\tau_Q) \tag{17}$$

$$\gamma_3^*(Q) = (-1)^{(a+c+ac)(b-1)/2}\gamma_3(\tau_Q) \tag{18}$$

$$
\mathfrak{f}^*(Q) = \begin{cases}
\zeta_{48}^{(a-c-ac^2)b}\mathfrak{f}(\tau_Q) & \text{if } 2 \mid a \text{ and } 2 \mid c \\
(-1)^{(\delta-1)/8}\, \zeta_{48}^{(a-c-ac^2)b}\mathfrak{f}_1(\tau_Q) & \text{if } 2 \mid a \text{ and } 2 \nmid c \\
(-1)^{(\delta-1)/8}\, \zeta_{48}^{(a-c+a^2c)b}\mathfrak{f}_2(\tau_Q) & \text{if } 2 \nmid a \text{ and } 2 \mid c
\end{cases}
\tag{19}
$$

In Table 1 we introduce some functions depending on various choices of u. In particular, Π_u denotes the (decimal) precision of our computations and is given by (22) below. Furthermore, ψ_u transforms a zero of $W_\delta[u]$ to a zero of $W_\delta[j]$.

Table 1. Choice of u

δ	u	u^*	$\psi_u(x)$	Π_u
$/$	j	j	x	Π_j
$2 \nmid \delta$	$\sqrt{\delta}\,\gamma_3$	$\sqrt{\delta}\,\gamma_3^*$	$x^2/\delta + 1728$	$(\Pi_j + h\log_{10}|\delta|)/2$
$3 \nmid \delta$	γ_2	γ_2^*	x^3	$\Pi_j/3$
$8\mid\delta-1,\,3\nmid\delta$	$(-1)^{(\delta-1)/8}\zeta_{48}\mathfrak{f}_2$	$\mathfrak{f}^*$	$(x^{24}-16)^3/x^{24}$	$1 + \Pi_j/47$

Theorem 9. *Let δ be the discriminant of an order $\mathcal{O}$ in $\mathbb{Q}(\sqrt{\delta})$ and choose δ, u, u^* and ψ_u according to Table 1. Then*

1. u^* *depends only on the* $SL_2(\mathbb{Z})$-*equivalence class of* Q, *so that we may write* $u^*([Q])$ *for* $u^*(Q)$.

2. $u^*([Q_0]) \in \mathbb{Q}(j([Q_0]))$ *where* $[Q_0]$ *is the class of the principal form* Q_0 *and hence is the identity in* $C(\delta)$.

3. $u^*(-[Q]) = \bar{u}^*([Q])$ *where* $\bar{}$ *denotes complex conjugation and* $-[Q]$ *stands for the inverse of* $[Q]$, *i.e.* $-[a,b,c] = [a,-b,c]$.

4. $$W_\delta[u](x) = \prod_{[Q]\in C(\delta)} (x - u^*([Q])) \tag{20}$$

5. $$W_\delta[u](x_0) = 0 \quad \Rightarrow \quad W_\delta[j](\psi_u(x_0)) = 0 \tag{21}$$

We summarize the steps required for computing $W_\delta[u]$:

1. Compute $C(\delta)$ by means of Theorem 8.
2. Compute approximations $\tilde{u}^*([Q])$ for the $h = h(\mathcal{O})$ values $u^*([Q])$. Apply Theorem 9.3 to speed up the computation.
3. Form the product (20) to obtain a polynomial $\tilde{W}_\delta[u] \in \mathbb{R}[x]$ corresponding to $\tilde{u}^*([Q])$.
4. $\tilde{W}_\delta[u]$ will be "close" to the desired reduced class equation $W_\delta[u] \in \mathbb{Z}[x]$, provided that the operations have been carried out with a sufficiently high precision Π_u.
5. Round $\tilde{W}_\delta[u]$ to obtain $W_\delta[u]$, i.e. round its coefficients.

The precision we use in the computations are given by

$$\Pi_j = 5 + h/4 + \frac{\pi\sqrt{-\delta}}{\ln 10} \sum_{[a,b,c]\in C(\delta)} a^{-1}. \tag{22}$$

Since u^* depends only on $[Q] \in C(\delta)$, we may represent the elements of $C(\delta)$ by the unique primitive reduced quadratic forms $Q = (a,b,c)$ of discriminant δ. Thus, we have $|b| \leqslant a \leqslant \sqrt{-\delta/3}$, and (6) converges at worst like a power series in $e^{-\pi\sqrt{3}}$.

4   A Defining Equation

The finite field $\mathbb{F}_q$ and the order of the elliptic curve E over $\mathbb{F}_q$ are connected by (2), giving us the imaginary quadratic field K. By Theorem 9.2, the class invariant u associated to the j-invariant of $\mathcal{E}$ by (21) is a primitive element of the ring class field $H_\mathcal{O}$ over K, that is,

$$H_\mathcal{O} = K(j(\mathcal{E})) = \left. K[x] \middle/ W_{\delta(\mathcal{O})}[u](x)\,K[x] \right. . \tag{23}$$

Note that, by the decomposition law for primes in an algebraic number field and by the algebraic properties of $H_\mathcal{O}$ and $\mathfrak{P} \subset H_\mathcal{O}$ (p splits completely in K and is relatively prime to the conductor of $\mathcal{O}$) it is easy to see that the reduced class equation $W_\delta[u]$ splits over $\mathbb{F}_p$ into irreducible factors of degree $f = f_{\mathfrak{P}|p}$. This is true because the $\mathfrak{P}$ in Theorem 4 has residue field $\mathcal{O}_{H_\mathcal{O}}/\mathfrak{P} \simeq \mathbb{F}_q$, where $q = p^f$. Observe also that, for each root x_0 of $W_\delta[u] \bmod \mathfrak{P}$, $\psi_u(x_0)$ yields the image of the j-invariant under the reduction modulo $\mathfrak{P}$ of an elliptic curve $\mathcal{E}$ over $H_\mathcal{O}$ with endomorphism ring $\operatorname{End}\mathcal{E} = \mathcal{O}$, and hence yields the j-invariant of an elliptic curve E over $\mathbb{F}_q$ with group order $N_K(1 - \zeta\pi)$ for $\zeta \in \mathcal{O}_K^*$. Therefore, the computation of a defining equation consists in the following steps:

1. Compute/choose δ and a prime p that splits completely in $K = \mathbb{Q}(\sqrt{\delta}\,)$.
2. Choose the class invariant u of Table 1 that requires the lowest precision Π_u.
3. Compute $W_\delta[u]$.
4. Find a root x_0 of $W_\delta[u]$ over $\mathbb{F}_q$.
5. Put $j_0 = \psi_u(x_0)$.
6. Compute an elliptic curve over $\mathbb{F}_q$ with j-invariant j_0 having the desired group order (4).

It remains to explain the last step. We need a relation between the j-invariant and the curve. Instead of using the usual (long) Weierstraß equation for solving this problem, we prefer to employ other normal forms which are more appropriate with respect to our problem.

Theorem 10.   *1. Let $p > 3$ be a prime and $j_0 \in \mathbb{F}_p$ be given. Then the elliptic curve over $\mathbb{F}_p$*

$$
\begin{aligned}
E : y^2 &= x^3 + 3\kappa x + 2\kappa && \text{with} && \kappa = \frac{j_0}{1728 - j_0} && \text{if} && j_0 \neq 0, 1728 \\
E : y^2 &= x^3 + ax && \text{with} && a \in \mathbb{F}_p^* && \text{if} && j_0 = 1728 \\
E : y^2 &= x^3 + b && \text{with} && b \in \mathbb{F}_p^* && \text{if} && j_0 = 0
\end{aligned}
\tag{24}
$$

has j-invariant j_0.

2. Let $\gamma \in \mathbb{F}_{2^n}$ be given with absolute trace $\operatorname{Tr}\gamma = 1$. Then a complete set of isomorphism classes of ordinary (which is equivalent to $j \neq 0$ in characteristic 2) elliptic curves over $\mathbb{F}_{2^n}$ is given by

$$y^2 + xy = x^3 + a_2 x^2 + j^{-1} \quad \text{with} \quad a_2 \in \{0, \gamma\}. \tag{25}$$

Unfortunately, by the reduction process of section 2, every isomorphism class of elliptic curves over H_O splits into several isomorphism classes of elliptic curves over $\mathbb{F}_q$. That is, an isomorphism class of elliptic curves over $\mathbb{F}_q$ is *not* uniquely determined by its j–invariant. The number of these isomorphism classes (for a fixed invariant) is given by the number $\#\mathcal{O}_K^*$ of units in K. We shall need the following

Theorem 11. *Let E and E' be two elliptic curves over $\mathbb{F}_q$. If E is ordinary, then E and E' are isomorphic iff $j(E) = j(E')$ and $\#E(\mathbb{F}_q) = \#E'(\mathbb{F}_q)$.*

Let the elliptic curve E over $\mathbb{F}_p$, $p > 3$, be given by a short Weierstraß equation

$$E : y^2 = x^3 + ax + b. \tag{26}$$

Define the c–twist $\hat{E}$ of E by

$$\hat{E} : y^2 = x^3 + \hat{a}x + \hat{b} \quad \text{with} \quad \hat{a} = ac^2 \text{ and } \hat{b} = bc^3 \tag{27}$$

for any fixed non-square $c \in \mathbb{F}_p^*$. In characteristic 2, we define the γ–twist of

$$E : y^2 + xy = x^3 + a_2 x^2 + a_6 \tag{28}$$

by

$$\hat{E} : y^2 + xy = x^3 + \hat{a}_2 x^2 + a_6 \quad \text{with} \quad \hat{a}_2 = a_2 + \gamma \text{ and } \mathrm{Tr}\,\gamma = 1. \tag{29}$$

We have (cf. [2])

Theorem 12. *Let E be an ordinary elliptic curve over $\mathbb{F}_q$ and $\hat{E}$ be a twist. Then*

1. $j(E) = j(\hat{E})$
2. $\#E(\mathbb{F}_q) + \#\hat{E}(\mathbb{F}_q) = 2q + 2$
3. $E(\mathbb{F}_{q^2}) \simeq \hat{E}(\mathbb{F}_{q^2})$

In the case of $\delta(K) < -4$, there are only two isomorphism classes. The defining equation of a curve satisfying (4) is then given by E or its twist $\hat{E}$, where $j(E) = j(\hat{E}) = j_0$. In any case, the right choice between E and $\hat{E}$ is made by trial and error for $p > 3$. If $p = 2$, we have (cf. [14])

Theorem 13. *Let E be an ordinary elliptic curve over $\mathbb{F}_{2^n}$ in the normal form (28). Then*

$$\#E(\mathbb{F}_{2^n}) \equiv 2\mathrm{Tr}\,a_2 \bmod 4. \tag{30}$$

5 Solving the Norm Equation

We want to compute all representations of a positive integer m as a norm in the imaginary quadratic field $K = \mathbb{Q}(\sqrt{\delta})$ of discriminant $\delta = \delta_K$. Equivalently, we must find the generators of all principal ideals of $\mathcal{O}_K$ with norm m. We write

$$m = \prod_{p|m} p^{e_p} \tag{31}$$

and

$$p\mathcal{O}_K = \begin{cases} \mathfrak{p}^2 \text{ with } \mathcal{N}(\mathfrak{p}) = p & \text{if } (\delta \mid p) = 0 \\ \mathfrak{p}\bar{\mathfrak{p}} \text{ with } \mathcal{N}(\mathfrak{p}) = p & \text{if } (\delta \mid p) = 1 \\ \mathfrak{p} \text{ with } \mathcal{N}(\mathfrak{p}) = p^2 & \text{if } (\delta \mid p) = -1 , \end{cases} \tag{32}$$

where $(\cdot \mid p)$ denotes the Legendre symbol for $p \neq 2$ and the Kronecker symbol for $p = 2$, i.e.

$$(\delta \mid 2) = \begin{cases} -1 & \text{if } \delta \equiv 5 \bmod 8 \\ 0 & \text{if } \delta \equiv 0 \bmod 4 \\ 1 & \text{if } \delta \equiv 1 \bmod 8 . \end{cases} \tag{33}$$

If e_p is odd for one p with $(\delta \mid p) = -1$, then there is no ideal of norm m in $\mathcal{O}_K$. Otherwise, the ideals $\mathfrak{a}$ of norm m are obviously given by

$$\mathfrak{a} = \prod_{\substack{p|p|m \\ (\delta|p)=0}} \mathfrak{p}^{e_p} \prod_{\substack{p|p|m \\ (\delta|p)=1}} \mathfrak{p}^{e_p - k_p} \bar{\mathfrak{p}}^{k_p} \prod_{\substack{p|p|m \\ (\delta|p)=-1}} \mathfrak{p}^{e_p/2} , \text{ with } 0 \leqslant k_p \leqslant e_p . \tag{34}$$

In order to decide whether or not an ideal $\mathfrak{a}$ is principal, we consider $\mathfrak{a}$ as a lattice in $\mathbb{C}$ and look for a minimal (with respect to the norm N_K) element π of $\mathfrak{a}\backslash\{0\}$. Since $\mathfrak{a}$ is principal iff $N_K(\pi) = \mathcal{N}(\mathfrak{a}) = m$ and π is unique up to the known units of K, we find all solutions to $m = N_K(\pi)$ with $\pi \in \mathcal{O}_K$.

To get an explicit representation of a prime ideal $\mathfrak{p}$, we use the decomposition law. The prime ideal $\mathfrak{p}$ is given (up to conjugation) by

$$\mathfrak{p} = \begin{cases} p\mathcal{O}_K & \text{if } (\delta \mid p) = -1 \\ p\mathcal{O}_K + (w_K - w_p)\mathcal{O}_K & \text{if } (\delta \mid p) \in \{0,1\} , \end{cases} \tag{35}$$

where $w_p \in \mathbb{Z}$ is any solution of $w_p \equiv w_K \bmod p$. Note that $\mathcal{O}_K = \mathbb{Z} + w_K\mathbb{Z}$. ¿From this it is easy to derive a $\mathbb{Z}$–basis for $\mathfrak{p}$. In particular, we have

$$\mathfrak{p} = p\mathbb{Z} + \frac{r + \sqrt{\delta}}{2}\mathbb{Z} \quad \text{with} \quad \begin{cases} r^2 \equiv \delta \bmod 4p & \text{if } p \neq 2 , (\delta \mid p) = 1 \\ r = p\delta & \text{if } p \neq 2 , (\delta \mid p) = 0 \\ r = \delta & \text{if } p = 2 , (\delta \mid p) = 1 \\ r = \delta/2 & \text{if } p = 2 , (\delta \mid p) = 0 . \end{cases} \tag{36}$$

6 Examples

We wish to construct an elliptic curve whose group has order $2q'$, where q' denotes the smallest prime greater than 10^{50}, i.e. $q' = 10^{50} + 151$. We want to find an imaginary quadratic field K which has class number as small as possible in order to reduce the running time. Starting with $h_K = 1$, we proceed until the norm equation $2q' = N_K(\pi)$ has an integral solution and $N_K(1-\pi)$ or $N_K(1+\pi)$ is a prime. We find $\delta_K = -5492$ and compute

$$\pi = 970666237845870942790637 + 27756708272407783694639 1 \cdot w_K$$
$$p = N_K(1 - \pi) = 2 \cdot 10^{50} - 19413324756917418855812431$$
$$j = 1750833118659272949909588776514668254512465143545756$$
$$a = 14848662876205547969080296514941739410907101516048$$
$$b = 9179194201470774410490417697865330623492717105946 \,.$$

Then for $E : y^2 = x^3 + ax + b$ we have $\#E(\mathbb{F}_p) = 2q'$. The running times using the computer algebra system SIMATH were as indicated in Table 2. The class number of K is $h_K = 18$. The required floating point precision to compute $W_{\delta_K}[\gamma_2]$ was $\Pi_{\gamma_2} = 126$. We used a dictionary that enabled us to get the fields K of given h_K.

Table 2. Running times for $E(\mathbb{F}_p)$ on a SPARC 2

Task	min	sec
find q'		8
prove the primality of q'		35
find K		50
compute $\mathcal{C}(\delta_K)$		1
compute $W_{\delta_K}[\gamma_2]$		4
factor $W_{\delta_K}[\gamma_2]$ over $\mathbb{F}_p$	4	30
compute a defining equation for E		7
prove the primality of p (using E)		3

The second example concerns the construction of an elliptic curve over $\mathbb{F}_{2^n}$ that fulfills the requirements of a cryptosystem. We chose n and proceed as follows. We look for an imaginary quadratic field K of class number n such that 2 splits completely in K, i.e. $\delta_K \equiv 1 \bmod 8$, and solve $2^n = N_K(\pi)$ for $\pi \in \mathcal{O}_K \backslash 2\mathcal{O}_K$. For $m = N_K(1-\pi)$ of the form $m = c \cdot q_*$ with q_* a prime (* will indicate the number of decimal digits of q_*) and $c \leqslant c_{\max} = 100$, we succeed. If we took K to have a divisor of n as class number, then m would not be of the desired form $m = c \cdot q_*$ because we then would have a representation $m = N_K(1 - \pi'^{n/h(K)})$ and m

could not have a large prime divisor q_*. We only look for K's with $\delta_K \not\equiv 0 \bmod 3$ which enables us to use the Yui-Zagier[3] reduced class equation $W = W_{\delta_K}[\mathfrak{f}^*]$. The polynomial W is clearly irreducible over $\mathbb{F}_2$ and we use it to generate $\mathbb{F}_{2^n}$. A root ϱ of $W \bmod 2$ over $\mathbb{F}_{2^n}$ is then trivially computed and from the last row of Table 1 we get $j(E) = \varrho^{48}$. Let

$$\mathbb{F}_{2^n} \simeq \mathbb{F}_2[X] \big/ W \, \mathbb{F}_2[X] \simeq \mathbb{F}_2(\varrho), \quad \text{where} \quad W(\varrho) = 0.$$

For

$$\alpha = \sum_{i=0}^{n-1} \alpha_i \varrho^i \in \mathbb{F}_2(\varrho), \qquad \alpha_i \in \mathbb{F}_2,$$

we list below the α_i in decreasing order omitting the ϱ^i. For $W \bmod 2 \in \mathbb{F}_2[X]$ we use the same notation. We set $h_K = n = 191$, find $\delta_K = -87887$ and compute

$$\pi = -3657017902810971643402695260 5 - 28548322031594433106650252 7 \cdot w_K$$

$$m = \#E(\mathbb{F}_{2^n}) = 313855086769334038191789471153040736677464234503289684871 2$$
$$= 2^3 \cdot 3 \cdot q_{58}$$

with a 58-digit prime cofactor q_{58} of m. The elliptic curve is

$$E : y^2 + xy = x^3 + \varrho^{-48}.$$

We have

$W \bmod 2 =$
$\qquad\qquad$ 11000111001101011 011010100010011100111 1010

10010111001001011101 00101 100111001100001010010 0101000 110010000100010000 1010111

101011011110110001 1011010 110010111011011000111 1110 1101010101001000101 000001

The group $E(\mathbb{F}_{2^n})$ is cyclic, generated by the point $P = (x_P, y_P)$ with coordinates

$x_P =$
$\qquad\qquad$ 1001010100011 101100111001101111 00100111

01111000101011100001 01101 0001110010101101100 011110 010101001101100011 0001110

00101000101010010101 11000 100101001101111001 1111011 010101100000010111 1101110

and

$y_P =$
$\qquad\qquad$ 110011111101101 0101010001110000010 011000

00010011000000000000 101111 1010110010101111110 10000 0 11011110101001011101 10011

10100000100100000100 10111 0000001100001011110 000000 1000100010110001000 010010

The leading ϱ-terms of x_P and y_P have respective degrees 187 and 189. It took 40 minutes to get an m with a large prime factor q_*, 60 seconds to show that

---

[3] We wish to thank Noriko Yui for making [23] available to us when she was visiting Saarbrücken in summer 1993.

q_* is actually a prime, 42 seconds to compute W and j and 80 seconds to find a generator P of E. In order to find a suitable m, i.e. an m which has a large prime factor q_* satisfying $m/q_* \leqslant c_{\max} = 100$, we had to search among $\nu = 6683$ different fields K. If we can benefit from a dictionary that gives us a list of δ_K for given h_K, then ν reduces to 11, the first running time to 6 seconds and the total running time to roughly 3 minutes. The algorithm also performs well for larger inputs.

Table 3. Data for $E(\mathbb{F}_{2^n})$

	$n = 191$	$n = 293$	$n = 300$	$n = 307$	$n = 311$
δ_K	-87887	-67559	-116087	-316759	-281959
h_K	191	293	300	307	311
$\Pi_{\mathfrak{f}^*}$	81	99	108	126	135
$c_{\max}$	100	100	100	100	100
q_*	q_{58}	q_{87}	q_{90}	q_{93}	q_{92}
$c = \#E(\mathbb{F}_{2^n})/q_*$	$2^3 \cdot 3$	$2 \cdot 3 \cdot 5$	2^2	2	$2 \cdot 5^2$
$\nu_n(\delta_K)$	10	2	22	24	23
$\nu_n(-1.6 \cdot 10^6)$	184	127	2216	145	128
memory in use (Kbyte)	130	295	307	330	330

In Table 3 and Table 4 we collect data and running times of the example just worked out together with $n = 300$ and another three $n \in \mathbb{P}$ in the size of about 300. We do not include the extensive outputs of these computations. The value $\nu_n(d)$ in Table 3 denotes the number of imaginary quadratic fields K' with $h_{K'} = n$ and $\delta_{K'} \geqslant d$.

Table 4. Running times for $E(\mathbb{F}_{2^n})$ on a SPARC 2

task	$n = 191$		$n = 293$		$n = 300$		$n = 307$		$n = 311$	
	min	sec	min	sec	min	sec	min	sec	min	sec
find K (with dictionary)		6		4		29		35		35
prove the primality of q_*	1	00	3	30	6	40	10	00	10	00
compute $W_{\delta_K}[\mathfrak{f}^*]$		42	1	40	2	00	2	40	3	00
compute a defining equation for E		1		1		1		1		1
find a generator P of E	1	20	3	30	4	00	4	40	6	00

The next example concerns the determination of a non-cyclic curve. We look for the smallest integer $n \geqslant 10^{50}$ for which there is an elliptic curve with $E \simeq n \times n$. We find $n = 10^{50} + 4$, $p = n^2 + n + 1$ and

$$E : y^2 = x^3 + b$$

where b has to satisfy $(4b)^{(p-1)/6} \equiv n+1 \bmod p$. Such a b is easily found by trial and error. It took about two minutes to work out this example.

If we want E to be of the form

$$E : y^2 = x^3 + ax \,,$$

then we take $n = 10^{50} + 206$, $p = n^2 + 1$ and $a^{(p-1)/4} \equiv -1 \bmod p$.

The times in Table 2 and Table 4 are slightly rounded up. All computations were done using the computer algebra system SIMATH [20].

References

1. H. Cohn: *Introduction to the construction of class fields.* Number 6 in Cambridge studies in advanced mathematics. Cambridge University Press, 1985.

2. D.A. Cox: *Primes of the form $x^2 + ny^2$.* John Wiley & Sons, New York 1989.

3. M. Deuring: *Die Klassenkörper der komplexen Multiplikation.* Enzyklopädie der mathematischen Wissenschaften mit Einschluß ihrer Anwendungen, Band 1, Heft 10, Teil 2. Teubner, Stuttgart 1958.

4. M. Deuring: *Die Typen der Multiplikatorenringe elliptischer Funktionenkörper.* Abhandlungen aus dem Mathematischen Seminar der Universität Hamburg **14** (1941), 197–272.

5. R. Fricke: *Lehrbuch der Algebra III.* Vieweg Braunschweig, 1928.

6. S. Goldwasser and J. Kilian: *Almost all primes can be quickly certified.* Proceedings of the 18th Annual ACM Symposium on Theory of Computation (STOC), Berkeley 1986, 316–329.

7. H. Hasse: *Abstrakte Begründung der komplexen Multiplikation und Riemannsche Vermutung in Funktionenkörpern.* Abhandlungen aus dem Mathematischen Seminar der Universität Hamburg **10** (1934), 235–348.

8. H. Hasse: *Beweis des Analogons der Riemannschen Vermutung für die Artinschen und F.K. Schmidtschen Kongruenzzetafunktionen in gewissen elliptischen Fällen.* Nachrichten der Akademie der Wissenschaften Göttingen, Mathematisch–Physikalische Klasse 1933, 253–262.

9. D. Husemöller: *Elliptic Curves.* Graduate Texts in Mathematics **111**, Springer, 1986.

10. E. Kaltofen and N. Yui: *Explicit construction of the Hilbert class fields of imaginary quadratic fields by integer lattice reduction.* Preliminary version 1989.

11. G.-J. Lay: *Konstruktion elliptischer Kurven mit vorgegebener Gruppenordnung über endlichen Primkörpern.* Diplomarbeit, Saarbrücken 1994.

12. A. Menezes, T. Okamoto and S. Vanstone: *Reducing elliptic curve logarithms to logarithms in a finite field.* preprint.

13. A. Menezes and S. Vanstone: *Isomorphism Classes of Elliptic Curves over Finite Fields.* Research Report 90–01, University of Waterloo 1990.

14. A. Menezes and S. Vanstone: *Isomorphism Classes of Elliptic Curves over Finite Fields of Characteristic 2.* Utilitas Mathematica **38** (1990), 135–153.

15. F. Morain: *Building Cyclic Elliptic Curves Modulo Large Primes.* Lecture Notes in Computer Science **547** (1991), 328–336.

16. F. Morain: *Courbes elliptiques et preuve de primalité.* Thèse. Lyon 1990.

17. F. Morain: *Implementation of the Goldwasser–Kilian–Atkin primality testing algorithm.* DRAFT 1988.

18. R. Scherz: *Die singulären Werte der Weberschen Funktionen* $\mathfrak{f}$, $\mathfrak{f}_1$, $\mathfrak{f}_2$, γ_2, γ_3. Journal für die reine und angewandte Mathematik **268–287** (1976), 46–74.

19. R. Schoof: *Elliptic curves over finite fields and the computation of square roots modulo p.* Mathematics of Computation **44** (1985), 483–494.

20. *SIMATH reference manual.* E–mail: simath@math.uni–sb.de

21. A.M. Spallek: *Konstruktion elliptischer Kurven über endlichen Körpern zu gegebener Punktegruppe.* Diplomarbeit, Essen 1993.

22. H. Weber: *Lehrbuch der Algebra III.* Chelsea Publishing Company, New York 1902.

23. N. Yui and D. Zagier: *Observation on the singular values of Weber modular functions.* Unpublished manuscript (1993).

Computing $\pi(x)$, $M(x)$ and $\Psi(x)$

Marc Deléglise and Joël Rivat

Université Lyon 1, Mathématiques, F-69622 Villeurbanne cedex, France

Computing $\pi(x)$

The ancient problem of computing $\pi(x)$ can be solved by at least three completely different methods:

- the sieve of *Eratosthenes* which finds all primes less than x and therefore cannot be achieved with less than about $\frac{x}{\log x}$ operations according to the Prime Number Theorem.
- the *Meissel-Lehmer* combinatorial method which uses sieve identities and is known to compute $\pi(x)$ in $O(\frac{x^{2/3}}{\log x})$ time and $O(x^{2/3} \log^2 x)$ space using the improvements introduced by *Lagarias*, *Miller* and *Odlyzko* [1].
- the *Lagarias-Odlyzko* analytic method [2], based on numerical integration of certain integral transforms of the *Riemann* ζ-function, for computing $\pi(x)$ using $O(x^{1/2+\epsilon})$ time and $O(x^{1/4+\epsilon})$ space for each $\epsilon > 0$.

We modified the combinatorial algorithm presented in [1] to compute $\pi(x)$ using $O(\frac{x^{2/3}}{\log^2 x})$ time and $O(x^{2/3} \log^3 x \log\log x)$ space. We then computed new large values of $\pi(x)$ up to 10^{18} using this method.

This work is submitted to Mathematics of Computation.

Computing $M(x)$ and $\Psi(x)$

The computation of $M(x) = \sum_{n \leq x} \mu(n)$ and $\Psi(x) = \sum_{n \leq x} \Lambda(n)$ appears to be much easier than the computation of $\pi(x)$. The method for computing $M(x)$ is based on the following identity:

$$M(x) = M(u) - \sum_{m \leq u} \mu(m) \sum_{\frac{u}{m} < n \leq \frac{x}{m}} M(\frac{x}{mn})$$

and can be achieved using $O(x^{2/3})$ time and $O(x^{1/3})$ space.

A similar identity is also available for computing $\Psi(x)$ with the same time and space complexity.

References

1. Lagarias, Miller, and Odlyzko. Computing $\pi(x)$: The Meissel Lehmer Method. *Mathematics of Computation*, 44(170):537–560, April 1985.
2. J.C. Lagarias and A.M. Odlyzko. Computing $\pi(x)$: An Analytic Method. *Journal of Algorithms*, 8:173–191, 1987.

On Some Applications of Finitely Generated Semi-Groups

Igor E. Shparlinski

School of MPCE, Macquarie University
NSW 2109, Australia
e-mail: igor@mpce.mq.edu.au

Abstract. Let V be a finitely generated multiplicative semi-group with r generators in the ring of integers $\mathbb{Z}_{\mathbb{K}}$ of an algebraic number field $\mathbb{K}$ of degree n over $\mathbb{Q}$. We use various bounds for character sums to obtain results on the distribution of the residues of elements of V modulo an integer ideal $\mathfrak{q}$. In the simplest case, when $\mathbb{K} = \mathbb{Q}$ and $r = 1$ this is a classical question on the distribution of residues of an exponential function, which may be interpreted as concerning the quality of the linear congruential pseudo-random number generator. Besides this well known application we consider several other problems from algebraic number theory, the theory of function fields over a finite field, complexity theory, cryptography, and coding theory where results on the distribution of some group V modulo $\mathfrak{q}$ play a central role.

1 Introduction

Let $\mathbb{K}$ be an algebraic number field of degree n over the field of rational numbers $\mathbb{Q}$, and let $\mathbb{Z}_{\mathbb{K}}$ be its ring of integers. For an integer ideal $\mathfrak{q}$, we denote by $\Lambda_{\mathfrak{q}}$ the residue ring modulo $\mathfrak{q}$ and by $\Lambda_{\mathfrak{q}}^*$ the multiplicative group of units of this ring. It is well known that $|\Lambda_{\mathfrak{q}}| = \mathrm{Nm}\,(\mathfrak{q})$ and $|\Lambda_{\mathfrak{q}}^*| = \phi(\mathfrak{q})$, where $\phi(\mathfrak{q})$ is the Euler function in $\mathbb{Z}_{\mathbb{K}}$.

Given a finitely generated multiplicative semi-group V of $\mathbb{Z}_{\mathbb{K}}$

$$V = \{\lambda_1^{x_1} \ldots \lambda_r^{x_r} \mid x_1, \ldots, x_r \geq 0\},$$

we denote its reduction modulo $\mathfrak{q}$ by $V_{\mathfrak{q}}$. We shall always suppose that the generators $\lambda_1, \ldots, \lambda_r$ are multiplicatively independent.

There are a great many results on the behavior of semi-groups V in $\mathbb{Z}_{\mathbb{K}}$ [8, 9]. Here we concentrate on their reductions $V_{\mathfrak{q}}$. In the simplest, but probably the most important case, when $\mathbb{K} = \mathbb{Q}$ and $r = 1$, this is a classical question about the distribution of residues of an exponential function equivalent to considering the quality of the linear congruential pseudo-random number generator [11, 21, 23]. We shall consider this and other applications which rely on not so widely known results concerning the distribution of $V_{\mathfrak{q}}$ in $\Lambda_{\mathfrak{q}}$. Such applications include:

Egami's question about minimal norm representatives of the residue classes modulo $\mathfrak{q}$ [5, 27];

The $1/M$ pseudo–random number generator of Blum, Blum and Shub [2];

Kodama's question about supersingular hyperelliptic curves [22, 32, 33];
Tompa's question about lower bounds for the QuickSort algorithm using a
congruential generator [31];
Lenstra's constants modulo q and Győry's arithmetical graphs [8, 9, 16, 24];
Estimating the dimension of BCH–codes [1, 18];
Niederreiter's problem about the multiplier of a linear congruential generator [21, 23].

Some of these questions deal only with the particular case $\mathbb{K} = \mathbb{Q}$, $r = 1$.
Moreover, the general case can often be reduced to this case.

It should not be a surprise that our main tool is various bounds for character sums. Specifically, given a principial additive character χ of the ring Λ_q, we set

$$\sigma_k(q, V) = \sum_{\alpha \in \Lambda_q^*} \Big[\sum_{v \in V_q} \chi(\alpha v)\Big]^k, \quad S(q, V) = \max_{\alpha \in \Lambda_q^*} \Big|\sum_{v \in V_q} \chi(\alpha v)\Big|,$$

noting that this definition does not of course depend on the choice of χ. Then for any integer ideal q prime to $\lambda_1 \ldots \lambda_r$,

$$S(q, V) \le \mathrm{Nm}\,(q)^{1/2}; \tag{1}$$

and for any prime ideal $\mathfrak{p}$ of first degree prime to $\lambda_1 \ldots \lambda_r$,

$$S(\mathfrak{p}, V) \le 2|V_{\mathfrak{p}}|^{5/12}\mathrm{Nm}\,(\mathfrak{p})^{1/4}. \tag{2}$$

These bounds are slight modifications and generalizations of bounds in [13]
and [29] respectively.

If V_q is sufficiently large the first bound is quite strong. In particular it is
non-trivial for $|V_q| > \mathrm{Nm}\,(q)^{1/2}$. Unfortunately, we know no more than that

$$|V_{\mathfrak{p}}| > \Big(\frac{\mathrm{Nm}\,(\mathfrak{p})}{\log\log \mathrm{Nm}\,(\mathfrak{p})}\Big)^{r/(r+1)} \tag{3}$$

for some set of prime ideals $\mathfrak{p}$ of asymptotic density 1 (for $\mathbb{K} = \mathbb{Q}$ it can be
improved, see [25]). This gives nothing if $r = 1$. For sets of almost all integer
ideals one obtains an even weaker result. Plainly, to get nontrivial results for
some infinite sequence of ideals we do need something nontrivial better than
the 'square-root' bound. The bound (2) provides such a result. It is nontrivial
for $|V_{\mathfrak{p}}| > 2^{12/7}\mathrm{Nm}\,(\mathfrak{p})^{3/7}$. This bound relies on results of [7] (for related results
see [14, 17, 30]).

The bounds (1) and (2) rely on estimates of the average values σ_k whereby
for any integer ideal q,

$$\sigma_2(q, V) = \mathrm{Nm}\,(q)|V_q|, \tag{4}$$

and for any prime ideal $\mathfrak{p}$ of first degree,

$$\sigma_4(\mathfrak{p}, V) \le 15\mathrm{Nm}\,(\mathfrak{p})|V_{\mathfrak{p}}|^{8/3}\mathrm{Nm}\,(\mathfrak{p})^{1/4}. \tag{5}$$

Indeed, for any $u \in V_q$

$$\sum_{v \in V_q} \chi(\alpha v) = \sum_{v \in V_q} \chi(\alpha u v),$$

so

$$|V_q|(S(q, V))^k \leq \sigma_k(q, V),$$

and after simple evaluations one can derive (1) and (2) from (4) and (5) respectively.

Now, in case when q is a large power k of a fixed prime ideal of first degree $\mathfrak{p}$, the papers [13] provides the following much stronger estimate

$$S(\mathfrak{p}^k, V) \leq c(\mathfrak{p}) \mathrm{Nm}\,(\mathfrak{p}^k) |V_{\mathfrak{p}^k}|^{-1} \tag{6}$$

where $c(\mathfrak{p})$ is a constant depending upon $\mathfrak{p}$ only. In [26], this bound was generalized on prime ideals $\mathfrak{p}$ of an arbitrary degree d as follows, for any $\varepsilon > 0$,

$$S(\mathfrak{p}^k, V) \leq c(\mathfrak{p}, \varepsilon) \mathrm{Nm}\,(\mathfrak{p}^k) |V_{\mathfrak{p}^k}|^{-1/d + \varepsilon}.$$

2 Representatives of Residue Classes

For $\alpha \in \Lambda_q$ denote by $N_q(\alpha)$ the minimal norm of all elements of α,

$$N_q(\alpha) = \min_{a \in \alpha} |\mathrm{Nm}\,(a)|,$$

and let $L(\mathbb{K}, q)$ and $A(\mathbb{K}, q)$ be the largest and average values of $N_q(\alpha)$ over all residue classes of Λ_q^*. That is, we set

$$L(\mathbb{K}, q) = \max_{\alpha \in \Lambda_q^*} N_q(\alpha), \quad A(\mathbb{K}, q) = \frac{1}{\phi(q)} \sum_{\alpha \in \Lambda_q^*} N_q(\alpha).$$

It is easy to see that $A(\mathbb{K}, q) \leq L(\mathbb{K}, q) = O(\mathrm{Nm}\,(q))$. Upper bounds for the implied constant in this estimate are obtained in [4] where some connections of this problem with the theory of diophantine approximation are displayed.

Let us also note that the inequality $L(\mathbb{K}, q) < \mathrm{Nm}\,(q)$ would mean that $\mathbb{K}$ is Euclidean with respect to its norm (that, is very untipical for algebraic number fields, see [16, 24],).

Now suppose that $\mathbb{K}$ has $r \geq 1$ principal units; that is, that $\mathbb{K}$ is neither the field of rationals $\mathbb{Q}$ nor an imaginary quadratic extension of $\mathbb{Q}$.

It is shown in [5] that $A(\mathbb{K}, p) = o(p^n)$ for almost all p in the sense of the asymptotic density of rational prime numbers p. From hereon we identify integer algebraic numbers and the corresponding principal ideals. Thus,

$$A(\mathbb{K}, q) = o(\mathrm{Nm}\,(q)) \tag{7}$$

for some infinite sequence of integer ideals q. In [5] this result is formulated in a slightly different but equivalent form.

Thus in fields with $r \geq 1$ the behavior of $A(\mathbb{K}, q)$ differs from that of $A(\mathbb{Q}, q)$ with integers $q \geq 2$, for there one has $A(\mathbb{Q}, q) = q/4 + O(1)$.

The results of [5] are improved in [27]. There it is shown that $A(\mathbb{K}, p) = O(p^n (\log p)^{-1/3})$ and $A(\mathbb{K}, p) = O(p^{n-1/6}(\log p)^{1/3})$ for all and almost all, respectively, rational prime numbers p. Thus in place of (7), we may get that there is an infinite sequence of integer ideals q with

$$A(\mathbb{K}, q) = O\left(\mathrm{Nm}\,(q)^{1-1/6n}(\log \mathrm{Nm}\,(q))^{1/3}\right). \tag{8}$$

In this paper we consider prime ideals $\mathfrak{p}$ and their powers $\mathfrak{p}^k$ instead of just principal ideals (p). This allows us to get much stronger bounds than (7) and (8). Moreover, for almost all prime ideals $\mathfrak{p}$ we obtain a nontrivial upper bound for $L(\mathbb{K}, \mathfrak{p})$ as well.

We begin by fixing an integral basis $\omega_1, \ldots, \omega_n$ of $\mathbb{Z}_\mathbb{K}$ over $\mathbb{Z}$ (see Theorem 2.5 of [20]), and for $h > 0$ we let $\mathfrak{B}_h$ denote the box

$$\mathfrak{B}_h = \{z = z_1\omega_1 + \ldots + z_n\omega_n \in \mathbb{Z}_\mathbb{K}; \; |z_i| \leq h, \; i = 1, \ldots, n\}.$$

Let U be the multiplicative group of units of $\mathbb{K}$ and denote by $R(U, q, h)$ the set of $\alpha \in \Lambda_q$ such that the congruence $\alpha u \equiv x - y \pmod{q}$, $u \in U_q$, $x, y \in \mathfrak{B}_h$, has no solution.

Further, let $H(\theta)$ be the height of $\theta \in \mathbb{Z}_\mathbb{K}$, that is, the maximal absolute value of coordinates in the representation of θ with respect to the basis $\omega_1, \ldots, \omega_n$. We have $|\mathrm{Nm}\,(\theta)| \leq (n\omega H(\theta))^n$, where ω is the maximum of the absolute values of $\omega_1, \ldots, \omega_n$ and of their conjugates over $\mathbb{Q}$. So, for $\alpha \in \Lambda_q \backslash R(U, q, h)$, $N_q\,(\alpha) \leq (n\omega h)^n$.

To get an upper bound for $L(\mathbb{K}, q)$ it is enough to find a nontrivial estimate for $H(U, q)$ that is the minimal h such that $R(U, q, h) = 0$; just as an upper bound of $|R(U, q, h)|$ is needed for estimating $A(\mathbb{K}, q)$. This reduces our original question to a question on the distribution of U_q.

To treat this problem, we use estimates for character sums ('individual' as well as 'in average'). In particular, we exploit the bounds (1), (2) and equation (4).

These bounds enable us to show that for any prime ideal $\mathfrak{p}$

$$|R(U, \mathfrak{p}, h)| = O\left(\mathrm{Nm}\,(\mathfrak{p})|U_\mathfrak{p}|^{-1}(1 + \mathrm{Nm}\,(\mathfrak{p})^2 h^{-2n})\right)$$

and

$$H(U, \mathfrak{p}) = O\left(\mathrm{Nm}\,(\mathfrak{p})^{3/2n}|U_\mathfrak{p}|^{-1/n}\right).$$

Accordingly,

$$A(\mathbb{K}, \mathfrak{p}) = O(\mathrm{Nm}\,(\mathfrak{p})|U_\mathfrak{p}|^{-1/2}), \qquad L(\mathbb{K}, \mathfrak{p}) = O(\mathrm{Nm}\,(\mathfrak{p})^{3/2}|U_\mathfrak{p}|^{-1}). \tag{9}$$

Also, for a prime ideal of first degree, we also have the upper bound

$$H(U, \mathfrak{p}) = O\left(\mathrm{Nm}\,(\mathfrak{p})^{5/4n}|U_\mathfrak{p}|^{-7/12n}\right).$$

Thus, in this case,

$$L(\mathbb{K}, \mathfrak{p}) = O(\mathrm{Nm}\,(\mathfrak{p})^{5/4}|U_{\mathfrak{p}}|^{-7/12}). \tag{10}$$

Finally, for a sufficiently large power $\mathfrak{p}^k$ of a fixed prime ideal $\mathfrak{p}$ of first degree, results of [13] (the bound (6) in particular) imply the very strong bound

$$H(U, \mathfrak{p}) = O(1). \tag{11}$$

Noting the following very simple bound $|U_{\mathfrak{q}}| \gg (\log \mathrm{Nm}\,(\mathfrak{q}))^r$, which holds for any integer ideal $\mathfrak{q}$, (9) yields

Theorem 1. $A(\mathbb{K}, \mathfrak{p}) = O\Big(Nm\,(\mathfrak{p})(\log Nm\,(\mathfrak{p}))^{-r/2}\Big).$

Recalling (3), from (9), (10) (and results on the distribution of prime ideals of first degree, for which see Corollary 2 of Proposition 7.10 of [20]) we also obtain

Theorem 2. *For a sequence of prime ideals $\mathfrak{p}$ of asymptotic density 1,*

$$L(\mathbb{K}, \mathfrak{p}) = \begin{cases} O\Big(Nm\,(\mathfrak{p})^{1/2+1/(r+1)}(\log\log Nm\,(\mathfrak{p}))^{r/(r+1)}\Big), & \text{if } r \geq 2, \\ O\Big(Nm\,(\mathfrak{p})^{23/24}(\log\log Nm\,(\mathfrak{p}))^{7/24}\Big), & \text{if } r = 1, \end{cases}$$

and

$$A(\mathbb{K}, \mathfrak{p}) = O\Big(Nm\,(\mathfrak{p})^{1/2+1/2(r+1)}(\log\log Nm\,(\mathfrak{p}))^{r/2(r+1)}\Big).$$

As $r \geq n/2 - 1$, we can rewrite our estimates in terms of n only. Finally, (11) implies

Theorem 3. *For any fixed prime ideal $\mathfrak{p}$ of first degree, $L(\mathbb{K}, \mathfrak{p}^k) = O(1)$.*

Now let us make a few comments.

First of all we note that perhaps an appropriate version of (9) (and therefore Theorem 1) is valid for arbitrary integer ideals $\mathfrak{q}$. Actually, some troubles arise with zero-divisors of $\Lambda_{\mathfrak{q}}$. The author hopes that it can be done nevertheless.

Then, let us stress links of problems considered here with an appropriate version of Artin's conjecture for algebraic number fields. Indeed, if $U_{\mathfrak{p}} = \Lambda_{\mathfrak{p}}^*$ for an infinite sequence of prime ideals $\mathfrak{p}$ (that can be considered as a modification of Artin's conjecture) then $L(\mathbb{K}, \mathfrak{p}) = 1$ for this sequence.

Note that quite simple elementary considerations give the lower bound

$$L(\mathbb{K}, \mathfrak{q}) \geq A(\mathbb{K}, \mathfrak{q}) \gg \mathrm{Nm}\,(\mathfrak{q})/|U_{\mathfrak{q}}|(\log\log \mathrm{Nm}\,(\mathfrak{q}))^{n-1}, \tag{12}$$

showing (together with (9) that the behavior of $L(\mathbb{K}, \mathfrak{q})$ and $A(\mathbb{K}, \mathfrak{q})$ does depend on the size of $U_{\mathfrak{q}}$. In particular, it follows from this bound that, for $r = 1$ (i.e. $n \leq 4$), $A(\mathbb{K}, \mathfrak{q})$ can be large enough. Indeed, let us consider the following sequence of principal ideals $\mathfrak{q}_k = (\rho^k - 1)$, $k = 1, 2, \ldots$, where ρ is a principal unit of $\mathbb{K}$. It is easy to prove that $|U_{q_k}| = O(k) = O(\log \mathrm{Nm}\,(q_k))$ for such ideals, hence

$$A(\mathbb{K}, q) \gg \mathrm{Nm}\,(q) / \log \mathrm{Nm}\,(q) \left(\log \log \mathrm{Nm}\,(q)\right)^n$$

for an infinite sequence of integer ideals.

The author does not know any other nontrivial general lower bounds for $L(\mathbb{K}, q)$ and $A(\mathbb{K}, q)$ (in terms of $\mathrm{Nm}\,(q)$ only) and any of them would be very interesting.

Of course there is a gap between (9) and (12), but (12) seems to be more precise.

Question 4. Does the bound $A(\mathbb{K}, q) = O\big(\mathrm{Nm}\,(q)|U_q|^{-1+\delta}\big)$ hold for any integer ideal q and any $\delta > 0$ (or at least for a prime ideal $\mathfrak{p}$) ?

As we have mentioned, we cannot hope to get even the following "weak" estimate $L(\mathbb{K}, q) < \mathrm{Nm}\,(q)$ for all integer ideals but what can we say about almost all integer ideals.

Question 5. Does the bound $L(\mathbb{K}, q) = o\big(\mathrm{Nm}\,(q)\big)$ hold for almost all integer ideal q ?

If the answer is "Yes", it would mean that any algebraic number field is "almost Euclidean" and moreover the analog of Euclid's algorithm would use a sub-logarithmic number of steps (at least for the majority of inputs).

The following question is a generalization of the previous two ones and seems to be realy hard.

Question 6. What is the exact behavior of $A(\mathbb{K}, q)$ and $L(\mathbb{K}, q)$ for "almost all" integer ideals q ?

Finally we mention that if we are really going to apply these results to Euclid's algorithm then for any (or "almost any") integer ideal q and any (or "almost any") $\alpha \in \Lambda_q^*$ we have to know how to compute $a \in \alpha$ with $|\mathrm{Nm}\,(a)| = N_q(\alpha)$.

3 The $1/M$-Pseudo-Random Number Generator

Let $g \geq 2$ be a fixed integer and let M be an integer positive number that can be written with at most L g-adic digits (that is, $M < g^L$).

It is proved in [2] that given $k = 2L + 3$ successive digits of the g-adic expansion of $1/M$ one can find M in polynomial time $L^{O(1)}$.

Then, once again in [2], we find that, under Artin's conjecture, $k = L - 1$ digits are not enough to determine M unambiguously.

Here we prove the weaker but unconditional statement that, for an arbitrary $\varepsilon > 0$, any string of $k = \lfloor (1/24 - \varepsilon)L \rfloor$ consecutive digits provides no information about M. Roughly speaking, we see without any unproven conjectures that M may take almost any value among all prime numbers $p < g^L$.

Theorem 7. *Given any string of $k = \lfloor(1/24 - \varepsilon)L\rfloor$ consecutive g-digits, there are at least $(1+o(1))\pi(g^L)$ prime numbers $p < g^L$ such that the g-adic expansion of $1/p$ contains the string.*

The main tool of the proof is the bound (2) (see Section 9.2 of [30]).

Further, using results of [13] (the bound (6) in particular) one can show (see Section 9.2 of [30]) that for an arbitrary $\varepsilon > 0$ any string of $k \leq (1-\varepsilon)L$ consecutive digits appears in the g-adic expansion of $1/M$ for at least $C(g)g^{\varepsilon L/2}$ values of $M < g^L$. Here $C(g)$ is some constant depending on g only.

Theorem 8. *There is a constant $c(g) > 0$, depending only on g, such that for every element M of the set $\mathfrak{M} = \{M = p^\alpha\mu \mid 1 \leq \mu \leq Q, (\mu, g) = 1\}$, where $Q = c(g)g^{\varepsilon L/2}$, p is the smallest odd prime number with $(p, g) = 1$, and $\alpha = \lfloor\log(g^l/Q)/\log p\rfloor$, and any string of $k = \lfloor(1-\varepsilon)L\rfloor$ consecutive g-digits, the g-adic expansion of $1/M$ contains the string.*

Since the set $\mathfrak{M}$ is exponentially large, this result means that even as many as $k = \lfloor(1-\varepsilon)L\rfloor$ digits give us very little information on the possible values of M. Applications of results of this type to cryptography are pointed out in [2].

4 Supersingular Hyperelliptic Curves

Let $\mathbb{P}$ be the set of prime numbers and for $l \in \mathbb{P}$ let $T(l)$ be the largest t with the property that there exists some integer g, $\gcd(g, l) = 1$, having the exponent t modulo l such that all the smallest positive residues of g^x, with $x = 1, \ldots, t$, belong to the interval $[1, (l-1)/2]$.

The matter is related to the question considered in [32, 33] on supersingular hyperelliptic curves of the form $y^2 = x^l + a$, $a \neq 0$, over $\mathbb{F}_p$ with l an odd prime. If this curve is supersingular then the exponent p modulo l does not exceed $T(l)$.

The function $T(l)$ is considered in [22] for the first time. There the upper bound of the shape $T(l) = O(l^{1/2} \log l)$ is stated. This result is improved in [29, 30] as follows:

Theorem 9. $T(l) \leq 100 l^{3/7}$.

Of course one can get a much better constant than 100.

Using Theorem 9, it is easy to show that the density $\delta(l)$ of primes p for which the corresponding curve is suppersingular over $\mathbb{F}_p$ is small.

Theorem 10. $\delta(l) = O(l^{-4/7+\varepsilon})$.

More precisely,

$$\delta(l) = (l-1)^{-1} \sum_{\substack{d|l-1,\\ d \leq T(l)}} \phi(d) = O(T(l)l^{-1+\varepsilon})$$

where $\phi(d)$ is the Euler function. In fact, the previous bound from [22] also produces a non-trivial estimate. On the other hand, for the following dual question it cannot give anything because we do need something below the 'square-root' bound for $T(l)$).

Given some prime p, let $N_p(L)$ be the number of primes $l \leq L$ which correspond to supersingular curves. Then

Theorem 11. $N_p(L) = O(L^{6/7} \log p)$.

Indeed, let $\tau(L) = \max\{T(l) \mid l \in \mathbb{P},\ 3 \leq l \leq L\}$. Then all the l with $T(l) \leq \tau(L)$ must divide the product

$$W_p(L) = \prod_{t=1}^{\tau(L)} (p^t - 1) = \exp\big(O(\tau(L)^2 \log p)\big).$$

On applying a known upper bound for the number of the prime divisors of an integer we get the theorem.

In particular, the curve $y^2 = x^l + a$, $a \in \mathbb{F}_p^*$ is not supersingular over $\mathbb{F}_p$ for almost all prime numbers l in the segment $[1, L]$ with

$$L > (\log p)^7 (\log \log p)^8.$$

This is plain because $N_p(L) = o(\pi(L))$ for such values of L.

5 Distribution of Powers of Primitive Roots

Let g be a fixed primitive root modulo p, and let $T(N, M, H)$ be the number of solutions of the congruence $g^x \equiv M + u \pmod{p}$ where $x = 1, \ldots, N$, $u = 0, \ldots, H - 1$.

It is proved in [19] that $T(N, M, H) = NH/p + O(p^{1/2} \log^2 p)$. Indeed this bound had been known for a long time (see [13, 14, 17, 21, 23, 30]). But, in [19], a new method providing a new upper bound for average value

$$R(N, H) = p^{-1} \sum_{M=0}^{p-1} \big(T(N, M, H) - NH/p\big)^2,$$

is proposed. These results are motivated by applications to some sorting algorithms [10, 31]. For example, if $HN \sim p$ (indeed, for the application mentioned we need only such pairs N and H) and $p^{5/7+\varepsilon} \leq N \leq p - 1$, the bound can be presented in the nice form

$$R(N, H) = O(NH/p). \tag{13}$$

Also, it is mentioned that one may expect the same bound for $p^\varepsilon \leq N \leq p - 1$. Of course this bound is much stronger than the considerably more easily proved bound

$$R(N, H) = O(Hp^\varepsilon) \tag{14}$$

holding for every $N \leq p - 1$ and $H \leq p$.

Actually, only a good lower bound is needed for the number $I(N, H)$ of i with $0 \leq i \leq p/H - 1$ such that $T(N, iH, H) > 0$. Of course, trivially,

$$\min\{p/H, N\} \geq I(N, H) \geq \min\{N/H, 1\}. \tag{15}$$

We shall show how bounds for the average value $R(N, H)$, like the two ones above, allow us to get better lower bounds for $I(N, H)$.

First of all, we mention the asymptotic result

$$I(N, H) = \frac{p}{H}\left(1 + O(p^2 R(N, \lfloor H/2 \rfloor)N^{-2}H^{-2})\right)$$

communicated to the author by Hugh Montgomery.

Indeed, since $T(N, iN, H) = 0$ for at least $p/H - 1 - I(N, H)$ values of i with $0 \leq i \leq p/H - 1$, then evidently $T(N, M, \lfloor H/2 \rfloor) = 0$ for at least $0.5(p/H - 1 - I(N, H))H$ values of M. Hence

$$0.5(p/H - 1 - I(N, H))H(NH/p)^2 \leq pR(N, \lfloor H/2 \rfloor).$$

Taking into account the trivial upper bound $I(N, H) \leq p/H$, we get the desired result.

Substituting the bounds (14) gives the asymptotic result $I(N, H) \sim p/H$ for $N^2 H \geq p^{2+\varepsilon}$. Unfortunately, when $NH \sim p$ even (13) does not give anything.

We shall now obtain another lower bound for $I(N, H)$ in terms of $R(N, H)$ which is nontrivial in a wider range of parameters (including $NH \sim p$). This bound was communicated to author by Sergey Konyagin.

For simplicity, we suppose that H is even.

Let $I(N, H, t)$ be the number of i, $0 \leq i \leq p/H - 1$ with $T(N, iH, H) = t$. If $I(N, H) < pH^{-1}/5$ then

$$W_1 = \sum_{t > 4NH/p} tI(N, H, t) \geq N - 4NH/p \sum_{0 < t \leq 4NH/p} I(N, H, t)$$
$$\geq N - 4NHI(N, H)/p > N/5.$$

If $T(N, iH, H) = t$ then either $T(N, iH, H/2) \geq t/2$ or $T(N, iH + H/2, H/2) > t/2$. In the first case we set $\Delta_i = [iH - H/2, iH - 1]$, in the second case, $\Delta_i = [iH, iH + H/2 - 1]$. Note that the intervals Δ_i do not intersect each other. Further, for $T(N, iH, H) = t > 4NH/p$ and $M \in \Delta_i$ we have

$$T(N, M, H) - NH/p \geq t/2 - NH/p > t/4.$$

Summing over all $t > 4NH/p$ and $M \in \Delta_i$ with $T(N, iH, H) = t$, we get

$$W_2 = \sum_{t > 4NH/p} t^2 I(N, H, t) \leq 32R(N, H)p/H.$$

Therefore

$$I(N, H) \geq \sum_{t > 4NH/p} I(N, H, t) \geq W_1^2/W_2 \gg N^2 H R(N, H)^{-1} p^{-1}.$$

Thus, in any case, $I(N, H) \gg \min(p/H, N^2 HR(N, H)^{-1}p^{-1})$. Now, from (13) and (14) we get

$$I(N, H) \gg \begin{cases} pH^{-1}, & \text{if } N \geq p^{5/7+\varepsilon}, \ NH \sim p; \\ \min\{N^2 p^{-1-\varepsilon}, pH^{-1}\}, & \text{if } 1 \leq N \leq p-1, \ 1 \leq H \leq p. \end{cases}$$

The first bound is best possible and improves the trivial lower bound (15) for all allowed values of N and H. The second bound does the same if $NH > p^{1+\varepsilon}$.

For applications of these bounds to the sorting algorithm of Theorem 4 of [10] (or of Theorem 12.1 of [31]) we have to consider the case $NH \sim p$. So the second bound cannot be useful, but the first one allows us to get an improvement.

Also, Sergey Konyagin proposed a different method to deal with $I(N, H)$ that for any integer $k \geq 2$ gives the estimate

$$I(N, H) \gg Np^{-\varepsilon} \frac{\min\{p^{1/k}H^{-1}, 1\}}{\min\{N^{1/k}, N^{1/k}Hp^{-1/k} + p^{1/2k}\}}.$$

In many cases, it is better then the previous estimates. Say if again $NH \sim p$ then it improves the trivial bound (15) (and therefore gives an improvement of the mentioned sorting algorithm) for $p^{1/3+\varepsilon} \leq N \leq p^{3/4+\varepsilon}$ (we use it for $k = 2$ and $k = 3$ only). Thus in this case we have an improvement for all $N \geq p^{1/3+\varepsilon}$.

Theorem 12.

$$I(N, \lfloor p/N \rfloor) \gg \begin{cases} N, & \text{if } N \geq p^{5/7+\varepsilon}; \\ N^{5/3}p^{-2/3-\varepsilon}, & \text{if } p^{5/8} \leq N < p^{5/7+\varepsilon}; \\ Np^{-1/4-\varepsilon}, & \text{if } p^{1/2} \leq N < p^{5/8}; \\ N^{3/2}p^{-1/2-\varepsilon}, & \text{if } p^{1/3+\varepsilon} \leq N < p^{1/2}. \end{cases}$$

6   Difference Sets in $V_{\mathfrak{p}}$

Let $M(\mathbb{K})$ denote the largest integer M such that there exists a set $\Omega = \{\omega_1, \ldots, \omega_M\} \subset \mathbb{Z}_{\mathbb{K}}$ with $\omega_i - \omega_j \in U$, $1 \leq i < j \leq M$, where U is the unit group of $\mathbb{K}$. If $M(\mathbb{K})$ is large enough then $\mathbb{K}$ is Euclidean, [16]. Some upper and lower bounds for $M(\mathbb{K})$ were obtained in [16] too.

The same functions $M(\mathbb{K}, V)$ can be defined with respect to any finitely generated multiplicative semigroup V of $\mathbb{Z}_{\mathbb{K}}$. But in this more general setting it is not quite clear how to apply the method of [16] (see also [24]) to obtain upper and lower bounds for $M(\mathbb{K}, V)$.

Here we consider an analog of this function modulo a prime ideal $\mathfrak{p}$. Define $M(\mathbb{K}, V, \mathfrak{p})$ as the largest integer M such that there exists a set $\Omega = \{\omega_1, \ldots, \omega_M\} \subset \Lambda_{\mathfrak{p}}$ with

$$\omega_i - \omega_j \in V_{\mathfrak{p}}, \quad 1 \leq i < j \leq M. \tag{16}$$

We will obtain an upper bound for $M(\mathbb{K}, V, \mathfrak{p})$. Using this bound and the trivial inequality $M(\mathbb{K}, V) \leq M(\mathbb{K}, V, \mathfrak{p})$, for an appropriate prime ideal $\mathfrak{p}$, we get an upper bound for $M(\mathbb{K}, V)$.

Theorem 13. *Let rank $V \geq 1$. If $V_\mathfrak{p} = \Lambda_\mathfrak{p}^*$ then $M(\mathbb{K}, \mathfrak{p}) = Nm(\mathfrak{p})$. Otherwise*

$$M(\mathbb{K}, V, \mathfrak{p}) \leq \begin{cases} 2Nm(\mathfrak{p})^{1/2} + 2, & \text{for any prime ideal } \mathfrak{p}, \\ 6|V_\mathfrak{p}|^{5/12} Nm(\mathfrak{p})^{1/4} + 2, & \text{for a prime ideal } \mathfrak{p} \text{ of first degree.} \end{cases}$$

Proof. The statement is trivial if $V_\mathfrak{p} = \Lambda_\mathfrak{p}^*$. In the other case we have $|V_\mathfrak{p}| \leq |Nm(\mathfrak{p})|/2$, because $V_\mathfrak{p}$ is a subgroup of $\Lambda_\mathfrak{p}^*$. Furthermore, suppose that for some set $\Omega = \{\omega_1, \ldots, \omega_M\} \subset \Lambda_\mathfrak{p}$ we have (16). Then the congruence $\omega_i - \omega_j \equiv u$ (mod $\mathfrak{p}$), $u \in V_\mathfrak{p}$, $1 \leq i, j \leq M$, has exactly $N = M(M-1)$ solutions.

On the other hand, from (1) and (2) one sees that

$$|N - M^2|V_\mathfrak{p}|/Nm(\mathfrak{p})| \leq \begin{cases} M Nm(\mathfrak{p})^{1/2}, & \text{for any } \mathfrak{p}, \\ 3M|V_\mathfrak{p}|^{5/12} Nm(\mathfrak{p})^{1/4}, & \text{if } \mathfrak{p} \text{ is of first degree,} \end{cases}$$

and the result follows. $\qquad\square$

More generally, for a set $\Omega = \{\omega_1, \ldots, \omega_M\} \subset \mathbb{Z}_\mathbb{K}$ suppose we consider the directed graphs $G(\Omega, V)$ and $G(\Omega, V, \mathfrak{p})$ which have M vertices labeled by $\omega_1, \ldots, \omega_M$, such that vertices ω_i and ω_j are connected iff $\omega_i - \omega_j \in V$ and $\omega_i - \omega_j \in V_\mathfrak{p}$, respectively. Thus $M(\mathbb{K}, V)$ and $M(\mathbb{K}, V, \mathfrak{p})$ are just the maximal values of M such that there exist sets Ω for which the graphs $G(\Omega, V)$ and $G(\Omega, V, \mathfrak{p})$ are each M-vertex complete graphs. Various other problems concerning graphs $G(\Omega, V)$ are considered in [8, 9].

Question 14. Generalize results of [8, 9] on graphs $G(\Omega, V, \mathfrak{p})$.

7 Dimension of BCH Codes

For integers n and d with $1 < d \leq n$ and a prime power q with $\gcd(n, q) = 1$, denote by $J(q, n, d)$ the maximal dimension of q-ary generalized BCH codes of length n and of designed distance d (see [1, 18]).

We do not give an exact definition of $J(q, n, d)$ (referring to [1, 18]) and note only that for some integer b, $J(q, n, d)$ equals the number of $j = 0, 1, \ldots, n-1$ for which the congruence $jq^x \equiv b + y$ (mod n), $x = 1, \ldots, t$, $y = 1, \ldots, d$, is unsolvable. Here t is the exponent of q modulo n.

We show that the bound $J(q, n, d) \leq 3n^3/(d-1)^2 t^{1/2}$, stated in [28], can be slightly improved.

Theorem 15. $J(q, n, d) \leq 24n^5/d^4 t$.

Proof. Let $D = \lfloor d \rfloor$ and let R_j denote the number of solutions of the congruence

$$jq^x \equiv b + D + u - v \pmod{n}, \quad x = 1, \ldots, t, \ u, v = 1, \ldots, D.$$

It is evident that $J(q, n, d) \leq |\mathfrak{J}(q, n, d)|$ where $\mathfrak{J}(q, n, d)$ is the set of $j = 0, 1, \ldots, n-1$ for which this congruence is unsolvable, i.e. $R_j = 0$.

Set

$$S(a) = \sum_{x=1}^{t} \exp(2\pi i a q^x/n), \quad V(a) = \sum_{u=1}^{D} \exp(2\pi i a u/n).$$

We have

$$\sum_{j=0}^{n-1} |S(aj)|^2 \le nt \gcd(a,n), \qquad \sum_{a=1}^{n/\delta-1} |V(a\delta)|^2 \le n^2/4\delta^2 \qquad (17)$$

for any $\delta \,|\, n$ (see Lemma 4 and Lemma 5 of [28], respectively). Clearly,

$$R_j = n^{-1} \sum_{a=0}^{n-1} S(aj)|V(a)|^2 \exp(2\pi i a(b+D)/p).$$

Separating the term tD^2/n corresponding $a = 0$ and summing over all $j \in \mathfrak{I}(q,n,d)$ we obtain from (17)

$$|\mathfrak{I}(q,n,d)|tD^2/n \le n^{-1} \sum_{j \in \mathfrak{I}(q,n,d)} \sum_{a=1}^{n-1} |S(aj)||V(a)|^2$$

$$= n^{-1} \sum_{\delta|n} \sum_{\gcd(a,n)=\delta} |V(a)|^2 \sum_{j \in \mathfrak{I}(q,n,d)} |S(aj)| \le \sum_{\delta|n} |\mathfrak{I}(q,n,d)|^{1/2}(nt\delta)^{1/2}n^2/4\delta^2$$

$$= n^{3/2}t^{1/2}|\mathfrak{I}(q,n,d)|^{1/2} \sum_{\delta|n} \delta^{-3/2} \le \zeta(3/2)n^{3/2}t^{1/2}|\mathfrak{I}(q,n,d)|^{1/2},$$

where $\zeta(s)$ is the Riemann Zeta-function. Hence, $|\mathfrak{I}(q,n,d)| \le \zeta(3/2)n^5/D^4t$. Since $D \ge (d-1)/2$ and $\zeta(3/2) < 1.5$, we get the result. $\qquad \square$

Since $q^t \ge n+1$ then $t > \log_q n$ and for a fixed q and linearly growing designed distance $\liminf_{n\to\infty} d/n > 0$ we obtain $J(q,n,d) = O(n/\log n)$. Earlier such bounds were known for primitive BCH-codes only [1, 18].

Question 16. Obtain an analog of the results above in terms of the actual code distance rather then the designed distance.

8 Optimal Coefficients and Congruential Pseudo-Random Numbers

For integers $a_1, \ldots, a_s, M$ we set

$$\rho(a_1,\ldots,a_s;M) = \min \overline{m}_1 \ldots \overline{m}_s, \qquad \omega(a_1,\ldots,a_s;M) = \min(m_1^2+\ldots+m_s^2)^{1/2},$$

where the minima are taken over all nontrivial solutions of the congruence

$$a_1 m_1 + \ldots + a_s m_s \equiv 0 \pmod{M},$$

and where $\overline{m} = \max(1, |m|)$.

The integers $a_1, \ldots, a_s$ are called optimal coefficients modulo M (and can be used for building up formulas for approximate integration) if and only if $\rho(a_1, \ldots, a_s; M)$ is large enough; say, of order $M^{1-\varepsilon}$ [12, 21, 23]. It is also known [12, 21, 23] that 'almost all' vectors $(a_1, \ldots, a_s)$ are optimal coefficients for a prime modulus $M = p$ with

$$\rho(a_1, \ldots, a_s; M) \gg M / \log^{s-1} M.$$

Moreover, it is easy to prove that an analogous result holds for arbitrary composite M. Unfortunately, these results are nonconstructive. Some algorithms for finding optimal coefficients are presented in [12]. The best for a special M has computing time $O(M^{4/3+\varepsilon})$. Recently [14], an algorithm with computing time $O(M^{1+\varepsilon})$ was given for $M = 2^m$; the algorithm uses Hensel lifting. A different algorithm, but with the same computing time, is designed in [3]. These papers deal with arithmetical complexity, but it is not difficult to obtain the estimates given above for the Boolean-complexity.

For the case $s = 2$ we can set $a_1 = 1$; $a_2 = F_k$; $M = F_{k+1}$, where $k = 1, 2, \ldots$, and $\{F_k\}$ is the Fibonacci sequence (see [12, 21]).

The theory of diophantine approximation in $\mathbb{R}$ allows one to obtain generalizations of these results [21]. It produces $a_1, \ldots, a_s$ and M with

$$\rho(a_1, \ldots, a_s; M) \gg M^{s/2(s-1)}.$$

We shall consider linear congruential pseudo-random number generators, that is, sequences of rational numbers $\alpha_n = u_n / M$, where

$$u_n \equiv \lambda u_{n-1} \pmod{M}, \qquad 0 < u_n < M, \qquad n = 1, 2, \ldots.$$

Here the *initial value* $u_0 = a$ and the *multiplier* λ are integers with $(a\lambda, M) = 1$.

We set $r_s(\lambda; M) = \rho(1, \lambda, \ldots, \lambda^{s-1}; M)$, $w_s(\lambda; M) = \omega(1, \lambda, \ldots, \lambda^{s-1}; M)$.

It is known that the s-dimensional discrepancy of the sequence α_n depends drastically on $r_s(\lambda; M)$, and that its lattice structure depends on $w_s(\lambda; M)$.

In the case $M = p \in \mathbb{P}$ it is shown in [12] that there exists a λ with

$$r_s(\lambda, p) \gg p / \log^{s-1} p.$$

The method for obtaining this estimate is based on a bound for the number of zeros of polynomials modulo p and does not generalize to an arbitrary modulus M. In general we can do no better than to get λ with $r_s(\lambda, M) \gg M^{1/s}$.

The case $s = 2$ can be dealt with by using the results on 2-dimensional optimal coefficients (we may set $\lambda \equiv a_1/a_2 \pmod{M}$).

The first nontrivial case is $s = 3$. This case is considered in [15] for fixed prime power moduli M (these are the most interesting moduli).It is shown that for $M = 2^m$ there exists a $\lambda \equiv 5 \pmod{8}$ with $r_3(\lambda, M) \gg M / \log^2 M$.

In the general case, if we define λ by the congruence $\lambda \tau \equiv \vartheta \pmod{M}$, with $(\vartheta \tau, M) = 1$ and $\vartheta \sim \tau \sim M^{1/(s+1)}$, then

$$r_s(\lambda, M) \gg M^{2/(s+1)}. \tag{18}$$

If in the previous construction one takes $\vartheta \sim \tau \sim (M/2s)^{1/s}$ then

$$w_s(\lambda, M) \geq 2^{1/2}(2s)^{-1/2s} M^{1/s} + o(M^{1/s}).$$

This bound meets the upper bound $w_s(\lambda, M) \leq \gamma_s M^{1/s}$, where γ_s is the Hermite constant (see Section 3.3.4 of [11]).

Similar problems also arise in cryptography. Let $F_s(\mu)$ be the size of the set of λ, $1 \leq \lambda \leq M$, with $w_s(\lambda, M) \leq M^\mu$. Amongst other very interesting and important results, the bound $F_3(\mu) = O(M^{1/2+3\mu/2+\varepsilon})$ is proved for $s = 3$ in [6].

For the case most important for applications, namely $M = 2^m$, $m = 1, 2, \ldots$, the bound (18) can be improved as follows (see Section 9.1 of [30]).

Theorem 17. *For $M = 2^m$, set $\lambda \equiv (2^r + 3)(2^r + 1)^{-1}$ (mod 2^m), where $r = 2 \lfloor m/(2s+1) \rfloor$. Then $r_s(\lambda, M) \gg M^{4/(2s+1)}$.*

It seems that methods from the theory of diophantine approximation in p-adic fields can be applied to this problem. Indeed, let $M = p^m$. Then, in the p-adic metric, $\left| m_1 + m_2\lambda + \ldots + m_s\lambda^{s-1} \right| \geq p^{-m}$ for $\overline{m}_1, \ldots, \overline{m}_s < r_s(\lambda, M)$. Roughly speaking, this means that the λ we are looking for should not be well p-adically approximated by roots of polynomials with small integer coefficients.

Acknowledgment. The author is grateful to Alf van der Poorten for much valuable advice and for the final translation from the author's 'Russian–English'. The author also wishes to thank Sergey Konyagin, Hugh Montgomery and Martin Tompa for helpful discussion of the problems considered in Section 5.

References

1. E. R. Berlekamp, *Algebraic Coding Theory*, McGraw-Hill, NY, 1968.
2. L. Blum, M. Blum and M. Shub, 'A Simple Unpredictable Pseudo-Random Number Generator', *SIAM J. Comp.* , **15** (1986), p.364–383.
3. L. P. Bocharova, V. S. Van'kova and N. M. Dobrovolski, 'On the Computation of Optimal Coefficients', *Matem. Zametki*, **49** (1991), 23–28 (in Russian).
4. H. Davenport, 'Linear Forms Associated with an Algebraic Number Field', *Quart Journ. Math.*, **3** (1952), 32–41.
5. S. Egami, 'The Distribution of Residue Classes Modulo a in an Algebraic Number Field', *Tsucuba J. of Math.*, 4 (1980), 9–13.
6. A. M. Frieze, J. Hastad, R. Kannan, J. C. Lagarias and A. Shamir, 'Reconstructing Truncated Integer Variables Satisfying Linear Congruence', *SIAM J. Comp.*, **17** (1988), 262–280.
7. A. Garcia and J. F. Voloch, 'Fermat Curves Over Finite Fields', *J. Number Theory*, **30** (1988), 345–356.
8. K. Győry, 'On Arithmetic graphs Associated with Integral Domains', *A Tribute To Paul Erdős*, Cambrige Univ. Press, 1990, 207–222.
9. K. Győry, 'Some Recent Applications of 'S-unit Equations', *Asterisque*, **209** (1992), 17–38.
10. H.J. Karloff and P. Raghavan, 'Randomized Algorithms and Pseudorandom Numbers', *J. ACM* , 40(1993), 454–476.

11. D. E. Knuth, *The Art of Computer Programming, vol.2*, Addison-Wesley, Massachusetts, 1981.

12. N. M. Korobov, *Number-Theoretic Methods in Approximate Analysis*, Moscow, 1963.

13. N. M. Korobov, 'On the Distribution of Signs in Periodic Fractions', *Matem. Sbornik*, 89 (1972), 654–670 (in Russian).

14. N. M. Korobov, *Exponential Sums and Their Applications*, Kluwer Acad. Publ., North-Holland, 1992.

15. G. Larcher and H. Niederreiter, 'Optimal Coefficients Modulo Prime Powers in the Three-Dimensional Case', *Annali di Mathematica pura ed applicata*, 155 (1989), 299–315.

16. H. W. Lenstra Jr., 'Euclidean Number Fields of Large Degree', *Invent. Math.*, 38 (1977), 237–254.

17. R. Lidl and H. Niederreiter, *Finite Fields*, Addison-Wesley, 1983.

18. F. J. MacWilliams and N. J. A. Sloane, *The Theory of Error-Correcting Codes*, North-Holland Publ. Comp., 1977.

19. H. L. Montgomery, 'Distribution of Small Powers of a Primitive Root', *Advances in Number Theory*, Clarendon Press, Oxford, 1993, 137–149

20. W. Narkiewicz, *Elementary and Analytic Theory of Algebraic Numbers*, Polish Sci. Publ., Warszawa, 1990.

21. H. Niederreiter, 'Quasi-Monte Carlo Methods and Pseudo-Random Numbers', *Bull. Amer. Math. Soc.*, 84 (1978), 957–1041.

22. H. Niederreiter, 'On a Problem of Kodama Concerning the Hasse-Witt Matrix and Distribution of Residues', *Proc. Japan Acad.*, Ser.A, 63 (1987), 367–369.

23. H. Niederreiter, *Random Number Generation and Quasi-Monte Carlo Methods*, SIAM Press, 1992.

24. G. Niklash and R. Queme, 'An Improvement of Lenstra's Criterion for Euclidean Number Fields: The Totally Real Case', *Acta Arithm.*, 58 (1991), 157–168.

25. F. Pappalardi, Ph.D. Thesis, McGill Univ., 1993.

26. I. E. Shparlinski, 'Bounds for exponential sums with recurring sequences and their applications', *Proc. Voronezh State Pedagogical Inst.*, 197 (1978), 74–85 (in Russian).

27. I. E. Shparlinski, 'On Residue Classes Modulo a Prime Number in an Algebraic Number Field', *Matem. Zametki*, 43 (1988), 433–437 (in Russian).

28. I. E. Shparlinski, 'On the Dimension of BCH Codes', *Problemy Peredachi Inform.*, 25(1) (1988), 100–103 (in Russian).

29. I. E. Shparlinski, 'On Gaussian Sums for Finite Fields and Elliptic Curves', *Proc. 1-st French-Soviet Workshop on Algebraic Coding., Paris, 1991*, Lect. Notes in Computer Sci., 537 (1992), 5–15.

30. I. E. Shparlinski, *Computational Problems in Finite Fields*, Kluwer Acad. Publ., North-Holland, 1992.

31. M. Tompa, 'Lecture Notes on Probabilistic Algorithms and Pseudorandom Generators', *Technical Report 91-07-05, Dept. of Comp. Sci. and Engin.*, Univ. of Washington, Seatle, 1991.

32. T. Washio and T. Kodama, 'Hasse-Witt Matrices of Hyperelliptic Function Fields', *Sci. Bull. Fac. Education Nagasaki Univ.*, 37 (1986), 9–15.

33. T. Washio and T. Kodama, 'A Note on a Supersingular Function Fields', *Sci. Bull. Fac. Education Nagasaki Univ.*, 37 (1986), 9–15.

Improved Incremental Prime Number Sieves

Paul Pritchard

Griffith University, School of Computing and Information Technology, Queensland, Australia 4111

Abstract. An algorithm due to Bengalloun that continuously enumerates the primes is adapted to give the first prime number sieve that is simultaneously sublinear, additive, and smoothly incremental:

- it employs only $\Theta(n/\log\log n)$ additions of numbers of size $O(n)$ to enumerate the primes up to n, equalling the performance of the fastest known algorithms for fixed n;
- the transition from n to $n+1$ takes only $O(1)$ additions of numbers of size $O(n)$. (On average, of course, $O(1)$ such additions increase the limit up to which all primes are known from n to $n + \Theta(\log\log n)$).

1 Introduction

A so-called "formula" for the i'th prime has been a long-lived concern, if not quite the Holy Grail, of Elementary Number Theory. This concern seems poorly motivated, as evidenced by the extraordinary freak-show of solutions proffered over the ages. The natural setting is Algorithmic Number Theory, and what is desired is much better cast as an algorithm to compute the i'th prime. Given that approaches involving (all) smaller primes have been deemed acceptable, the problem can perhaps best be formulated as that of finding an algorithm to enumerate the primes, with efficiency and (as we shall see) incrementality the desirable properties.

The problem of enumerating the prime numbers is one of the most venerable of algorithmic problems. It received a deceptively simple and efficient solution from Eratosthenes of Alexandria in the 3rd century B.C. His insight was to recast the problem as that of removing the non-primes up to a fixed limit N. Eratosthenes' sieve requires $\Theta(N\log\log N)$ additions of numbers of size $O(N)$.

To paraphrase a remark of C.A.R. Hoare about Algol 60, Eratosthenes' sieve was an improvement on most of its successors. In this connection, note that the asymptotically fastest known multiplication algorithm for a RAM, that of Schönhage and Strassen (see [2]), has a complexity equal to that of $\Theta(\log\log N\log\log\log N)$ additions. So the bit-complexity of Eratosthenes' sieve is lower than one of $\Theta(N)$ multiplications. The latter is characteristic of many proposed *parallel* algorithms (let alone some sequential ones).

Nevertheless, progress was eventually made. The fastest known (sequential) algorithm is now our wheel sieve [9, 10], which requires $\Theta(N/\log\log N)$ additions. It enjoys the properties of being *sublinear*, i.e., $o(N)$, and *additive*, i.e., not requiring multiplications.

Bengalloun [3] promoted another desideratum, that of being *incremental*. This means that N need not be fixed, so that the algorithm can in theory be run indefinitely to enumerate the primes. Bengalloun's incremental sieve requires $\Theta(N)$ multiplications (and additions) to find the primes up to N. Any sieve algorithm can be made incremental by repeatedly running it with a doubled limit. But Bengalloun's is incremental in a much more natural way: it takes bounded time to progress from N to $N + 1$, for all N. We shall describe it as *smoothly incremental*.

Bengalloun also outlined how our technique of using wheels could be incorporated in his basic algorithm to reduce its complexity to $\Theta(N/\log\log N)$ multiplications, but observed that our method in [9] of avoiding multiplications would require a prohibitively large number of multiplication tables.

The main contribution of this paper is to exhibit an algorithm that simultaneously enjoys all these desirable properties: it is a sublinear, additive, and smoothly incremental prime number sieve. The construction of this algorithm proceeds in two phases.

The first phase modifies Bengalloun's linear (i.e., $\Theta(N)$) smoothly incremental but multiplicative sieve to avoid multiplications. It does so by calculating differences using an incrementally grown multiplication table. Theorems by Heath-Brown and Iwaniec show that the table does not grow too quickly.

The second phase shows how dynamic wheels, i.e., reduced residue systems of the product of the primes up to a limit that grows with N, can be used to reduce the complexity while still avoiding multiplications and retaining the property of being smoothly incremental.

2 Notation

We employ a consistent notation, due to E. W. Dijkstra, wherever a variable is bound by a quantifier such as $\forall$, $\exists$, $\sum$, $\prod$, **Max**, **Min** and $\{\}$. The set-forming quantifier $\{\}$ is our invention; it was first used in [13]. The general form for a set-constructor is $(\{\}x : D(x) : t(x))$ which denotes the set of values $t(x)$ when x ranges over the domain characterized by $D(x)$. However, $t(x)$ is commonly x, and in such cases we omit the second colon and term, giving a form very close to traditional mathematical notation. Such an abbreviation is also used with $\sum$, $\prod$, **Max** and **Min**.

The notion of a *wheel* is central to the algorithms to be discussed. The k'th wheel W_k ($k > 0$) is a particular reduced residue system of the product of the first k primes. In more prosaic terms, W_k is the set of all natural numbers no greater than the product of the first k primes and not divisible by one of the first k primes. W_k^* is the set of all natural numbers not divisible by one of the first k primes. g_k is the maximum gap between successive natural numbers not divisible by one of the first k primes. We use the following notation (introduced in [11]):

$\emptyset$: the empty set;

$|S|$: the cardinality of set S;

next(S, x): $(\mathbf{Min}\ y : y \in S \wedge y > x)$;

$a..b$: $(\{\}x : a \leq x \leq b)$;

(x, y): the g.c.d. of x and y;

$x \mid y$: x divides y;

p_i: the i'th prime number;

Primes(S): $(\{\}x : x \in S \wedge x$ is a prime number$)$;

$\pi(n)$: $|$Primes$(2..n)|$;

d_n: $(\mathbf{Max}\ i : p_i \leq n : p_i - p_{i-1})$, $n > 2$;

Π_k: $(\prod i : 1 \leq i \leq k : p_i)$, with $\Pi_0 = 1$;

W_k: $(\{\}x : 1 \leq x \leq \Pi_k \wedge (x, \Pi_k) = 1)$;

$W_k^{(i)}$: the i'th greatest member of W_k;

W_k^* : $(\{\}x : 1 \leq x \wedge x \bmod \Pi_k \in W_k)$;

g_k: $(\mathbf{Max}\ x : x \in W_k^* : \text{next}(W_k^*, x) - x)$.

Our algorithms are expressed in the language of guarded commands used in [4], extended with **if-** and **forall-**commands. The command

$$\textbf{if } b \textbf{ then } S \textbf{ fi}$$

is an abbreviation of

$$\textbf{if } b \rightarrow S \,[]\, \neg b \rightarrow skip \textbf{ fi}.$$

The **forall-**command denotes iteration over a fixed finite set in unspecified order (see [12]).

3 Bengalloun's Sieve

The following discussion of Bengalloun's sieve recapitulates our presentation in [12].

Bengalloun's basic sieve is based on the following normal form for composites c:

$$c = p \cdot f \text{ where } p = \text{lpf}(c) \text{ and } f > 1 \tag{1}$$

"lpf" denotes the least prime factor function. Note that in (1) we have $\text{lpf}(f) \geq p$ and $\text{lpf}(c) \leq \sqrt{c}$.

Bengalloun's sieve tabulates the function lpf on a superset of an incrementally increasing segment $2..n$ of the natural numbers. $\text{lpf}(n)$ is tabulated when n is processed if n is even, otherwise it is tabulated when processing the largest composite $< n$ with the same value of f in its normal form. Changing perspective, when composite $n = p_i \cdot f$ is processed, lpf is tabulated at $p_{i+1} \cdot f$, provided $p_i < \text{lpf}(f)$. The primes are gathered in an array as they are discovered.

In our presentation of the algorithm, $f|_S$ denotes the restriction of the function f to the sub-domain S.

Bengalloun's Sieve:

$n, P, LPF := 2, \{2\}, \{(2, 2)\};$

$\{P = \text{Primes}(2..n) \wedge LPF = \text{lpf}|_{\text{domain}(LPF)} \wedge \text{domain}(LPF) = (2..n) \cup$
$(\{\}p, f : f > 1 \wedge p \in \text{Primes}(3..\text{lpf}(f)) \wedge \text{prev}(P, p) \cdot f \le n : p \cdot f)\}$

do *true* →

    $n := n + 1;$

    if $2 \mid n \to LPF := LPF \cup \{(n, 2)\}$

    $[] \neg(2 \mid n) \to skip$

    fi;

    if $\neg(n \in \text{domain}(LPF)) \to P, LPF := P \cup \{n\}, LPF \cup \{(n, n)\}$

    $[] n \in \text{domain}(LPF) \to$

      $p := LPF(n);$

      $f := n \div p;$

      if $p < LPF(f) \to p' := \text{next}(P, p);$

                       $LPF := LPF \cup \{(p' \cdot f, p')\}$

      $[] p = LPF(f) \to skip$

      fi

    fi

od

A neat implementation uses a single array *lpf* with the representation invariant

$$lastp = \max(P) \wedge lpf[x] = \begin{cases} 0, & \text{if } 2 \le x \le 2n \wedge \\ & \neg(x \in \text{domain}(LPF)), \\ LPF(x), & \text{if } x \text{ is composite } \wedge \\ & x \in \text{domain}(LPF), \\ \text{next}(P, x), & \text{if } x \in \text{Primes}(2..(lastp - 1)). \end{cases}$$

Bertrand's theorem — there is always a prime between n and $2n$ — justifies only defining array *lpf* up to index $2n$.

The algorithm clearly requires $\Theta(N)$ multiplications and additions to reach $n = N$.

4 Avoiding Multiplications in Bengalloun's Sieve

Multiplicative operations occur in two places in Bengalloun's sieve. The first is the computation $f := n \div p$ after p has been looked-up in $lpf[n]$. This is avoided by the simple expedient of also tabulating the co-factor f along with the least prime factor. Another array, *cf* (for co-factor) is used for this purpose. So the multiplicative operation is replaced with $f := cf[n]$.

The other multiplicative operation occurs in the abstract operation

$$LPF := LPF \cup \{(p' \cdot f, p')\} \tag{2}$$

which is now implemented as

$$n' := p' \cdot f;$$
$$lpf[n'], cf[n'] := p', f$$

Since $p' \cdot f = p \cdot f + (p' - p) \cdot f$, and $p \cdot f = n$, the problem of computing $p' \cdot f$ reduces to that of computing $(p' - p) \cdot f$. Suppose, after it is computed, this latter product and the value f are recorded abstractly as $\Delta(p)$ and $\mathrm{lastf}(p)$ respectively. (We shall postpone implementation details.)

Now when a value n is processed, and $p' \cdot f$ must be computed, the product may be written as

$$n + (p' - p) \cdot \mathrm{lastf}(p) + (p' - p) \cdot (f - \mathrm{lastf}(p)), \tag{3}$$

provided $\mathrm{lastf}(p)$ is defined, in which case the second summand is $\Delta(p)$, which may be looked up, and the third involves very small multiplicands.

Under this provisional arrangement $\mathrm{lastf}(p_i)$ and $\Delta(p_i)$ are initialised when processing $n = p_i \cdot p_{i+1}$, so only in this case is $\mathrm{lastf}(p_i)$ undefined. Then $p' \cdot f$ is p_{i+1}^2. It is possible to compute this value incrementally without using multiplications, but we shall adopt a more straightforward approach.

When $p = f$, so that $n = p^2$, $\mathrm{lastf}(p)$ is initialised to p and $\Delta(p)$ to $(p' - p) \cdot p$. The effect is to extend the invariant with the conjunct

$$(\forall i : p_i^2 \le n : \mathrm{lastf}(p_i) = (\mathbf{Max}\, f : f = p_i \vee (\mathrm{lpf}(f) > p_i \wedge p_i \cdot f \le n)) \wedge$$
$$\Delta(p_i) = (p_{i+1} - p_i) \cdot \mathrm{lastf}(p_i))$$

Now the product $p' \cdot f$ may always be written as (3) above.

We shall arrange for a multiplication table to be incrementally tabulated (together with row-offsets) so that the third summand $(p' - p) \cdot (f - \mathrm{lastf}(p))$ may be looked-up (without a multiplication).

The first multiplicand is $p' - p$. Since $n = p \cdot f$ and $f \ge p'$, $n \ge p \cdot p'$, so $p < \sqrt{n}$ and hence $p' < 2\sqrt{n}$ by Bertrand's Theorem. Therefore

$$p' - p < d_{2\sqrt{n}}.$$

The best known upper bound on prime gaps is the very conservative

$$d_n = O(n^{0.55 + \epsilon}) \text{ for any } \epsilon > 0$$

[6]. Thus

$$p' - p = O(n^{0.275 + \epsilon}) \text{ for any } \epsilon > 0.$$

Now consider the second multiplicand. $\mathrm{lastf}(p_i)$ and f are either p_i and p_{i+1} respectively, or successive members of W_{i+1}^*. In the former case,

$$f - \mathrm{lastf}(p_i) = O(n^{0.275 + \epsilon})$$

as above. In the latter,

$$f - \mathrm{lastf}(p_i) = O(g_{i+1}) = O(p_{i+1}^2)$$

by a result of Iwaniec [7]. Since $f = n \div p_i$, the difference is also bounded by

$$d_{n \div p_i} = O((n \div p_i)^{0.55 + \epsilon}).$$

We may approximately balance these two known bounds as follows. If $p_i = O(n^{0.216})$, the difference is $O(p_{i+1}^2) = O(n^{0.432})$. Otherwise, $p_i = \Omega(n^{0.216})$, and the difference is $O((n \div p_i)^{0.55+\epsilon}) = O(n^{0.432})$.

We are thus able to prove that an $O(n^{0.275+\epsilon})$ by $O(n^{0.432})$ multiplication table suffices for the product $(p' - p) \cdot (f - \text{lastf}(p))$.

In passing, we note that this is certainly a gross over-estimate. According to [1], Cramér's conjecture that $d_n = \Theta(\log^2 n)$ "has been called into question", but the slightly weaker

$$d_n = O(\log^{2+\epsilon} n) \text{ for any } \epsilon > 0$$

is "still probably true". Suppose this latter claim is true. Then when $p_i = O(\log n)$, the difference is

$$O(p_{i+1}^2) = O(\log^2 n).$$

And when $p_i = \Omega(\log n)$, the difference is

$$O(\log^{2+\epsilon}(n \div p_i)) = O(\log^{2+\epsilon} n).$$

So it is probably the case that a square multiplication table of side $O(\log^{2+\epsilon} n)$ suffices!

For simplicity we construct a square multiplication table. Because of symmetry, we only construct the upper triangle, going down each column in order. A one-dimensional array is used, so that $i \cdot j$ is stored at index $j(j - 1)/2 + i$, for $j \geq i$. In order to avoid a multiplication when accessing the table, the values j^2 are incrementally computed and stored as well. With this scheme, a full m by m table is available after $m(m + 1)/2$ entries have been made. c_1 entries are added at each iteration (i.e., for each value of n), for an appropriate constant c_1, which is guaranteed to exist because each side is $O(\sqrt{n})$. It is very likely that $c_1 = 1$ suffices. Each entry may be computed in $O(1)$ additions.

It remains to implement the abstract functions Δ and lastf. These are merely tabulated in arrays, indexed either by their prime argument (which is wasteful), or its index (in which case the prime indices of the least prime factors would need to be stored in array lpf). Because a bounded number of one-dimensional arrays is used, they may be interleaved in a single incrementally growing one-dimensional array with $O(n)$ elements, without requiring multiplications for access.

The resulting algorithm is not only incremental, linear and additive, but it is optimally smooth at the bit-complexity level — each value of n is processed in $O(1)$ additions of numbers of size $O(n)$. Note that performing just one multiplication would vitiate the latter property.

5 Using Wheels to Add Sublinearity

As was observed by its inventor in [3], Bengalloun's Sieve may be sped up by exploiting the same simple idea powering our wheel sieve (see [10]): a number n

exceeding Π_k can only be prime if $n \in W_k^*$, for otherwise it is divisible by one of the first k primes.

To derive maximum benefit, k is maximised. In order to do so, the wheel that is maintained must be updated to W_{k+1} when n passes Π_{k+1}. This is accomplished by incrementally building W_{k+1} while the complete wheel W_k is used to generate candidates n.

So rather than simply incrementing n, our new algorithm updates n to next(W_k^*, n), where k is maximal such that $\Pi_k < n$. Since n will be odd, the first if-command is no longer needed.

Now suppose composite $n = p \cdot f$ is processed by this modified algorithm, so $p \geq p_{k+1}$. If $p < \mathrm{lpf}(f)$, then $p' \cdot f \in W_{k+1}^*$, so it too will later be processed. The present code caters for this by tabulating lpf at $p' \cdot f$. The only concern, therefore, is that the least composite number n with a given co-factor f that is processed by the new algorithm is factored.

These numbers are just those of the form

$$p_{k+1} \cdot f \text{ where } k \geq 1 \text{ and } f \in W_k - W_{k-1}.$$

The first of them, 25, is treated specially, and the factorisation of each of the others is recorded when the previous one n in numerical order is processed. Let $n = p_{k+1} \cdot f$. There are two cases, depending on whether f is the greatest member of W_k.

If not, then with f' the succeeding member,

$$p_{k+1} \cdot f' = n + p_{k+1} \cdot (f' - f). \tag{4}$$

Consider the product in (4). The multiplicand $f' - f$ falls within the bounds of the multiplication table. To bound the other multiplicand, note that

$$\Pi_k < n < \Pi_{k+1}$$

by definition of k. Since

$$\sum_{p \leq x} \log p \sim x$$

(see [5, theorems 420 and 434]),

$$p_{k+1} = \Theta(\log n).$$

Hence the product may be looked up in the multiplication table.

In the case that f is the greatest member $\Pi_k - 1$, the next number is $p_{k+2} \cdot (\Pi_k + 1)$, which differs from n by

$$(p_{k+2} - p_{k+1}) \cdot \Pi_k + p_{k+2} + p_{k+1}. \tag{5}$$

Unfortunately, Π_k falls outside the bounds of the multiplication table. However, the product in (5) may be incrementally computed by repeated additions of Π_k. Since

$$p_{k+2} - p_{k+1} = O(d_{p_{k+2}}) = \Theta(\log^{0.55+\epsilon} n),$$

there is ample time to do the calculation after n reaches Π_k.

The incremental construction of W_{k+1} is straightforward. The numbers $n > \Pi_k$ with $\mathrm{lpf}(n) > p_{k+1}$ may simply be linked as encountered, and those less than Π_k linked in when used for the last time. The otherwise unused even-numbered elements of array lpf may be used to construct the linked list. Alternatively, two new arrays may be used: one to hold the current wheel W_k, while the next wheel W_{k+1} is built in the other.

Both new incremental sub-computations — building the next wheel and the product of Π_k — may be carried out in $O(1)$ additions of numbers of size $O(n)$ for each of the $O(n/\log\log n)$ numbers up to n that are processed.

We have arrived at an algorithm that

- employs only $\Theta(n/\log\log n)$ additions of numbers of size $O(n)$ to enumerate the primes up to n, equalling the performance of the fastest known algorithms for fixed n;
- moves from n to $n + 1$ in only $O(1)$ additions of numbers of size $O(n)$. (On average, of course, $O(1)$ such additions increase the limit up to which all primes are known from n to $n + \Theta(\log\log n)$).

A remark on the machine model is called for. We have been implicitly assuming a RAM with a word size that grows with the computation, so that the running time to find all primes up to n is $O(n/\log\log n)$ arithmetic operations on numbers of size $O(n)$. The reader may be more comfortable with a model positing a fixed size word, in which case the numbers used may be represented by linked lists of the appropriate length, with no change to the underlying bit-complexity.

6 Closing Remarks

Since the author's discovery of the *wheel sieve* in 1979, no algorithm for finding the primes up to even fixed N has been discovered with a complexity of $o(N/\log\log N)$ arithmetic operations. We conjecture that $O(N/\log\log N)$ additions of numbers of size $O(N)$ is best possible. We have achieved this with a smoothly incremental algorithm.

Determining the status of the above conjecture, or more generally, giving good lower bounds on the time-complexity of the problem of enumerating the primes up to N (whether or not by incremental algorithms), are daunting open problems. $\Omega(N/\log N)$ additive operations on numbers of size $O(N)$ are needed to list the primes up to N. The same lower bound of $\Omega(N)$ bit operations applies if output as a bit-vector is permitted. But if output in the form of prime gaps is allowed (and why not?), even this lower bound has yet to be established, since we showed in [11] that the primes up to N may be stored in $O(N\log\log N/\log N)$ bits and still recovered in order in $O(1)$ additive operations per prime.

Our new algorithm lacks one important property: it is not *compact*. As we showed in [8, 11], Eratosthenes' sieve and some linear wheel-based sieves can be implemented to run in $O(N^{0.5})$ bits, whereas our new algorithm and Bengalloun's sieve require $O(N \log N)$ bits (to enumerate the primes up to N).

The space requirement of our new algorithm may be reduced by a factor of $1/\Theta(\log \log n)$, while preserving its sublinear, additive and smoothly incremental properties, by adapting the invertible mapping technique introduced in section 7 of [9], but the other techniques presented therein rely on a bounded problem size, and do not readily adapt to an incremental setting.

Little is known about lower bounds for space for sublinear algorithms for finding the primes, whether or not the algorithms are incremental. The known upper bounds exhibit a great disparity in moving from a linear to a sublinear algorithm. Is there a sublinear algorithm with a space complexity of $o(N^{1-\epsilon})$ bits for some $\epsilon > 0$?

As has been mentioned, great reductions in space requirements for nonincremental algorithms may be obtained by trading in sublinearity. We plan to investigate the extent to which this is possible for incremental algorithms in a future paper.

References

1. Adleman, L.M. and McCurley, K.S.: Open problems in number theoretic complexity, II. Proceedings of the First Algorithmic Number Theory Symposium. This volume.
2. Aho, A., Hopcroft, J., Ullman., J.: The Design and Analysis of Computer Algorithms. Addison-Wesley, Reading, Massachusetts, 1974.
3. Bengalloun, S.: An incremental primal sieve. Acta Informatica 23 (1986) 119–125
4. Gries, D.: The Science of Programming. Springer-Verlag, New York, 1981.
5. Hardy, G.H. and Wright, E.M.: An Introduction to the Theory of Numbers, 5th ed. London Univ. Press, London, 1979.
6. Heath-Brown, D.R., Iwaniec, H.: On the difference between consecutive primes. Inventiones Mathematicae 55 (1979) 49–69
7. Iwaniec, H.: On the problem of Jacobsthal. Demonstratio Math. 11 (1978) 225–231
8. Pritchard, P.: On the prime example of programming. In Tobias, J. (ed.): Language Design and Programming Methodology. Lecture Notes in Computer Science 79, (1980) 85–94
9. Pritchard, P.: A sublinear additive sieve for finding prime numbers. Comm. ACM 24 (1981) 18–23
10. Pritchard, P.: Explaining the wheel sieve. Acta Informatica 17 (1982) 477–485
11. Pritchard, P.: Fast compact prime number sieves (among others). J. Algorithms 4 (1983) 332–344.
12. Pritchard, P.: Linear prime number sieves: a family tree. Sci. Comp. Prog. 9 (1987) 17–35
13. Pritchard, P.: Opportunistic algorithms for eliminating supersets. Acta Informatica 28 (1991) 733–754

Polynomial Time Algorithms for Discrete Logarithms and Factoring on a Quantum Computer

Peter W. Shor

AT&T Bell Labs, Room 2D-149
600 Mountain Ave.
Murray Hill, NJ 07974 USA

A computer is generally considered to be a universal computational device, in other words, it is able to simulate any physical computational device with a cost of at most a polynomial factor in the computation time. It is not clear that this is still true if quantum mechanics is taken into account. Feynman seems to have been the first to ask what effect the non-local properties of quantum mechanics have on computation [7, 8]. He gave arguments as to why these properties might make it intrinsically computationally expensive to simulate quantum mechanics on a classical (Von Neumann) computer. He also asked the converse question: whether these properties permit more powerful computation. Several researchers have since developed models for quantum mechanical computers and investigated their computational properties. [4, 5, 6, 2, 3, 1, 10, 9] We give Las Vegas algorithms for the discrete logarithm and integer factoring problems that take random polynomial time on a quantum computer.

References

1. E. Bernstein and U. Vazirani, "Quantum complexity theory," in *Proc. 25th ACM Symp. on Theory of Computation*, pp. 11–20 (1993).
2. A. Berthiaume and G. Brassard, "The quantum challenge to structural complexity theory," in *Proc. 7th IEEE Conference on Structure in Complexity Theory*, pp. 132–137 (1992).
3. A. Berthiaume and G. Brassard, "Oracle quantum computing," in *Proc. Workshop on Physics of Computation*, pp. 195–199, IEEE Press (1992).
4. D. Deutsch, "Quantum theory, the Church–Turing principle and the universal quantum computer," *Proc. R. Soc. Lond.* Vol. A400, pp. 96–117 (1985).
5. D. Deutsch, "Quantum computational networks," *Proc. R. Soc. Lond.* Vol. A425, pp. 73–90 (1989).
6. D. Deutsch and R. Jozsa, "Rapid solution of problems by quantum computation," *Proc. R. Soc. Lond.* Vol. A439, pp. 553–558 (1992).
7. R. Feynman, "Simulating physics with computers," *International Journal of Theoretical Physics*, Vol. 21, nos. 6/7, pp. 467–488 (1982).
8. R. Feynman, "Quantum mechanical computers," *Foundations of Physics*, Vol. 16, pp. 507–531 (1986). (Originally appeared in *Optics News*, February 1985.)
9. D. Simon, "On the power of quantum computation," manuscript (1993).
10. A. Yao, "Quantum circuit complexity," in *Proc. 34th Symp. on Foundations of Computer Science*, pp. 352–361, IEEE Press (1993).

On Dispersion and Markov Constants

Guangheng Ji* and Hongwen Lu**

Computer Center*, and Department of Mathematics
University of Science and Technology of China
Hefei, Anhui 230026, P.R. China

We proved some results on the dispersion of the real quadratic irrational numbers, and use LEO 386/25 to compute some numerical results when the dscriminant $d < 200$. The details are as follows.

Let $\{x_n\}$ be a sequence of numbers, $0 \le x_n \le 1$. H. Niederreiter introduced a measure of denseness of such a sequence as follows. For each N, let

$$d_N = \sup_{0 \le x \le 1} \min_{1 \le n \le N} |x - x_n|$$

and define

$$D(\{x_n\}) = \limsup_{N \to \infty} N d_N.$$

In particular, for irrational α, the *dispersion constant* $D(\alpha)$ is defined by $D(\{n\alpha \mod 1\})$. Then we proved

Theorem *Let $d \equiv b \pmod 4$ be a positive discriminant which is not a perfect square of a rational integer, and $b = 0$ or 1. Then we have*

$$D\left(\frac{b + \sqrt{d}}{2}\right) = \frac{2d + (d + b)\sqrt{d}}{4d}.$$

We proved a general but not so simple result which is used to compute some results for the discriminants $d < 200$ on a pc computer LEO 386/25 and recorded in a table. The detail of this note will be published elsewhere.

This project was supported by a grant from NNSF of P.R. China

Open Problems
in Number Theoretic Complexity, II

Leonard M. Adleman[1] and Kevin S. McCurley[2]

[1] Department of Computer Science, University of Southern California, Los Angeles, CA 90089-0782, USA, adleman@cs.usc.edu

[2] Organization 1423, MS 1110, Sandia National Laboratories, Albuquerque, NM 87185, USA, mccurley@cs.sandia.gov

Introduction.

This conference (ANTS-1) marks the beginning of what we hope will be a long series of international conferences on algorithmic number theory. It seems appropropriate, at the beginning, to state some of the central open problems in the field. Accordingly, this paper contains a list of 36 open problems in number-theoretic complexity. We expect that none of these problems are easy; we are sure that many of them are hard.

This list of problems reflects our own interests and should not be viewed as definitive. As the field changes and becomes deeper, new problems will emerge and old problems will lose favor. Ideally there will be other 'open problems' papers in future ANTS proceedings to help guide the field.

It is likely that some of the problems presented here will remain open for the forseeable future. However, it is possible in some cases to make progress by solving subproblems, or by establishing reductions between problems, or by settling problems under the assumption of one or more well known hypotheses (e.g. the various extended Riemann hypotheses, $\mathcal{NP} \neq \mathcal{P}, \mathcal{NP} \neq \text{co}\mathcal{NP}$).

For the sake of clarity we have often chosen to state a specific version of a problem rather than a general one. For example, questions about the integers modulo a prime often have natural generalizations to arbitrary finite fields, to arbitrary cyclic groups, or to problems with a composite modulus. Questions about the integers often have natural generalizations to the ring of integers in an algebraic number field, and questions about elliptic curves often generalize to arbitrary curves or abelian varieties.

The problems presented here arose from many different places and times. To those whose research has generated these problems or has contributed to our present understanding of them but to whom inadequate acknowledgement is given here, we apologize.

Our list of open problems is derived from an earlier 'open problems' paper we wrote in 1986 [AM86]. When we wrote the first version of this paper, we feared that the problems presented were so difficult that young researchers reading the list might be discouraged rather than inspired. Happily, despite the difficulties, eight years has brought considerable progress on a number of these problems. Even for the two most central problems in the field, primality testing and factoring, there has been impressive progress: the primes are now known to be decidable in random polynomial time and the 'number field sieve' has given us the most powerful factoring algorithms yet. To emphasize the progress that has been made, the statement of each problems is followed by the original 1986 remarks and then the remarks which now seem appropriate.

The authors would appreciate your comments, particularly with regard to further progress on these problems.

Definitions, notation, and conventions.

In this paper:

- **R** denotes the set of real numbers,
- **Z** denotes the set of integers,
- **N** denotes the set of positive integers,
- *Primes* denotes the set of primes in **N**,
- *Squarefrees* denotes the set of squarefree numbers in **N**,
- **Q** denotes the set of rationals.
- ERH refers to the extended Riemann hypothesis.

For $a, b \in \mathbf{Z}$,

- we write $a \mid b$ if there exists $k \in \mathbf{Z}$ with $b = ka$,
- we write $a \nmid b$ if there does not exist $k \in \mathbf{Z}$ with $b = ka$,
- $\gcd(a, b)$ denotes the greatest common divisor of a and b,
- $\left(\frac{a}{b}\right)$ denotes the Jacobi symbol if b is odd and $\gcd(a, b) = 1$,
- $\langle a, b \rangle$ denotes the ordered pair.

For $n \in \mathbf{N}$,

- $\mathbf{Z}/n\mathbf{Z}$ denotes the ring of integers modulo n,
- $(\mathbf{Z}/n\mathbf{Z})^*$ denotes the corresponding multiplicative group,
- $\phi(n)$ denotes the number of elements in $(\mathbf{Z}/n\mathbf{Z})^*$,
- $L(n)$ represents any function of the form

$$\exp((1 + o(1))(\log n \log \log n)^{1/2}) .$$

- For $n \in \mathbf{N}$, $\alpha, \beta \in \mathbf{R}$, $\alpha, \beta > 0$, $L_n[\alpha, \beta]$ represents any function of the form

$$\exp((\beta + o(1))((\log n)^\alpha (\log \log n)^{1-\alpha})) .$$

If R is a ring, then we write $R[x]$ for the ring of polynomials with coefficients in R. The set of finite strings composed of the letters a and b is denoted $\{a, b\}^*$. For $n, a, b \in \mathbf{N}$ with $\gcd(n, 4a^3 + 27b^2) = 1$, let

$$S_{n,a,b} = \{\langle x, y \rangle \mid x, y \in \mathbf{Z}/n\mathbf{Z} \ \& \ y^2 \equiv x^3 + ax + b \pmod{n}\} \cup \{0\} .$$

When $p \in \textit{Primes}$, $S_{p,a,b}$ is well known to be endowed with a group structure. We denote this group by $E_{p,a,b}$ and use $\#E_{p,a,b}$ for the number of elements of this group. More generally, if S is a set, we write $\#S$ for the cardinality of S.

In stating open problems we have decided to continue the ad hoc notation from [AM86]. For example, we label the first computational problem as **C1**, the corresponding open problem as **O1** (or **O1a** and **O1b** if there are two), and the original 1986 remarks concerning **C1** and **O1** we label as $\mathbf{Rem1}_{86}$. Any new remarks we label as $\mathbf{Rem1}_{94}$. Any additional references are given in **Ref1**. Computational problems **C2** and **C6** are stated in terms of a parameter S which is an arbitrary subset of N. Computational problem **C30** is stated in terms of a parameter $c \in \mathsf{N}$.

While it seems inappropriate to spend a great deal of time giving rigorous definitions of the complexity-theoretic notions used in this paper, it seems worthwhile to provide some guidance. More information on these notions may be found in [Gil77], [AHU74], [AM77], and [GJ79]. We assume the concept of a polynomial time computable function is understood. A computational problem $\mathbf{C}$ is thought of as a set of pairs $\langle x, S_x \rangle$, where x is an input for which an output is desired and S_x is the set of possible 'correct' outputs on input x. For example

$$\mathbf{C1} = \{\langle n, S_n \rangle \mid n \in Primes \Rightarrow S_n = \{1\} \ \& \ n \notin Primes \Rightarrow S_n = \{0\}\}$$

$$\mathbf{C17} = \{\langle \langle p, d \rangle, S_{\langle p,d \rangle} \rangle \mid d \in \mathsf{N} \ \& \ p \in Primes \ \& $$
$$S_{\langle p,d \rangle} = \{f \mid f \in (\mathsf{Z}/p\mathsf{Z})[x] \mid \deg(f) = d \ \& \ f \text{ irreducible}\}\}.$$

$$\mathbf{C19} = \{\langle p, S_p \rangle \mid p \in Primes \ \& \ S_p = \{g \mid g \in \mathsf{N}, 1 \le g \le p-1 \ \& $$
$$g \text{ generates } (\mathsf{Z}/p\mathsf{Z})^*\}\}$$

$$\mathbf{C28} = \{\langle \langle a, b, p, P, Q \rangle, S_{\langle a,b,p,P,Q \rangle} \rangle \mid a, b \in \mathsf{N}, p \in Primes, P, Q \in E_{p,a,b},$$
$$(\exists n \in \mathsf{N})[nP = Q] \ \& \ S_{\langle a,b,p,P,Q \rangle} = \{n \mid n \in \mathsf{N} \ \& \ nP = Q\}\}$$

Definition 1 *If* $\mathbf{C} = \{\langle x, S_x \rangle\}$ *is a computational problem then we let* $\pi(\mathbf{C}) = \{x \mid \langle x, S_x \rangle \in \mathbf{C}\}$.

We use $|x|$ to denote the length of an object x, where we hope that the meaning of 'length' will be clear from the context.

Definition 2 $\mathbf{C}$ *is in* $\mathcal{P}$ *iff there exists a polynomial time computable function* f *such that* $(\forall x \in \pi(\mathbf{C}))[f(x) \in S_x]$.

Thus for example, in **O18** below we ask if **C18** is in $\mathcal{P}$. Any deterministic algorithm which runs in polynomial time with input-output behaviour consistent with that described in **C18** would provide an affirmative answer to **O18**. In particular how that algorithm behaves on an input $p \notin Primes$ is irrelevant.

Definition 3 $\mathbf{C}$ *is in* $\mathcal{R}$ *iff there exists a* c *in* N *and a polynomial time computable function* f *such that*

i. $(\forall x \in \pi(\mathbf{C}))(\forall |r| \le |x|^c)[f(x,r) \in S_x \text{ or } f(x,r) = "?"]$

ii. $(\forall x \in \pi(\mathbf{C}))\left[\dfrac{\#\{r \mid |r| \le |x|^c \ \& \ f(x,r) \in S_x\}}{\#\{r \mid |r| \le |x|^c\}} \ge \frac{1}{2}\right]$

Definition 4 $\mathbf{C}$ *is in* $\mathcal{NP}$ *iff there exists a* c *in* N *and a polynomial time computable function* f *such that*

i. $(\forall x \in \pi(C))(\forall |r| \le |x|^c)[f(x,r) \in S_x$ or $f(x,r) = "?"]$.

ii. $(\forall x \in \pi(C))(\exists y \in S_x)(\exists |r| \le |x|^c)[f(x,r) = y]$.

Definition 5 **C** *is recognized in* $\mathcal{R}$ *iff*

i. $(\forall x \in \pi(C))[S_x = \{1\} \Rightarrow (\forall |r| \le |x|^c)[f(x,r) = \{1\}$ or $f(x,r) = "?"]]$

ii. $(\forall x \in \pi(C))\left[S_x = \{1\} \Rightarrow \frac{\#\{r| |r| \le |x|^c \ \& \ f(x,r)=1\}}{\#\{r| |r| \le |x|^c\}} \ge \frac{1}{2}\right]$

iii. $(\forall x \in \pi(C))[S_x \ne \{1\} \Rightarrow (\forall |r| \le |x|^c)[f(x,r) = "?"]]$.

Definition 6 **C** *is recognized in* $\mathcal{NP}$ *iff there exists a c in* **N** *and a polynomial time computable function f such that*

i. $(\forall x \in \pi(C))[S_x = \{1\} \Rightarrow (\forall |r| \le |x|^c)[f(x,r) = \{1\}$ or $f(x,r) = "?"]]$

ii. $(\forall x \in \pi(C))[S_x = \{1\} \Rightarrow (\exists |r| \le |x|^c)[f(x,r) = 1]]$

iii. $(\forall x \in \pi(C))[S_x \ne \{1\} \Rightarrow (\forall |r| \le |x|^c)[f(x,r) = "?"]]$.

For notions involving the reduction of one problem to another we will be even less formal.

Definition 7 f *is a deterministic solution to* **C** *iff* $(\forall x \in \pi(C))[f(x) \in S_x]$.

Let $D(C) = \{f \mid f$ is a deterministic solution to C$\}$. For all deterministic algorithms $\mathcal{A}$ and functions f and g, we say that $\mathcal{A}$ translates f into g iff when given a subroutine for f, $\mathcal{A}$ computes g in polynomial time (where the time used in the subroutine for f is not counted). We remark that calls to the subroutine may be 'dovetailed' but the algorithm A cannot know if the absence of a response on a particular call means that no response is forthcoming or that a response has just not arrived yet. See **C18** for an example.

Definition 8 **C1** $\le_P$ **C2** *iff there exists a deterministic algorithm* $\mathcal{A}$ *such that for all* $f \in D(\textbf{C2})$, *there exists a* $g \in D(\textbf{C1})$ *such that* $\mathcal{A}$ *translates* f *into* g *in polynomial time.*

Definition 9 **C** *is* $\mathcal{NP}$-*hard with respect to* $\mathcal{P}$ *iff for all* **C'**, (**C'** *is in* $\mathcal{NP}$) $\Rightarrow$ (**C'** $\le_P$ **C**).

We will follow the convention of using $\mathcal{NP}$-hard to denote $\mathcal{NP}$-hard with respect to $\mathcal{P}$.

Definition 10 f *is a random solution to* **C** *iff there exists a c in* **N** *such that*

i. $(\forall x \in \pi(C))(\forall |r| \le |x|^c)[f(x,r) \in S_x$ or $f(x,r) = "?"]$

ii. $(\forall x \in \pi(C))\left[\frac{\#\{r| |r| \le |x|^c \ \& \ f(x,r) \in S_x\}}{\#\{r| |r| \le |x|^c\}} \ge \frac{1}{2}\right]$

Let $R(C) = \{f \mid f$ is a random solution to C$\}$.

Definition 11 **C1** $\le_{\mathcal{R}}$ **C2** *iff there exists a deterministic algorithm* $\mathcal{A}$ *such that for all* $f \in D(\textbf{C2})$, *there exists a* $g \in R(\textbf{C1})$ *such that* $\mathcal{A}$ *translates* f *into* g *in polynomial time.*

Definition 12 **C** *is* $\mathcal{NP}$-*hard with respect to* $\mathcal{R}$ *iff for all* **C'**,

$$(\textbf{C'} \text{ is in } \mathcal{NP}) \Rightarrow \textbf{C'} \le_{\mathcal{R}} \textbf{C} .$$

1 Primality testing

C1   Input $n \in \mathbf{N}$
                 Output   1 if $n \in Primes$,
                                0 otherwise.

O1a   Is **C1** in $\mathcal{P}$?

O1b   Is **C1** recognized in $\mathcal{R}$?

Rem1$_{86}$ A classical problem. The following quote appears in art. 329 of Gauss' *Disquisitiones Arithmeticæ*:(translation from [Knu81, page 398])

> The problem of distinguishing prime numbers from composites, and of resolving composite numbers into their prime factors, is one of the most important and useful in all of arithmetic. ... The dignity of science seems to demand that every aid to the solution of such an elegant and celebrated problem be zealously cultivated.

It is known that the set of composites is recognized in $\mathcal{R}$ [SS77]. If the extended Riemann hypothesis for Dirichlet L-functions is true, then **C1** is in $\mathcal{P}$ [Mil76]. There exists a constant $c \in \mathbf{N}$ and a deterministic algorithm for **C1** with running time $O((\log n)^{c \log \log \log n})$ [APR83]. If Cramér's conjecture on the gaps between consecutive primes is true, then **C1** is recognized in $\mathcal{R}$ [GK86]. **C1** is recognized in $\mathcal{NP}$ [Pra75]. Fürer [Für85] has shown that the problem of distinguishing between products of two primes that are $\not\equiv 1 \pmod{24}$ and primes that are $\not\equiv 1 \pmod{24}$ is in $\mathcal{R}$.

Rem1$_{94}$ Problem **O1b** has been settled in the affirmative by Adleman and Huang [AH92]. As a result of the work of H. Maier on gaps between consecutive primes, the exact formulation of Cramér's conjecture has now been called into question, however the conjecture required for [GK86] is unaffected.

Ref1   [Guy77], [Knu81], [Len81], [CL84], [Pom81], [Rab80a], [Rie85b], [Rie85a], [Wil78].

2 Testing an infinite set of primes

Let $S \subset \mathbf{N}$.

C2   Input $n \in \mathbf{N}$.
                 Output   1 if $n \in S$,
                                0 otherwise.

O2   Does there exist an infinite set $S \subset Primes$ such that **C2** is in $\mathcal{P}$?

Rem2$_{86}$ In light of **Rem1$_{86}$** it is remarkable that **O2** remains unsettled. The related problem of the existence of an infinite set $S \subset Primes$ such that **C2** is recognized in $\mathcal{R}$ is addressed in [GK86].

Rem2₉₄ | Problem **O2** been settled in the affirmative by Pintz, Steiger, and Szemerédi [PSS89]. One can now ask what the densest such set S is. In this direction, Konyagin and Pomerance [KP94] have proved that for every $\epsilon > 0$ there exists an algorithm that will prove primality in deterministic polynomial time for at least $x^{1-\epsilon}$ primes less than x.

Ref2 | [PSS88].

3 Prime greater than a given bound

C3 | Input $n \in \mathbb{N}$.
Output $p \in Primes$ with $p > n$.

O3 | Is **C3** in $\mathcal{P}$?

Rem3₈₆ | If Cramér's conjecture (see [Cra36]) on the gaps between consecutive primes is true, then **C3** $\leq_P$ **C1**. Since the density of primes between n and $2n$ is approximately $1/\log n$, it follows that **C3** $\leq_{\mathcal{R}}$ **C1**. This problem has cryptographic significance [DH76], [RSA78].

Rem3₉₄ | As we mentioned in **Rem1₉₄**, the exact formulation of Cramér's conjecture has now been called into question. It is still probably true that for every constant $c > 2$, there is a constant $d > 0$ such that there is a prime between x and $x + d(\log x)^c$. This hypothesis still implies that **C3** $\leq_P$ **C1**.

Note, since **C1** is recognized in $\mathcal{R}$ (see **Rem1₉₄**), it follows that **C3** is in $\mathcal{R}$. If anything, the importance of this problem has grown since 1986, since there have been numerous cryptosystems proposed since then that require the ability to construct large primes, sometimes with special properties. See [Pom90].

Ref3 | [Bac88], [Pla79]. See also **Ref1**.

4 Prime in an arithmetic progression

C4 | Input $a, n \in \mathbb{N}$.
Output $p \in Primes$ with $p \equiv a \pmod{n}$ if $\gcd(a, n) = 1$.

O4 | Is **C4** in $\mathcal{P}$?

Rem4₈₆ | It was conjectured by Heath-Brown [HB78] that if $\gcd(a, n) = 1$, then the least prime $p \equiv a \pmod{n}$ is $O(n \log^2 n)$, and this would imply that **C4** $\leq_P$ **C1**. If there are no Siegel zeroes, then the density of small primes in the arithmetic progression a modulo n is sufficient to conclude that **C4** $\leq_{\mathcal{R}}$ **C1** [Bom74]. Without hypothesis, it is known [EH71] that Heath-Brown's conjecture is true for almost all pairs a, n with $\gcd(a, n) = 1$. Hence if **C1** is in $\mathcal{P}$, then one can solve **C4** in deterministic polynomial time for almost all inputs. See also **Rem20₈₆**.

Rem4$_{94}$ Since **C1** is now known to be in $\mathcal{R}$ (see **Rem1$_{94}$**), it follows that **C4** is also in $\mathcal{R}$. **C4** also has cryptographic applications [Sch91], [BM92], [oC91].

Ref4   [AM77]

5   Integer factoring

C5   Input $n \in \mathbf{N}$.
Output $p_1, p_2, \ldots, p_k \in Primes$ and $e_1, e_2, \ldots, e_k \in \mathbf{N}$ such that

$$n = \prod_{i=1}^{k} p_i^{e_i} \text{ if } n > 1 .$$

O5a   Is **C5** in $\mathcal{P}$?

O5b   Is **C5** in $\mathcal{R}$?

Rem5$_{86}$ Another classical problem, mentioned by Gauss in his *Disquisitiones Arithmeticæ* (see **Rem1$_{86}$**). There are a large number of random algorithms for **C5** whose running time is believed to be $L(n)^c$ for varying constants $c \geq 1$ [Pom82], [Len87], [SL84]. The only random algorithm of this class whose running time has actually been proved to be $L(n)^c$ is due to Dixon [Dix81]. Dixon's algorithm is unfortunately not practical. A determination of the complexity of **C5** would have significance in cryptography [RSA78].

Rem5$_{94}$ A great deal of progress has been made in the area of factoring integers. Lenstra and Pomerance [LP92] proved the existence of a probabilistic algorithm for factoring integers with an expected running time of $L_n[1/2, 1]$, improving on Dixon's bound. Another interesting development was the discovery of the number field sieve. A heuristic analysis suggests that there exists a constant $c > 0$ such that the number field sieve factors an integer n in expected time $L_n[1/3, c]$. Contributions to the number field sieve were made by a number of researchers, including (but not limited to) Adleman, Buhler, Coppersmith, Couveignes, A.K. Lenstra, H.W. Lenstra, Manasse, Odlyzko, Pollard, Pomerance and Schroeppel. See [Adl91], [Cop90], [Cou93], [LL93], and the references cited therein.

In a very recent development Peter Shor [Shoar] has shown that factoring can be done in polynomial time on a "quantum computer". It is premature to judge the implications of this development.

Ref5   [Dix81], [Guy77], [Knu81], [Len87], [MB75], [Pom82], [Rie85b], [Rie85a], [Sha71], [Sch82], [SL84], [Wil84].

6 Factoring a set of positive density

Let $S \subset \mathbf{N}$.

C6   Input $n \in \mathbf{N}$.
          Output $p_1, p_2, \ldots, p_k \in Primes$ and $e_1, e_2, \ldots, e_k \in \mathbf{N}$ such that

$$n = \prod_{i=1}^{k} p_i^{e_i} \text{ if } n > 1 \text{ and } n \in S \ .$$

O6   Does there exist a set S such that

$$\liminf_{x \to \infty} \frac{\#\{n \mid n \leq x \ \& \ n \in S\}}{x} > 0$$

and **C6**(S) is in $\mathcal{P}$?

Rem6$_{86}$ Assuming the necessary hypotheses for the running time analysis for Lenstra's elliptic curve factoring method (see [Len87]), it is probably possible to prove that a set S satisfying

$$\liminf_{x \to \infty} \frac{\#\{n \mid n \leq x \ \& \ n \in S\}}{\frac{x \log \log^2 x}{\log x \log \log \log x}} > 0 \tag{1}$$

can be factored in random polynomial time. This set will still have density zero, however. A related question is whether factoring a set of positive density is random polynomial time equivalent to **C5**. The set *Squarefrees* has density $6/\pi^2$ however it is not even clear that **C5** $\leq_R$ **C6**(*Squarefrees*).

Rem6$_{94}$ Let A denote a deterministic algorithm for factoring integers, and define $F(x, t, A)$ to be the number of integers n with $1 \leq n \leq x$ such that A will factor n in at most t bit operations. **O6** can then be stated as asking whether there exists an algorithm A and a constant $c > 0$ such that

$$\liminf_{x \to \infty} \frac{F(x, \log^c x, A)}{x} > 0 \ .$$

This problem remains open, but Hafner and McCurley [HM89a] and later Sorenson [Sor90] proved several results about the behaviour of F for various factoring algorithms (including a generalization to cover probabilistic algorithms). The estimate (1) has still not been proved, and the best result known [HM89a] in this direction is

$$F(x, \log^c x, A) \gg_c \frac{x (\log \log x)^{\frac{6}{5} - \epsilon}}{\log x} \ ,$$

using a probabilistic algorithm. In this formulation, one may also ask for the slowest growing function $t(x)$ such that there exists an algorithm A with

$$\liminf_{x \to \infty} \frac{F(x, t(x), A)}{x} > 0 \ .$$

7 Squarefree part

C7   Input $n \in \mathbf{N}$.
          Output $r, s \in \mathbf{N}$ with $n = r^2 s$ and $s \in Squarefrees$.

O7a   Is **C7** in $\mathcal{P}$?

O7b   Is **C5** $\leq_{\mathcal{R}}$ **C7** ?

Rem7₈₆ See **Rem13₈₆**. Clearly **C7** $\leq_{\mathcal{P}}$ **C5**. The analogous question for $f \in$ $\mathbf{Q}[x]$ or $(\mathbf{Z}/p\mathbf{Z})[x]$ is solvable in polynomial time by performing calculations of the form $\gcd(f, f')$, where f' is the (formal) derivative of f. (see [Knu81, page 421]).

Rem7₉₄ Landau [Lan88] proved that **C7** $\leq_{\mathcal{P}}$ **C23**. According to [Len92], Chistov [Chi89] has shown that **C7** is polynomial time equivalent to determining the ring of integers in a number field.

8 Squarefreeness

C8   Input $n \in \mathbf{N}$.
          Output   1 if $n \in Squarefrees$,
                    0 otherwise.

O8   Is **C8** in $\mathcal{P}$?

Rem8₈₆ A generalization of this is, given n and $k \in \mathbf{N}$, to determine if n is divisible by the kth power of a prime. Another generalization is to output $\mu = \mu(n)$, where

$$\mu(n) = \begin{cases} 1 & \text{if } n = 1, \\ 0 & \text{if there exists a } p \in Primes \text{ with } p^2 \mid n, \\ (-1)^k & \text{if } n \text{ is a product of } k \text{ distinct primes.} \end{cases}$$

Shallit and Shamir have shown that this generalization is reducible to the problem of computing the function d mentioned in **Rem9₈₆**.

Rem8₉₄ We are unaware of any progress on this problem.

9 Number of distinct prime factors

C9   Input $n \in \mathbf{N}$.
          Output $\omega(n) = \#\{p \mid p \in Primes \ \& \ p \mid n\}$.

O9   Is **C9** in $\mathcal{P}$?

Rem9₈₆ Clearly **C1** $\leq_{\mathcal{P}}$ **C9**, since we can easily check to see if n is a perfect power. An interesting variant of **C9** is to output $\Omega(n) = e_1 + \ldots + e_k$, where $n = \prod_{i=1}^{k} p_i^{e_i}$ is the prime factorization of n. Another variant is to output $d(n) = \#\{k \mid k \in \mathbf{N} \ \& \ k \mid n\}$, and still another variant is to output the multiset $\{e_1, \ldots, e_k\}$. Shallit and Shamir [SS85] have proved that the last two variants are polynomial time equivalent to each other. As a consequence we have that **C9** is polynomial time reducible to the problem of computing the function $d(n)$ mentioned above.

Rem9$_{94}$ We are unaware of any progress on this problem. It is remarkable that one can decide if $\omega(n) = 1$ in random polynomial time [AH92], but there are no other partial results known on this problem.

10 Roots modulo a composite

C10   Input $e, a, n \in \mathsf{N}$.
Output $x \in \mathsf{N}$ such that $x^e \equiv a \pmod{n}$, if $\gcd(e, \phi(n)) = 1$ and $\gcd(a, n) = 1$.

O10   Is **C5** $\leq_{\mathcal{R}}$ **C10**?

Rem10$_{86}$ When the restriction that $\gcd(e, \phi(n)) = 1$ is dropped, it is known that **C5** $\leq_{\mathcal{R}}$ **C10** [Rab79]. A resolution of this problem would have important consequences in public-key cryptography [RSA78]. It is known that **C10** $\leq_{\mathcal{P}}$ **C23**.

Rem10$_{94}$ We are unaware of any progress on this problem.

11 Quadratic residuosity modulo a composite

C11   Input $a, n \in \mathsf{N}$.
Output   1   if there exists an $x \in \mathsf{N}$ such that $x^2 \equiv a \pmod{n}$ and $\gcd(a, n) = 1$,
0   otherwise.

O11a   Is **C11** in $\mathcal{P}$?

O11b   Is **C5** $\leq_{\mathcal{R}}$ **C11**?

Rem11$_{86}$ It is easy to show that **C11** $\leq_{\mathcal{P}}$ **C5**. There is an obvious generalization where the exponent 2 is replaced by another exponent k that is either fixed for the problem or supplied as an input. The presumed difficulty of **C11** has been used as a basis for cryptographic systems [GM82], [GM84], [Yao82], [BBS86]. **C11** is related to **C9** since the proportion of residues modulo n that are quadratic residues is $2^{-\omega(n)}$, where $\omega(n)$ is the number of distinct prime divisors of n. Therefore given an algorithm for **C11**, one can obtain a confidence interval for $\omega(n)$ by checking random values.

Rem11$_{94}$ We are unaware of any progress on this problem.

Ref11   [AM82].

12 Quadratic non-residue modulo a prime

C12 Input $p \in \mathbf{N}$.

Output $b \in \mathbf{N}$ such that there does not exist $c \in \mathbf{N}$ with $c^2 \equiv b$ (mod p), if $p \in Primes$.

O12 Is **C12** in $\mathcal{P}$?

Rem12$_{86}$ **C12** is easily seen to be in $\mathcal{R}$, since polynomial time algorithms for the corresponding problem of distinguishing quadratic residues from nonresidues can be based on the Jacobi symbol and the law of quadratic reciprocity, or else on Euler's criterion:

$$p \in Primes \text{ and } p \nmid a \Rightarrow a^{\frac{p-1}{2}} \equiv (\frac{a}{p}) \pmod{p} .$$

Curiously, Gauss was aware of Euler's criterion, but was apparently unimpressed by its efficiency [Gau86, art. 106]:

> Although it is of almost no practical use, it is worthy of mention because of its simplicity and generality ... But as soon as the numbers we are examining are even moderately large this criterion is practically useless because of the amount of calculation involved.

Under the extended Riemann hypothesis, **C12** is in $\mathcal{P}$ [Mil76]. It is also known that the least quadratic nonresidue is almost always small [Erd61], so **C12** can be solved in deterministic polynomial time for almost all inputs.

Rem12$_{94}$ On the problem of calculating kth power non-residues in $\mathrm{GF}(p^n)$, the following is known. On ERH, the algorithm of Huang [Hua85], generalized by Evdokimov [Evd89], constructs a kth power non-residue, in $\mathrm{GF}(p^n)$ in deterministic time $(kn \log p)^{O(1)}$. Buchmann and Shoup [BS91], on ERH, construct a kth power non-residue in $\mathrm{GF}(p^n)$ in deterministic time $(\log p)^{O(n)}$. Bach [Bac90], on ERH, has given explicit bounds for estimations of the least kth power non-residue. See also **Rem19$_{94}$**.

Ref12 [Ank52], [Bac85].

13 Quadratic signature

C13 Input $\sigma \in \{-1, 1\}^*$.

Output The least $p \in Primes$ such that for all i with $1 \leq i \leq |\sigma|, (\frac{p_i}{p}) = \epsilon_i$, where $|\sigma|$, the length of σ, is the number of symbols in σ, p_i is the ith prime, and ϵ_i is the ith symbol of σ.

O13 Is **C13** in $\mathcal{P}$?

Rem13$_{86}$ If n has the form $m^2 q$ with q an odd prime and m odd, then for any a with $\gcd(a, n) = 1$ we have $\left(\frac{a}{n}\right) = \left(\frac{a}{q}\right)$. It follows that if **C13** is in $\mathcal{P}$, then n could be partially factored since, assuming the extended Riemann hypothesis, q can be determined by a signature of length $O(\log^2 n)$ [Mil76], [Ank52]. The notion of quadratic signature can be generalized; see [AM82].

Rem13$_{94}$ The concept of quadratic signature has found application in the number field sieve [Adl91].

Ref13 [Ank52], [Bac85], [Bac90].

14 Square roots modulo a prime

C14   Input $a, p \in \mathbb{N}$.
          Output $x \in \mathbb{N}$ with $x^2 \equiv a \pmod{p}$ if $p \in \textit{Primes}$ and such an x exists.

O14   Is **C14** in $\mathcal{P}$?

Rem14$_{86}$ Among the researchers who have presented algorithms for **C14** are [Gau86, art. 319-322], [Ton91], [Leh69], [Sha72], [Ber67], [Rab80b], [AMM77]. It is now known that **C14** is in $\mathcal{R}$. It is also known that **C14** $\leq_{\mathcal{P}}$ **C12** and that on the extended Riemann hypothesis, **C14** is in $\mathcal{P}$. There is a natural generalization of **C14** where the exponent 2 is replaced by a fixed k. Another generalization has k as part of the input. For this version there is a random time $O((k \log p)^c)$ algorithm based on known algorithms for **C15**. One can also use a discrete logarithm algorithm (see **Rem21$_{86}$**) to solve this variant, resulting in a random time $O(L(p))$ algorithm, which for large k will be faster.

Rem14$_{94}$ It is an oversight that we did not mention the work of Schoof [Sch85] on this problem in our earlier manuscript. Schoof proved that for fixed a, there exists a deterministic algorithm with running time polynomial in $\log p$.

Ref14 Many additional references are given in [LN83, page 182]. See also **Ref16** and [Hua85], [Evd89], [BS91].

15 Polynomial roots modulo a prime

C15   Input $p \in \mathbb{N}$, $f \in (\mathbb{Z}/p\mathbb{Z})[x]$.
          Output $a \in \mathbb{Z}$ with $f(a) \equiv 0 \pmod{p}$ if $p \in \textit{Primes}$ and such an a exists.

O15   Is **C15** in $\mathcal{P}$?

Rem15₈₆ See **Rem14₈₆**. C15 is in $\mathcal{R}$ [Ber70], [CZ81], [Rab80b]. If the extended Riemann hypothesis is assumed and f has abelian Galois group over the rationals, then the problem is in $\mathcal{P}$ [Hua85].

Rem15₉₄ If f is fixed the problem appears to remain difficult; however, for certain f progress has been made. When f is linear the problem is trivial. When f is a quadratic there exists a deterministic polynomial time algorithm due to Schoof [Sch85]. When f is a cyclotomic polynomial, there exists a deterministic polynomial time algorithm due to Pila [Pil90].

Ref15 [Sho90b], [BS91]. See also **Ref16**.

16 Factoring polynomials modulo a prime

C16   Input $p \in \mathbb{N}, f \in (\mathbb{Z}/p\mathbb{Z})[x]$.
                Output   irreducible $g_1, \ldots, g_k \in (\mathbb{Z}/p\mathbb{Z})[x]$, and $e_1, \ldots, e_k \in \mathbb{N}$ such that $f = \prod_{i=1}^{k} g_i^{e_i}$, if $p \in Primes$.

O16   Is **C16** in $\mathcal{P}$?

Rem16₈₆ See **Rem15₈₆**. C16 is in $\mathcal{R}$ [Ber70], [CZ81], [Rab80b]. The corresponding problem over $\mathbb{Q}$ is in $\mathcal{P}$ [LLL82].

Rem16₉₄ Let n denote the degree of f. Rónyai [Rón88] on ERH gives a deterministic algorithm with running time $(n^n \log p)^{O(1)}$. Evdokimov [Evdar] on ERH gives a deterministic algorithm with running time $(n^{\log n} \log p)^{O(1)}$. In particular, both algorithms are polynomial time if the degree is bounded. For the case $f \in Z[x]$, f irreducible and $Q[x]/(f)$ Abelian over Q, Huang [Hua91] on ERH gives a deterministic polynomial time algorithm. For the case $f \in Z[x]$, f irreducible and $Q[x]/(f)$ Galois over Q, Rónyai on ERH gives a deterministic polynomial time algorithm [Rón89]. For the case $f \in Z[x]$ solvable, Evdokimov [Evd89] on ERH gives a deterministic polynomial time algorithm.

Lenstra [Len90] has shown in many cases the assumption of ERH above may be removed if irreducible polynomials of appropriate degree can be found in deterministic polynomial time.

Buchmann and Shoup [BS91] proved, under ERH, that for all $n \in \mathbb{N}$, there exists a deterministic algorithm for **C16** with running time $\sqrt{k}$ times a polynomial in the input size, where k is the largest prime dividing $\phi_n(p)$ and ϕ_n is the n-th cyclotomic polynomial.

Ref16 [Ber67], [Ber68], [Knu81, pages 420–441], [LN83, pages 147-185].

17 Irreducible polynomials

C17   Input $d, p \in \mathbb{N}$.

Output   irreducible $f \in (\mathbb{Z}/p\mathbb{Z})[x]$ with $degree(f) = d$, if $p \in$ *Primes*.

O17   Is **C17** in $\mathcal{P}$?

Rem17₈₆ **C17** is in $\mathcal{R}$ [Ber68], [Rab80b]. **C17** is in $\mathcal{P}$ if the extended Riemann hypothesis is true [AL86]. There is a $c \in \mathbb{N}$ and a deterministic polynomial time algorithm which on input d, p with $p \in$ *Primes* outputs an irreducible $f \in (\mathbb{Z}/p\mathbb{Z})[x]$ of degree greater than $cd/\log p$ and less than or equal to d [AL86]. Since irreducible quadratics yield quadratic nonresidues, it is clear that **C12** $\leq_{\mathcal{P}}$**C17**, and also from the results on **C14** that **C14** $\leq_{\mathcal{P}}$**C17**.

Rem17₉₄ The result of [AL86] was discovered independently by Evdokimov [Evd89]. Shoup [Sho90a] proved **C17** $\leq_{\mathcal{P}}$**C16**, and gave a deterministic algorithm for finding an irreducible polynomial of degree d over $\mathbb{Z}/p\mathbb{Z}$ in time $\sqrt{p}(d + \log p)^{O(1)}$.

Ref17   [Len92].

18 Recognition of a primitive root modulo a prime

C18   Input $b, p \in \mathbb{N}$.
             Output   1 if b is a generator of $(\mathbb{Z}/p\mathbb{Z})^*$ and $p \in$ *Primes*,
                      0 if b is not a generator of $(\mathbb{Z}/p\mathbb{Z})^*$ and $p \in$ *Primes*.

O18a   Is **C18** in $\mathcal{P}$?

O18b   Is **C18** recognized in $\mathcal{R}$?

Rem18₈₆ It is known that **C18** $\leq_{\mathcal{P}}$**C5**, since b is a primitive root modulo p if and only if $p \nmid b$ and

$$\forall q[[q \in Primes \ \& \ q \mid p - 1] \Rightarrow b^{(p-1)/q} \not\equiv 1 \pmod{p}] \ .$$

A generalization of **C18** where a third input $c \in \mathbb{N}$ is given and the output is 1 if and only if b has order c is also of interest.

Rem18₉₄ We are unaware of any progress on this problem. We would like to point out however that under ERH, **C18** $\leq_{\mathcal{P}}$**C21**. To see why, recall that under ERH, the least primitive root modulo p is $\leq c\log^6 p$ for some constant c [Sho90c]. Let g be a suspected primitive root modulo p. We dovetail the following procedures:

process A for $b = 1, 2, \ldots, c\log^6 p$: ask oracle for **C21** to compute an x with $g^x \equiv b \pmod{p}$. If the oracle returns an x keep it only if you confirm that $g^x \equiv b \pmod{p}$. If for all b an x is kept then output "primitive root".

process B for $b = 1, 2, \ldots, c\log^6 p$: ask oracle for **C21** to compute x such that $b^x \equiv g \pmod{p}$. If the oracle returns an x keep it only if you confirm that $b^x \equiv g \pmod{p}$. If for some b an x is kept with $\gcd(x, p-1) > 1$, then output "not a primitive root".

19 Finding a primitive root modulo a prime

C19   Input $p \in \mathbb{N}$.
            Output $g \in \mathbb{N}$ such that $1 \le g \le p-1$ and g generates $(\mathbb{Z}/p\mathbb{Z})^*$, if $p \in Primes$.

O19   Is C19 in $\mathcal{P}$?

Rem19$_{86}$ The density of generators is sufficient that it is easily shown that C19 $\le_{\mathcal{R}}$ C18. If the extended Riemann hypothesis is true, then the least generator is small [Wan61], and C19 $\le_{\mathcal{P}}$ C18. An interesting variant of C19 involves finding elements of $(\mathbb{Z}/p\mathbb{Z})^*$ of desired order. C19 has an obvious extension to an arbitrary finite field, or for that matter to any cyclic group.

Rem19$_{94}$ Shoup [Sho90c] proved several results related to this problem. Among other things, he proved under the assumption of the extended Riemann hypothesis that a primitive root for $GF(p^2)$ can be constructed in deterministic polynomial time. Buchmann and Shoup [BS91], on ERH, give a deterministic algorithm, which on input an irreducible f of degree n over $\mathbb{Z}/p\mathbb{Z}$, outputs a generating set for $\mathbb{Z}/p\mathbb{Z}[x]/(f)$ in time $(\log p)^{O(n)}$. As a consequence, if the factorization of $p^n - 1$ is known, then under the assumption of ERH, a primitive root of $GF(p^n)$ can be computed in deterministic polynomial time.

20 Calculation of orders modulo a prime

C20   Input $a, p \in \mathbb{N}$.
            Output $k = min\{x \mid x \in \mathbb{N}, a^x \equiv 1 \pmod{p}\}$, if $p \in Primes$ and $\gcd(a, p) = 1$.

O20   Is C20 in $\mathcal{P}$?

Rem20$_{86}$ The variant in which p is not required to be prime is random polynomial time equivalent to C5 [Mil76]. A related question: is the problem of factoring numbers of the form $p-1$, with p prime, polynomial time reducible to C20? If C6 is in $\mathcal{P}$, then the problem of factoring numbers of the form $p-1$ with p prime is polynomial time equivalent to factoring.

Rem20$_{94}$ We are unaware of any progress on this problem.

21 Discrete logarithm modulo a prime

C21   Input $g, b, p \in \mathbb{N}$.
            Output $x \in \mathbb{N}$ with $g^x \equiv b \pmod{p}$, if $p \in Primes$ and such an x exists.

O21   Is **C21** in $\mathcal{P}$?

Rem21$_{86}$ If the prime factors of $p-1$ are less than $\log^c p$ for some constant $c > 0$, then the problem is in $\mathcal{P}$ [PH78]. The fastest known algorithms for solving **C21** have running times of $L(p)$ [COS86]. The resolution of **O21** would have important consequences in cryptography [ElG85], [BM84]. There is an obvious generalization of **C21** to an arbitrary finite field. Bach [Bac84] has asked if the problem of factoring numbers of the form $p-1$, with p prime, is polynomial time reducible to **C21**.

Rem21$_{94}$ There has been considerable progress on this problem. Pomerance [Pom86] proved that there exists a probabilistic algorithm to compute discrete logarithms in $GF(q)$ with expected running time of $L_q[1/2, \sqrt{2}]$, for the case where q is prime or q is a power of 2. Gordon [Gor93] presented an adaptation of the number field sieve to computing discrete logarithms in $\mathbf{Z}/p\mathbf{Z}$, along with a heuristic argument to suggest an expected running time of $L_p[1/3, c]$ for some positive constant c.

For discrete logarithms over general finite fields, progress has also been made. At the time that we wrote our original paper, we neglected to mention the work of Coppersmith [Cop84], who had published an algorithm for $GF(2^n)$ with a heuristic expected running time bounded by $L_{2^n}[1/3, c]$ for some positive constant c. Lovorn [Lov92] proved a running time of $L_q[1/2, c]$ for some positive constant c when $q = p^n$ with $\log p \leq n^{0.98}$. Adleman and De-Marrais [AD93a] gave an algorithm for arbitrary finite fields whose heuristic expected running time is $L_q[1/2, c]$ for some positive constant c. Adleman's function field sieve [Adlar] gives a heuristic expected running time of $L_q[1/3, c]$ for some positive constant c when $q = p^n$ and $\log p \leq n^{g(n)}$, where g is any function such that $0 < g(n) < 0.98$ and $\lim_{n \to \infty} g(n) = 0$.

Surveys on the discrete logarithm problem have been published: [vO91], [McC90a], [Odl94].

Historically, advances in integer factoring algorithms have brought corresponding advances in discrete logarithm algorithms. The first author thinks it is an interesting research problem to establish whether reductions exist between **C5** and **C21**. The second author finds the evidence for the existence of such reductions to be unconvincing.

In a very recent development Peter Shor [Shoar] has shown that discrete logarithms can be computed in polynomial time on a "quantum computer". It is premature to judge the implications of this development.

Ref21   [Odl85], [Sch93], [AD93b].

22   Discrete logarithm modulo a composite

C22   Input $g, b, n \in \mathbb{N}$.
            Output $x \in \mathbb{N}$ with $g^x \equiv b \pmod{n}$, if such an x exists.

O22a   Is C22 in $\mathcal{P}$?

O22b   Is C5 $\leq_{\mathcal{P}}$ C22?

Rem22₈₆ Clearly C21 $\leq_{\mathcal{P}}$ C22. It is also known that C5 $\leq_{\mathcal{R}}$ C22 [Bac84].
            The resolution of **O22** would have consequences in public-key cryp-
            tography [McC88]. There is an obvious generalization to an arbitrary
            group (see also **C28**).

Rem22₉₄ We are unaware of any progress on this problem.

23   Calculation of $\phi(n)$

C23   Input $n \in \mathbb{N}$.
            Output $\phi(n)$.

O23   Is C5 $\leq_{\mathcal{P}}$ C23?

Rem23₈₆ It is known that **C5** $\leq_{\mathcal{R}}$ **C23** [Mil76], and it is obvious that **C23**
            $\leq_{\mathcal{P}}$ **C5**. **C5** is known to be random polynomial time equivalent to the
            problem of computing $\sigma(n)$, the sum of the positive integral divisors
            of n [BMS84].

Rem23₉₄ We are unaware of any progress on this problem. See **Rem7₉₄**.

24   Point on an elliptic curve

C24   Input $a, b, p \in \mathbb{N}$.
            Output $x, y \in \mathbb{N}$ with $y^2 \equiv x^3 + ax + b \pmod{p}$, if $p \in Primes$ and
            $p \nmid 4a^3 + 27b^2$.

O24   Is C24 in $\mathcal{P}$?

Rem24₈₆ One can show that **C24** is in $\mathcal{R}$, since there is an easy argument
            to show that **C24** $\leq_{\mathcal{R}}$ **C14**: choose random values of x, evaluate the
            right hand side, and use a random algorithm for **C14** to try to solve
            for y. A theorem of Hasse implies that the probability of choosing a
            successful x is approximately $\frac{1}{2}$.

Rem24₉₄ We are unaware of any progress on this problem. **C24** has applica-
            tions in cryptography [Kob87b, p. 162].

25   Binary quadratic congruences

C25   Input $k, m, n \in \mathbb{N}$.

Output $x, y \in \mathbb{N}$ with $x^2 - ky^2 \equiv m$ (mod n), if n is odd and $\gcd(km, n) = 1$.

O25   Is C25 in $\mathcal{P}$?

Rem25$_{86}$ C25 is in $\mathcal{R}$ [AEM87]. If the extended Riemann hypothesis and Heath-Brown's conjecture on the least prime in an arithmetic progression are true, then **C25** is in $\mathcal{P}$ [Sha84]. **C25** arose from cryptography [OSS84], [PS87]. In fact, **C25** is only one example of a wide range of questions concerning solutions of $f \equiv 0$ (mod n), where f is a multivariate polynomial with coefficients in $\mathbb{Z}/n\mathbb{Z}$. Such questions can vary greatly in their complexity as the form of the question changes. We may ask questions about determining if a solution exists, finding a solution, finding the least solution, or finding the number of solutions. We may vary the form of the polynomial or the properties of n (e.g. prime, composite, squarefree). As an example of the variation in complexity, even for the polynomial $f(x) = x^2 - a$ we have the following situation:

1. The problem of deciding from inputs $a, p \in \mathbb{N}$ whether $x^2 - a \equiv 0$ (mod p) has a solution when p is prime is in $\mathcal{P}$ (see **Rem12$_{86}$**.)

2. The problem of finding from inputs $a, p \in \mathbb{N}$ a solution of $x^2 - a \equiv 0$ (mod p) when p is prime is in $\mathcal{R}$ (see **Rem14$_{86}$**).

3. The problem of finding from inputs $a, n \in \mathbb{N}$ a solution of $x^2 - a \equiv 0$ (mod n) is random equivalent to the problem of factoring n (see **Rem10$_{86}$**).

4. The problem of finding from inputs $a, n \in \mathbb{N}$ the least positive integer solution of $x^2 - a \equiv 0$ (mod n) is $\mathcal{NP}$-hard [MA78].

We therefore view the problem of classifying all problems concerning solutions of $f \equiv 0$ (mod n) according to their complexity as an important metaproblem.

Rem25$_{94}$ We are unaware of any progress on this problem. There has been marginal progress on the "metaproblem". We regard this area as a very fruitful one for future investigations.

Ref25   [vzGKS93]. Some cryptographic problems related to the metaproblem are mentioned in [McC90b]. That paper also contains pointers to other unsolved number-theoretic problems relating to cryptology.

26   Key distribution

C26   Input $g, p, a, b \in \mathbb{N}$.
Output $c \in \mathbb{N}$, where $c \equiv g^{xy}$ (mod p), if $p \in Primes$, g is a primitive root modulo p, $a \equiv g^x$ (mod p), and $b \equiv g^y$ (mod p).

O26   Is C21 $\leq_{\mathcal{R}}$ C26?

Rem26$_{86}$ The motivation for this problem comes from cryptography [DH76]. It is obvious that **C26** $\leq_{\mathcal{P}}$ **C21**. There is a generalization where p is replaced by a composite n, and we ask only for an output c when a and b are powers of g. For this generalization is the problem equivalent to **C5** or **C22** (see [Bac84], [McC88])?

Rem26$_{94}$ Bert den Boer [dB90] proved that when all prime factors of $\phi(p-1)$ are small, the key distribution problem is as hard as computing discrete logarithms.

Ref26 [Odl85], [ElG85].

27 Construction of an elliptic curve group of a given order

C27 Input $p, n \in \mathbf{N}$.
Output $a, b \in \mathbf{N}$ with $\#E_{p,a,b} = n$, if $p \in Primes$ and such an a, b exist.

O27 Is C27 in $\mathcal{P}$?

Rem27$_{86}$ There is a polynomial time algorithm that, given p, a, and b with $p \nmid 4a^3 + 27b^2$ computes $\#E_{p,a,b}$ [Sch85].

Rem27$_{94}$ We are unaware of any progress on this problem, however it is known that for some primes p, supersingular curves of order $p + 1$ can be constructed efficiently (see [MOV94]).

Ref27 [Kob87b], [Kob87a], [Sch85], [Sil86], [Kob91], [Kob91], [Kob88].

28 Discrete logarithms in elliptic curve groups

C28 Input $a, b, p \in \mathbf{N}, P, Q \in S_{p,a,b}$
Output $n \in \mathbf{N}$ with $P = nQ$, if $p \in Primes$ and such an n exists.

O28 Is C28 in $\mathcal{P}$?

Rem28$_{86}$ The presumed difficulty of this problem has been used as the basis for a public key cryptosystem and digital signature scheme [Kob87b], [Mil86]. Whereas for the discrete logarithm problem in the multiplicative group modulo a prime there is a subexponential algorithm (see **Rem21$_{86}$**), no such algorithm is known to exist for **C28**. A related problem is given a, b, and p to construct a minimal set of generators for $E_{p,a,b}$.

Rem28$_{94}$ Menezes, Okamoto, and Vanstone [MOV94] used Weil pairing to prove that there exists a probabilistic reduction from **C28** to the problem of computing discrete logarithm in the multiplicative group of a (perhaps high degree) extension of GF(q). For supersingular curves, this reduction can be carried out in random polynomial time, with the result that a probabilistic subexponential algorithm is obtained for **C28** in this special case.

Koblitz [Kob90] has suggested cryptographic uses for the rational subgroups of the Jacobian of a hyperelliptic curve over a finite field. Adleman, Huang, and DeMarrais [AHDar] discovered a heuristic subexponential probabilistic algorithm for the discrete logarithm problem in these subgroups when the genus of the curve is large with respect to the size of the finite field.

29 Shortest vector in a lattice

C29   Input $b_1, \ldots, b_n \in \mathbf{Z}^n$
Output $v \in \Lambda$ with $\|v\|_2 = \min\{\|x\|_2 \mid x \in \Lambda, x \neq 0\}$, where $\Lambda = \mathbf{Z}b_1 \oplus \ldots \oplus \mathbf{Z}b_n$ if $b_1, \ldots, b_n$ span $\mathbf{R}^n$.

O29   Is **C29** $\mathcal{NP}$-hard?

Rem29$_{86}$ The corresponding problems with norms $\|\cdot\|_\infty$ and $\|\cdot\|_1$ are known to be $\mathcal{NP}$-hard [Lag85], [vEB81]. See also **Rem30$_{86}$**.

Rem29$_{94}$ It was an oversight that we did not mention the result of Lenstra [Len83], who proved that if the dimension n is fixed, the shortest vector in a lattice of dimension n can be found in polynomial time.

Ref29   [GLS88] and [Lov86] contain nice surveys of this and related topics.

30 Short vector in a lattice

Let $c \in \mathbf{N}$

C30   Input $b_1, \ldots, b_n \in \mathbf{Z}^n$
Output $v \in \Lambda$ with $\|v\|_2 \leq n^c \min\{\|x\|_2 \mid x \in \Lambda, x \neq 0\}$, where $\Lambda = \mathbf{Z}b_1 \oplus \ldots \oplus \mathbf{Z}b_n$ if $b_1, \ldots, b_n$ span $\mathbf{R}^n$.

O30   Does there exist a $c \in \mathbf{N}$ for which **C30** is in $\mathcal{P}$?

Rem30$_{86}$ In [LLL82] it was shown that there is a polynomial time algorithm that produces a vector $v \in \Lambda$ with

$$\|v\|_2 \leq 2^{\frac{n-1}{2}} \min\{\|x\|_2 \mid x \in \Lambda, x \neq 0\} \ ,$$

and in [Sey87] it was shown that for any $\epsilon > 0$ there is a polynomial time algorithm $\mathcal{A}_\epsilon$ that produces a vector $v \in \Lambda$ with

$$\|v\|_2 \leq (1 + \epsilon)^n \min\{\|x\|_2 \mid x \in \Lambda, x \neq 0\} \ .$$

A number of related problems in simultaneous diophantine approximation are discussed in [Lag85] and [Fru85].

Rem30$_{94}$ We are unaware of any progress on this problem.

Ref30   [LLS90], [GLS88], [Lov86].

31 Galois group of a polynomial

C31   Input $f \in \mathbb{Q}[x]$.

Output $n = [K : \mathbb{Q}]$, where K is the splitting field of f.

O31   Is **C31** in $\mathcal{P}$?

Rem31₈₆ n is the order of the Galois group associated with f. Polynomial time algorithms exist for determining if n is a power of 2 or if the Galois group is solvable [LM85]. Many other properties of the Galois group can also be determined in polynomial time [Kan85].

Rem31₉₄ Landau [Lan85] proved that the Galois group can be computed in deterministic time $O((\#G + \ell)^c)$ for some constant $c > 0$, where ℓ is the length of the input specification of f and K. Further results are discussed in [Len92], but the problem remains open.

32 Class numbers

C32   Input $d \in \mathbb{N}$.

Output $h(-d)$, the order of the group of equivalence classes of binary quadratic forms with discriminant $-d$ under composition.

O32   Is **C32** in $\mathcal{P}$?

Rem32₈₆ This is related to classical questions of Gauss [Gau86, art. 303]. It appears that the results of Shanks [Sha72], [Sha71], Schnorr & Lenstra [SL84], Seysen [Sey87], and Schoof [Sch82] establish that **C5** $\leq_{\mathcal{R}}$ **C32**, and that ERH implies **C5** $\leq_{\mathcal{P}}$ **C32**. It is remarked in [BMS84] that it is not even known if **C32** is in $\mathcal{NP}$. The best known algorithm for computing $h(-d)$ is due to Shanks [Sha71]. The question could also be stated in terms of the class number of orders in the field $\mathbb{Q}(\sqrt{-d})$.

Rem32₉₄ McCurley [McC89] proved under ERH that **C32** is in $\mathcal{NP}$. Hafner and McCurley [HM89b] proved under ERH that there exists a probabilistic algorithm with expected running time $L_d[1/2, \sqrt{2}]$ that will compute not only the class number $h(-d)$, but also the structure of the class group. These results were extended to the case of real quadratic fields by Buchmann and Williams [BW89]. Thiel [Thiar] has shown under ERH that verifying the class number belongs to $\mathcal{NP} \cap co\mathcal{NP}$.

The more general question of computing class numbers and class groups of arbitrary algebraic number fields is also of interest. According to Lenstra [Len92], Buchmann and Lenstra proved that there is a deterministic exponential time algorithm for computing the cardinality and structure of the class group. Buchmann [Buc90] gave a probabilistic subexponential algorithm for a special case of this problem. Lenstra [Len92] outlines an approach to obtaining a probabilistic subexponential algorithm in the general case.

Lenstra's paper [Len92] is an important source for information concerning algorithms and open problems concerning algebraic number fields.

Ref32   [Gol85], [Sha72], [Sch82], [Lag80b], [Buc90].

33   Solvability of binary quadratic diophantine equations

C33   Input $a, b, c, d, e, f \in \mathbf{Z}$.

Output   1   if there exists $x, y \in \mathbf{Z}$ with $ax^2 + bxy + cy^2 + dx + ey + f = 0$ and there does not exist a $g \in \mathbf{Z}$ with $b^2 - 4ac = g^2$,

0   otherwise.

O33a   Is **C33** $\mathcal{NP}$-hard?

O33b   Is **C33** $\mathcal{NP}$-hard with respect to $\mathcal{R}$?

Rem33$_{86}$ It is known that **C33** is recognized in $\mathcal{NP}$ [Lag79]. Without the constraint that $b^2 - 4ac$ is not a square, the problem is known to be $\mathcal{NP}$-hard [MA78]. Certain variants of **C33** are known to be $\mathcal{NP}$-hard with respect to $\mathcal{R}$ [AM77].

Rem33$_{94}$ We are unaware of any progress on this problem.

34   Solvability of anti-Pellian equation

C34   Input $d \in \mathbf{N}$.

Output   1 if there exist $x, y \in \mathbf{Z}$ with $x^2 - dy^2 = -1$,

0 otherwise.

O34   Is **C34** in $\mathcal{P}$?

Rem34$_{86}$ There exist choices of d for which the smallest solution of $x^2 - dy^2 = -1$ cannot be written down in polynomial space [Lag79]. It is known that **C34** is in $\mathcal{NP}$ [Lag80a]. If the factorization of d is provided as part of the input, then the problem is recognized in $\mathcal{R}$, and if in addition we assume the extended Riemann hypothesis, then the problem is in $\mathcal{P}$ [Lag80a].

Rem34$_{94}$ We are unaware of any progress on this problem.

35   Greatest common divisors in parallel

C35   Input $a, b \in \mathbf{N}$.

Output $\gcd(a, b)$.

O35   Is **C35** in $\mathcal{NC}$?

Rem35$_{86}$ The best known results for computing greatest common divisors in parallel are contained in [BK83], [CG] and [KMR87]. One may ask a similar question for the modular exponentiation problem: given $a, b, n \in \mathbf{N}$, compute a^b (mod n). For a definition of $\mathcal{NC}$ see [Coo85] or [Coo81].

Rem35$_{94}$ Polylog depth, subexponential size circuits for both integer GCD and modular exponentiation have been obtained by Adleman and Kompella [AK88].

Ref35 [KMR84].

36 Integer multiplication in linear time

C36 Input $\quad a, b \in \mathbf{N}$.
Output $\quad ab$.

O36 Does there exist an algorithm to solve **C36** that uses only $O(\log(ab))$ bit operations ?

Rem36$_{86}$ The best known algorithm is due to Schönhage and Strassen and uses $O(\log(ab) \cdot \log\log(ab) \cdot \log\log\log(ab))$ bit operations [SS71].

Rem36$_{94}$ We are unaware of any progress on this problem.

Ref36 [Knu81, pages 278-301]

Acknowledgments

During the course of writing this paper we have benefited from conversations with several people, and we would like especially to thank Neal Koblitz, Jeff Lagarias, Gary Miller, Jonathan DeMarrais, Ming-Deh Huang, and Andrew Granville for their contributions. The work of the first author was supported by NSF grant CCR-9214671.

References

[AD93a]   Leonard M. Adleman and Jonathan DeMarrais. A subexponential algorithm for discrete logarithms over all finite fields. In Douglas R. Stinson, editor, *Advances in Cryptology: Crypto '93*, volume 773 of *Lecture Notes in Computer Science*, pages 147–158, New York, 1993. Springer-Verlag.

[AD93b]   Leonard M. Adleman and Jonathan DeMarrais. A subexponential algorithm for discrete logarithms over all finite fields. *Mathematics of Computation*, 61:1–15, 1993. Extended abstract in [AD93a].

[Adl91]   Leonard M. Adleman. Factoring numbers using singular integers. In *Proceedings of the 23th Annual Symposium on Theory of Computing*, pages 64–71, 1991.

[Adlar]   Leonard M. Adleman. The function field sieve. In *Proceedings of the 1994 Algorithmic Number Theory Symposium*, Lecture Notes in Computer Science. Springer-Verlag, to appear.

[AEM87]   Leonard M. Adleman, Dennis Estes, and Kevin S. McCurley. Solving bi-
           variate quadratic congruences in random polynomial time. *Mathematics of
           Computation*, 48:17–28, 1987.

[AH92]   Leonard M. Adleman and Ming-Deh Huang. *Primality testing and two
           dimensional Abelian varieties over finite fields*, volume 1512 of *Lecture Notes
           in Mathematics*. Springer-Verlag, 1992.

[AHDar]   Leonard M. Adleman, Ming-Deh A. Huang, and Jonathan DeMarrais. A
           subexponential algorithm for discrete logarithms in the rational subgroup of
           the Jacobian of a hyperelliptic curve over a finite field. In *Proceedings of the
           1994 Algorithmic Number Theory Symposium*, Lecture Notes in Computer
           Science. Springer-Verlag, to appear.

[AHU74]   Alan Aho, John Hopcroft, and Jeffrey Ullman. *The Design and Analysis of
           Computer Algorithms*. Addison-Wesley, Reading, MA, 1974.

[AK88]   Leonard M. Adleman and Kireeti Kompella. Using smoothness to achieve
           parallelism. In *Proceedings of the 20th ACM Symposium on Theory of Com-
           puting*, pages 528–538, 1988.

[AL86]   Leonard M. Adleman and H. W. Lenstra, Jr. Finding irreducible poly-
           nomials over finite fields. In *Proceedings of the 18th Annual Symposium
           on Theory of Computing*, pages 350–355, New York, 1986. Association for
           Computing Machinery.

[AM77]   Leonard M. Adleman and K. Manders. Reducibility, randomness, and in-
           tractibility. In *Proc. 9th Annual ACM Symposium On Theory Of Comput-
           ing*, pages 151–163, New York, 1977. Association for Computing Machinery.

[AM82]   Leonard M. Adleman and R. McDonnell. An application of higher reci-
           procity to computational number theory. In *Proceedings of the 22nd An-
           nual Symposium on Foundations of Computer Science*, pages 100–106. IEEE
           Computer Society, 1982.

[AM86]   Leonard M. Adleman and Kevin S. McCurley. Open problems in number-
           theoretic complexity. In *Discrete Algorithms and Complexity (Proceedings
           of the Japan-US Joint Seminar on Discrete Algorithms and Complexity The-
           ory)*, pages 237–262. Academic Press, 1986.

[AMM77]   Leonard M. Adleman, K. Manders, and Gary L. Miller. On taking roots in
           finite fields. In *Proceedings of the 18th Annual Symposium on Foundations
           of Computer Science*, pages 175–178, Rhode Island, 1977. IEEE Computer
           Society.

[Ank52]   N. Ankeny. The least quadratic nonresidue. *Annals of Mathematics*, 55:65–
           72, 1952.

[APR83]   Leonard M. Adleman, Carl Pomerance, and Robert Rumely. On distin-
           guishing prime numbers from composite numbers. *Annals of Mathematics*,
           117:173–206, 1983.

[Bac84]   Eric Bach. Discrete logarithms and factoring. Technical Report UCB/CSD
           84/186, University of California, Computer Science Division (EECS), Uni-
           versity of California, Berkely, California, June 1984.

[Bac85]   Eric Bach. *Analytic Methods in the Analysis and Design of Number Theo-
           retic Algorithms*. MIT Press, Cambridge, 1985.

[Bac88]   Eric Bach. How to generate factored random numbers. *SIAM Journal of
           Computing*, 17:179–193, 1988.

[Bac90]   Eric Bach. Explicit bounds for primality testing and related problems.
           Mathematics of Computation, 55:355–380, 1990.

[BBS86]   Lenore Blum, Manuel Blum, and Michael Shub. A simple unpredictable pseudo-random number generator. *SIAM Journal of Computing*, 15:364–383, 1986.

[Ber67]   Elwyn Berlekamp. Factoring polynomials over finite fields. *Bell System Technical Journal*, 46:1853–1859, 1967.

[Ber68]   Elwyn Berlekamp. *Algebraic Coding Theory*. McGraw-Hill, New York, 1968.

[Ber70]   Elwyn Berlekamp. Factoring polynomials over large finite fields. *Mathematics of Computation*, 24:713–735, 1970.

[BK83]   Richard Brent and H. Kung. Systolic VLSI arrays for linear time gcd computation. In F. Anceau and E. Aas, editors, *VLSI 83*, pages 145–154. IFIP, Elsevevier, 1983.

[BM84]   Manuel Blum and Silvio Micali. How to generate cryptographically strong sequences of pseudorandom bits. *SIAM Journal of Computing*, 13:850–864, 1984.

[BM92]   Ernest F. Brickell and Kevin S. McCurley. An interactive identification scheme based on discrete logarithms and factoring. *Journal of Cryptology*, 5:29–40, 1992.

[BMS84]   Eric Bach, Gary L. Miller, and Jeffrey O. Shallit. Sums of divisors, perfect numbers, and factoring. In *Proceedings of the 16th Annual Symposium on Theory of Computing*, New York, 1984. Association for Computing Machinery.

[Bom74]   Enrico Bombieri. Le grand crible dans la théorie analytique des nombres. Avec une sommaire en anglais. *Astérisque*, 18, 1974.

[BS91]   Johannes Buchmann and Victor Shoup. Constructing non-residues in finite fields and the extended Riemann hypothesis. In *Proceedings of the 23th Annual Symposium on Theory of Computing*, pages 72–79, 1991.

[Buc90]   Johannes Buchmann. Complexity of algorithms in algebraic number theory. In R.A. Mollin, editor, *Proceedings of the First Conference of the Canadian Number Theory Association*, pages 37–53, Berlin, 1990. De Gruyter.

[BW89]   Johannes Buchmann and Hugh C. Williams. On the existence of a short proof for the value of the class number and regulator of a real quadratic field. In Richard A. Mollin, editor, *Proceedings of the NATO Advanced Study Institute on Number Theory and Applications*, pages 327–345, The Netherlands, 1989. Kluwer.

[CG]   B. Chor and O. Goldreich. An improved parallel algorithm for integer GCD. *Algorithmica*. To Appear.

[Chi89]   A. L. Chistov. The complexity of constructing the ring of integers of a global field. *Dokl. Akad. Nauk. SSSR*, 306:1063–1067, 1989. English translation: Soviet Math. Dokl. 39 (1989), 597-600.

[CL84]   H. Cohen and H. W. Lenstra, Jr. Primality testing and Jacobi sums. *Mathematics of Computation*, 42:297–330, 1984.

[Coo81]   Stephen Cook. Towards a complexity theory of synchronous parallel computation. *Enseignment Math.*, 27:99–124, 1981.

[Coo85]   Stephen Cook. A taxonomy of problems with fast parallel algorithms. *Information and Control*, 64:2–22, 1985.

[Cop84]   Don Coppersmith. Fast evaluation of discrete logarithms in fields of characteristic two. *IEEE Transactions on Information Theory*, 30:587–594, 1984.

[Cop90]   Don Coppersmith. Modifications to the number field sieve. Technical Report RC16264, IBM TJ Watson Research Center, Yorktown Heights, New York, 1990.

[COS86]   Don Coppersmith, Andrew Odlyzko, and Richard Schroeppel. Discrete logarithms in GF(p). *Algorithmica*, 1:1–15, 1986.

[Cou93]   Jean-Marc Couveignes. Computing a square root for the number field sieve. In A. K. Lenstra and H. W. Lenstra, Jr., editors, *The development of the number field sieve*, number 1554 in Lecture Notes in Mathematics, pages 95–102. Springer-Verlag, 1993.

[Cra36]   H. Cramér. On the order of magnitude of the difference between consecutive prime numbers. *Acta Arithmetica*, 2:23–46, 1936.

[CZ81]   David Cantor and Hans Zassenhaus. A new algorithm for factoring polynomials over finite fields. *Mathematics of Computation*, 36:587–592, 1981.

[dB90]   Bert den Boer. Diffie-Hellman is as strong as discrete log for certain primes. In *Advances in Cryptology: Proceedings of Crypto '88*, volume 403 of *Lecture Notes in Computer Science*, pages 530–539, New York, 1990. Springer-Verlag.

[DH76]   W. Diffie and M. E. Hellman. New directions in cryptography. *IEEE Transactions on Information Theory*, 22:644–654, 1976.

[Dix81]   John D. Dixon. Asymptotically fast factorization of integers. *Mathematics of Computation*, 36:255–260, 1981.

[EH71]   P. D. T. A. Elliot and H. Halberstam. The least prime in an arithmetic progression. In *Studies in Pure Mathematics*, pages 69–61. Academic Press, London, 1971.

[ElG85]   Taher ElGamal. A public key cryptosystem and a signature scheme based on discrete logarithms. *IEEE Transactions on Information Theory*, 31:469–472, 1985.

[Erd61]   P. Erdös. Remarks on number theory, I. *Mat. Lapok*, 12:10–17, 1961.

[Evd89]   S. A. Evdokimov. Factoring a solvable polynomial over a finite field and generalized Riemann hypothesis. *Zapiski Nauchn. Semin. Leningr. Otdel. Matem. Inst. Acad. Sci. USSR*, 176:104–117, 1989. In Russian.

[Evdar]   S. A. Evdokimov. Factorization of polynomials over finite fields. In *Proceedings of the 1994 Algorithmic Number Theory Symposium*, Lecture Notes in Computer Science, Berlin, to appear. Springer-Verlag.

[Fru85]   M. A. Frumkin. Complexity questions in number theory. *J. Soviet Math.*, 29:1502–1517, 1985. Translated from Zapiski Nauchnykh Seminarov Leningradskogo Otdeleniya Matematicheskogo Instituta im. V. A. Steklova AN SSSR, vol. 118 (1982), 188-210.

[Für85]   Martin Fürer. Deterministic and Las Vegas primality testing algorithms. In *Proceedings of ICALP 1985*, 1985.

[Gau86]   Karl Friedrich Gauss. *Disquisitiones Arithmeticæ*. Springer-Verlag, New York, 1986. Reprint of the 1966 English translation by Arthur A. Clarke, S.J., Yale University Press, revised by William C. Waterhouse. Original 1801 edition published by Fleischer, Leipzig.

[Gil77]   John Gill. Computational complexity of probabilistic Turing machines. *SIAM Journal of Computing*, 4:675–695, 1977.

[GJ79]   Michael Garey and David Johnson. *Computers and Intractibility: A Guide to the Theory of NP-Completeness*. W. H. Freeman, San Francisco, 1979.

[GK86]   Shafi Goldwasser and Joe Kilian. Almost all primes can be quickly certified. In *Proceedings of the 18th Annual Symposium on Theory of Computing*, pages 316–329, New York, 1986. Association for Computing Machinery.

[GLS88]   Martin Grötschel, László Lovász, and Alexander Schrijver. *Geometric Algorithms and Combinatorial Optimization.* Springer-Verlag, Berlin, 1988.

[GM82]   Shafi Goldwasser and Silvio Micali. Probabilistic encryption & how to play mental poker keeping secret all partial information. In *Proceedings of the 14th Annual Symposium on Theory of Computing*, pages 365–377, New York, 1982. Association for Computing Machinery.

[GM84]   Shafi Goldwasser and Silvio Micali. Probabilistic encryption. *Journal of Computer and System Science*, 28:270–299, 1984.

[Gol85]   Dorian Goldfeld. Gauss' class number problem for imaginary quadratic fields. *Bulletin of the American Mathematical Society*, 13:23–38, 1985.

[Gor93]   Daniel M. Gordon. Discrete logarithms in GF(p) using the number field sieve. *SIAM Journal of Discrete Mathematics*, 6:124–138, 1993.

[Guy77]   Richard Guy. How to factor a number. *Congressus Numeratium*, XXVII:49–89, 1977. Proceedings of the Fifth Manitoba Conference on Numerical Mathematics, University of Manitoba.

[HB78]   D. R. Heath-Brown. Almost-primes in arithmetic progressions and short intervals. *Mathematical Proceedings of the Cambridge Philosophical Society*, 83:357–375, 1978.

[HM89a]   James Lee Hafner and Kevin S. McCurley. On the distribution of running times of certain integer factoring algorithms. *Journal of Algorithms*, 10:531–556, 1989.

[HM89b]   James Lee Hafner and Kevin S. McCurley. A rigorous subexponential algorithm for computation of class groups. *Journal of the American Mathematical Society*, 2:837–850, 1989.

[Hua85]   Ming-Deh A. Huang. Riemann hypothesis and finding roots over finite fields. In *Proceedings of the 17th Annual Symposium on Theory of Computing*, pages 121–130, New York, 1985. Association for Computing Machinery.

[Hua91]   Ming-Deh A. Huang. Generalized riemann hypothesis and factoring polynomials over finite fields. *Journal of Algorithms*, 12:464–481, 1991.

[Kan85]   William Kantor. Polynomial-time algorithms for finding elements of prime order and Sylow subgroups. *Journal of Algorithms*, 4:478–514, 1985.

[KMR84]   Ravi Kannan, Gary L. Miller, and L. Rudolph. Sublinear parallel algorithm for computing the greatest common divisor of two integers. In *Proceedings of the 25th Annual Symposium on Foundations of Computer Science*, pages 7–11. IEEE Computer Society, 1984.

[KMR87]   Ravi Kannan, Gary L. Miller, and L. Rudolph. Sublinear parallel algorithm for computing the greatest common divisor of two integers. *SIAM Journal of Computing*, 16:7–16, 1987. Extended abstract in [KMR84].

[Knu81]   Donald E. Knuth. *Seminumerical Algorithms*, volume 2 of *The Art of Computer Programming*. Addison-Wesley, Reading, Massachusetts, second edition, January 1981.

[Kob87a]   Neal Koblitz. *A Course in Number Theory and Cryptography.* Number 114 in Graduate Texts in Mathematics. Springer-Verlag, New York, 1987.

[Kob87b]   Neal Koblitz. Elliptic curve cryptosystems. *Mathematics of Computation*, 48:203–209, 1987.

[Kob88]   Neal Koblitz. Primality of the number of points on an elliptic curve over a finite field. *Pacific Journal of Mathematics*, 131:157–165, 1988.

[Kob90]   Neal Koblitz. A family of Jacobians suitable for discrete log cryptosystems. In S. Goldwasser, editor, *Advances in Cryptology: Proceedings of Crypto*

'88, volume 403 of *Lecture Notes in Computer Science*, pages 94–99, Berlin, 1990. Springer-Verlag.

[Kob91] Neal Koblitz. Constructing elliptic curve cryptosystems in characteristic 2. In A. J. Menezes and S. A. Vanstone, editors, *Advances in Cryptology: Proceedings of Crypto '90*, volume 537 of *Lecture Notes in Computer Science*, pages 156–167, Berlin, 1991. Springer-Verlag.

[KP94] Sergei Konyagin and Carl Pomerance. On primes recognizable in deterministic polynomial time. preprint, May 1994.

[Lag79] Jeffrey C. Lagarias. Succinct certificates for the solvability of binary quadratic diophantine equations. In *Proceedings of the 20th Annual Symposium on Foundations of Computer Science*, pages 47–54. IEEE Computer Society, 1979.

[Lag80a] Jeffrey C. Lagarias. On the computational complexity of determining the solvability or unsolvability of the equation $x^2 - dy^2 = -1$. *Transactions of the American Mathematical Society*, 260:485–508, 1980.

[Lag80b] Jeffrey C. Lagarias. Worst-case complexity bounds for algorithms in the theory of integral quadratic forms. *Journal of Algorithms*, 1:142–186, 1980.

[Lag85] Jeffrey C. Lagarias. The computational complexity of simultaneous diophantine approximation problems. *SIAM Journal of Computing*, 14:196–209, 1985.

[Lan85] Susan Landau. Polynomial time algorithms for Galois groups. In J. Fitch, editor, *Proceedings of EUROSAM '84*, volume 174 of *Lecture Notes in Computer Science*, pages 225–236, New York, 1985. Springer-Verlag.

[Lan88] Susan Landau. Some remarks on computing the square parts of integers. *Information and Computation*, 78:246–253, 1988.

[Leh69] D. H. Lehmer. Computer technology applied to the theory of numbers. In *Studies in Number Theory*, pages 117–151. Mathematical Association of America, 1969. Distributed by Prentice Hall, Englewood Cliffs, NJ.

[Len81] H. W. Lenstra, Jr. Primality testing algorithms (after Adleman, Rumely, and Williams). In *Séminaire Bourbaki 1980/81, Exposé 576*, number 901 in Lecture Notes in Mathematics, pages 243–257. Springer-Verlag, Berlin, 1981.

[Len83] H. W. Lenstra, Jr. Integer programming with a fixed number of variables. *Mathematics of Operations Research*, 8:538–548, 1983.

[Len87] H. W. Lenstra, Jr. Factoring integers with elliptic curves. *Annals of Mathematics*, 126:649–673, 1987.

[Len90] H. W. Lenstra, Jr. Algorithms for finite fields. In *Number Theory and Cryptography*, volume 154 of *London Mathematical Society Lecture Note Series*, pages 76–85. Cambridge University Press, Cambridge, 1990.

[Len92] H. W. Lenstra, Jr. Algorithms in algebraic number theory. *Bulletin of the American Mathematical Society*, 26:211–244, 1992.

[LL93] Arjen K. Lenstra and H. W. Lenstra, Jr., editors. *The development of the number field sieve*. Number 1554 in Lecture Notes in Mathematics. Springer-Verlag, 1993.

[LLL82] Arjen K. Lenstra, H. W. Lenstra, Jr., and László Lovász. Factoring polynomials with rational coefficients. *Mathematische Annalen*, 261:515–534, 1982.

[LLS90] Jeffrey C. Lagarias, H. W. Lenstra, Jr., and Claus-Peter Schnorr. Korkine-Zolotarev bases and successive minima of a lattice and its reciprocal lattice. *Combinatorica*, 10:333–348, 1990.

[LM85]   Susan Landau and Gary L. Miller. Solvability by radicals is in polynomial time. *Journal of Computer and System Science*, 30:179–208, 1985.

[LN83]   Rudolf Lidl and Harald Niederreiter. *Finite Fields*. Addison-Wesley, Reading, MA, 1983.

[Lov86] László Lovász. *An Algorithmic Theory of Numbers, Graphs, and Convexity*. Number 50 in CBMS-NSF Regional Conference Series in Applied Mathematics. Society of Industrial and Applied Mathematicians, Philadelphia, PA, 1986.

[Lov92] Renet Lovorn. *Rigorous Subexponential Algorithms for Discrete Logarithms over Finite Fields*. PhD thesis, University of Georgia, May 1992.

[LP92]   H. W. Lenstra, Jr. and Carl Pomerance. A rigorous time bound for factoring integers. *Journal of the American Mathematical Society*, 5:483–516, 1992.

[MA78]   K. Manders and Leonard M. Adleman. NP-complete decision problems for binary quadratics. *Journal of Computer and System Science*, 16:168–184, 1978.

[MB75]   Michael Morrison and John Brillhart. A method of factoring and the factorization of F_7. *Mathematics of Computation*, 29:183–205, 1975.

[McC88] Kevin S. McCurley. A key distribution system equivalent to factoring. *Journal of Cryptology*, 1:95–105, 1988.

[McC89] Kevin S. McCurley. Cryptographic key distribution and computation in class groups. In Richard A. Mollin, editor, *Proceedings of the NATO Advanced Study Institute on Number Theory and Applications*, pages 459–479, The Netherlands, 1989. Kluwer.

[McC90a] Kevin S. McCurley. The discrete logarithm problem. In Pomerance [Pom90], pages 49–74.

[McC90b] Kevin S. McCurley. Odds and ends from cryptology and computational number theory. In Pomerance [Pom90], pages 145–166.

[Mil76] Gary L. Miller. Riemann's hypothesis and tests for primality. *Journal of Computer and System Science*, 13:300–317, 1976.

[Mil86] Victor Miller. Use of elliptic curves in cryptography. In *Advances in Cryptology: Proceedings of Crypto '85*, volume 218 of *Lecture Notes in Computer Science*, pages 417–426, Berlin, 1986. Springer-Verlag.

[MOV94] Alfred J. Menezes, Tatsuaki Okamoto, and Scott A. Vanstone. Reducing elliptic curve logarithms to logarithms in a finite field. *IEEE Transactions in Information Theory*, ???, 1994. Extended abstract in Proceedings of the 23rd ACM Symposium on Theory of Computing, 1991, ACM, pp. 80–89.

[oC91]   U.S. Department of Commerce. A proposed federal information processing standard for digital signature standard. In *Federal Register*, volume 56, no. 169, pages 42980–42982. U.S. GPO, August 1991.

[Odl85] Andrew Odlyzko. The discrete logarithm problem and its cryptographic significance. In *Advances in Cryptology: Proceedings of Eurocrypt '84*, volume 209 of *Lecture Notes in Computer Science*, pages 224–314, Berlin, 1985. Springer-Verlag.

[Odl94] Andrew Odlyzko. Discrete logarithms and smooth polynomials. In Gary L. Mullen and Peter Shiue, editors, *Finite Fields: Theory, Applications, and Algorithms*, Contemporary Mathematics Series, Providence, RI, 1994. American Mathematical Society.

[OSS84] H. Ong, Claus-Peter Schnorr, and Adi Shamir. An efficient signature scheme based on quadratic equations. In *Proceedings of the 16th Annual*

Symposium on Theory of Computing, pages 208–216, New York, 1984. Association for Computing Machinery.

[PH78]   Stephen Pohlig and Martin Hellman. An improved algorithm for computing discrete logarithms over GF(p) and its cryptographic significance. *IEEE Transactions on Information Theory*, 24:106–110, 1978.

[Pil90]   Jonathan Pila. Frobenius maps of abelian varieties and finding roots of unity in finite fields. *Mathematics of Computation*, 55:745–763, 1990.

[Pla79]   D. A. Plaisted. Fast verification, testing, and generation of large primes. *Theoretical Computer Science*, 9:1–16, 1979.

[Pom82]   Carl Pomerance. Analysis and comparison of some integer factoring methods. In H. W. Lenstra, Jr. and R. Tijdeman, editors, *Computational Methods in Number Theory, Part I*, number 154 in Math. Centre Tract, pages 89–139. Math. Centre, Amsterdam, 1982.

[Pom86]   Carl Pomerance. Fast, rigorous factorization and discrete logarithm algorithms. In *Discrete Algorithms and Complexity (Proceedings of the Japan-US Joint Seminar on Discrete Algorithms and Complexity Theory)*, pages 119–143. Academic Press, 1986.

[Pom90]   Carl Pomerance, editor. *Cryptography and Computational Number Theory*, volume 42 of *Proceedings of Symposia in Applied Mathematics*. American Mathematical Society, Providence, 1990.

[Pom81]   Carl Pomerance. Recent developments in primality testing. *Mathematical Intelligencer*, 3:97–105, 1980/81.

[Pra75]   Vaughn Pratt. Every prime has a succinct certificate. *SIAM Journal of Computing*, 4:214–220, 1975.

[PS87]   John Pollard and Claus-Peter Schnorr. Solution of $x^2 + ky^2 \equiv m \pmod{n}$, with application to digital signatures. *IEEE Transactions on Information Theory*, 22:702–709, 1987.

[PSS88]   Janos Pintz, William L. Steiger, and Endre Szemerédi. Two infinite sets of primes with fast primality tests. In *Proceedings of the 20th Annual Symposium on Theory of Computing*, pages 504–509. Association for Computing Machinery, 1988. Journal version in [PSS89].

[PSS89]   Janos Pintz, William L. Steiger, and Endre Szemerédi. Infinite sets of primes with fast primality tests and quick generation of large primes. *Mathematics of Computation*, 53:399–406, 1989. Extended abstract in [PSS88].

[Rab79]   Michael O. Rabin. Digitalized signatures and public-key functions as intractible as factorization. Technical Report TR-212, Massachussetts Institute of Technology, Laboratory for Computer Science, 1979.

[Rab80a]   Michael O. Rabin. Probabilistic algorithm for testing primality. *Journal of Number Theory*, 12:128–138, 1980.

[Rab80b]   Michael O. Rabin. Probabilistic algorithms in finite fields. *SIAM Journal of Computing*, 9:273–280, 1980.

[Rie85a]   Hans Riesel. Modern factorization methods. *BIT*, 25:205–222, 1985.

[Rie85b]   Hans Riesel. *Prime Numbers and Computer Methods for Factorization*. Birkhäuser, Boston, 1985.

[Rón88]   L. Rónyai. Factoring polynomials over finite fields. *Journal of Algorithms*, 9:391–400, 1988.

[Rón89]   L. Rónyai. Galois groups and factoring polynomials over finite fields. In *Proceedings of the 30th Annual Symposium on Foundations of Computer Science*, pages 99–104. IEEE Computer Society, 1989.

[RSA78]   Ronald Rivest, Adi Shamir, and Leonard M. Adleman. A method for obtaining digital signatures and public key cryptosystems. *Communications of the ACM*, 21:120–126, 1978.

[Sch82]   Rene Schoof. Quadratic fields and factorisation. In H. W. Lenstra, Jr. and R. Tijdeman, editors, *Computational Methods in Number Theory, Part I*, number 154 in Math. Centre Tract. Math. Centre, Amsterdam, 1982.

[Sch85]   R. Schoof. Elliptic curves over finite fields and the computation of square roots modulo p. *Mathematics of Computation*, 44:483–494, 1985.

[Sch91]   Claus-Peter Schnorr. Efficient signature generation by smart cards. *Journal of Cryptology*, 4:161–174, 1991.

[Sch93]   Oliver Schirokauer. Discrete logarithms and local units. *Philisophical Transactions of the Royal Society of London (A)*, 345:409–423, 1993.

[Sey87]   Martin Seysen. A probabilistic factorization algorithm with quadratic forms of negative discriminant. *Mathematics of Computation*, 48:757–780, 1987.

[Sha71]   Daniel Shanks. Class number, a theory of factorization, and genera. In *Analytic Number Theory*, volume 20 of *Proceedings of Symposia in Pure Mathematics*, pages 415–440. American Mathematical Society, 1971.

[Sha72]   Daniel Shanks. Five number-theoretic algorithms. In *Proceedings of the Second Manitoba Conference on Numerical Mathematics*, pages 51–70, 1972.

[Sha84]   Jeffrey O. Shallit. An exposition of Pollard's algorithm for quadratic congruences. Technical Report 84-006, University of Chicago, Department of Computer Science, December 1984.

[Sho90a] Victor Shoup. New algorithms for finding irreducible polynomials over finite fields. *Mathematics of Computation*, 54:435–447, 1990. Extended abstract in Proceedings of the 29th Annual IEEE Symposium on Foundations of Computer Science (1988), pp. 283–290.

[Sho90b] Victor Shoup. On the deterministic complexity of factoring polynomials over finite fields. *Information Processing Letters*, 33:261–267, 1990.

[Sho90c] Victor Shoup. Searching for primitive roots in finite fields. In *Proceedings of the 22th Annual Symposium on Theory of Computing*, pages 546–554, 1990.

[Shoar]   Peter W. Shor. Polynomial time algorithms for discrete logarithms and factoring on a quantum computer. In *Proceedings of the 1994 Algorithmic Number Theory Symposium*. Springer-Verlag, to appear.

[Sil86]   Joseph H. Silverman. *The Arithmetic of Elliptic Curves*, volume 106 of *Graduate Texts in Mathematics*. Springer-Verlag, 1986.

[SL84]   Claus-Peter Schnorr and H. W. Lenstra, Jr. A Monte Carlo factoring algorithm with linear storage. *Mathematics of Computation*, 43:289–311, 1984.

[Sor90]   Jonathan Sorenson. Counting the integers factorable via cyclotomic methods. Technical Report 919, University of Wisconsin, Deparment of Computer Sciences, 1990.

[SS71]   Alfred Schönhage and Volker Strassen. Schnelle Multiplikation grosser Zahlen. *Computing*, 7:281–292, 1971.

[SS77]   R. Solovay and Volker Strassen. A fast Monte-Carlo test for primality. *SIAM Journal of Computing*, 6:84–85, 1977.

[SS85]   Jeffrey O. Shallit and Adi Shamir. Number-theoretic functions which are equivalent to number of divisors. *Information Processing Letters*, 20:151–153, 1985.

[Thiar]   C. Thiel. Verifying the class number belongs to $\mathcal{NP} \cap co\mathcal{NP}$. In *Proceedings of the 1994 Algorithmic Number Theory Symposium*. Springer-Verlag, to appear.

[Ton91]   A. Tonelli. Bemerkung ber die aufl sung quadratischer Congruenzen. *Göttinger Nachrichten*, pages 314–346, 1891.

[vEB81]   P. van Emde Boas. Another NP-complete partition problem and the complexity of computing short vectors in lattices. Technical Report 81-04, Mathematics Department, University of Amsterdam, 1981.

[vO91]   Paul van Oorschot. A comparison of practical public key cryptosystems based on integer factorization and discrete logarithms. In Gustavus J. Simmons, editor, *Contemporary Cryptology: The Science of Information Integrity*, IEEE Proceedings, pages 280–322. IEEE, 1991.

[vzGKS93]   Joachim von zur Gathen, Marek Karpinski, and Igor Shparlinski. Counting curves and their projections. preprint, March 1993.

[Wan61]   Y. Wang. On the least primitive root of a prime. *Scientia Sinica*, 10:1–14, 1961.

[Wil78]   Hugh C. Williams. Primality testing on a computer. *Ars Combinatorica*, 5:127–185, 1978.

[Wil84]   Hugh C. Williams. Factoring on a computer. *Mathematical Intelligencer*, 6:29–36, 1984.

[Yao82]   Andrew C. Yao. Theory and applications of trapdoor functions. In *Proceedings of the 23rd Annual Symposium on Foundations of Computer Science*, pages 80–91. IEEE Computer Society, 1982.

Author Index

Lecture Notes in Computer Science

For information about Vols. 1–801 ᵛ
please contact your bookseller or Springer-Verlag

Vol. 838: C. MacNish, D. Pearce, L. Moniz Pereira (Eds.), Logics in Artificial Intelligence. Proceedings, 1994. IX, 413 pages. 1994. (Subseries LNAI).

Vol. 839: Y. G. Desmedt (Ed.), Advances in Cryptology - CRYPTO '94. Proceedings, 1994. XII, 439 pages. 1994.

Vol. 840: G. Reinelt, The Traveling Salesman. VIII, 223 pages. 1994.

Vol. 841: I. Prívara, B. Rovan, P. Ružička (Eds.), Mathematical Foundations of Computer Science 1994. Proceedings, 1994. X, 628 pages. 1994.

Vol. 842: T. Kloks, Treewidth. IX, 209 pages. 1994.

Vol. 843: A. Szepietowski, Turing Machines with Sublogarithmic Space. VIII, 115 pages. 1994.

Vol. 844: M. Hermenegildo, J. Penjam (Eds.), Programming Language Implementation and Logic Programming. Proceedings, 1994. XII, 469 pages. 1994.

Vol. 845: J.-P. Jouannaud (Ed.), Constraints in Computational Logics. Proceedings, 1994. VIII, 367 pages. 1994.

Vol. 846: D. Shepherd, G. Blair, G. Coulson, N. Davies, F. Garcia (Eds.), Network and Operating System Support for Digital Audio and Video. Proceedings, 1993. VIII, 269 pages. 1994.

Vol. 847: A. L. Ralescu (Ed.) Fuzzy Logic in Artificial Intelligence. Proceedings, 1993. VII, 128 pages. 1994. (Subseries LNAI).

Vol. 848: A. R. Krommer, C. W. Ueberhuber, Numerical Integration on Advanced Computer Systems. XIII, 341 pages. 1994.

Vol. 849: R. W. Hartenstein, M. Z. Servít (Eds.), Field-Programmable Logic. Proceedings, 1994. XI, 434 pages. 1994.

Vol. 850: G. Levi, M. Rodríguez-Artalejo (Eds.), Algebraic and Logic Programming. Proceedings, 1994. VIII, 304 pages. 1994.

Vol. 851: H.-J. Kugler, A. Mullery, N. Niebert (Eds.), Towards a Pan-European Telecommunication Service Infrastructure. Proceedings, 1994. XIII, 582 pages. 1994.

Vol. 852: K. Echtle, D. Hammer, D. Powell (Eds.), Dependable Computing – EDCC-1. Proceedings, 1994. XVII, 618 pages. 1994.

Vol. 853: K. Bolding, L. Snyder (Eds.), Parallel Computer Routing and Communication. Proceedings, 1994. IX, 317 pages. 1994.

Vol. 854: B. Buchberger, J. Volkert (Eds.), Parallel Processing: CONPAR 94 – VAPP VI. Proceedings, 1994. XVI, 893 pages. 1994.

Vol. 855: J. van Leeuwen (Ed.), Algorithms – ESA '94. Proceedings, 1994. X, 510 pages. 1994.

Vol. 856: D. Karagiannis (Ed.), Database and Expert Systems Applications. Proceedings, 1994. XVII, 807 pages. 1994.

Vol. 857: G. Tel, P. Vitányi (Eds.), Distributed Algorithms. Proceedings, 1994. X, 370 pages. 1994.

Vol. 858: E. Bertino, S. Urban (Eds.), Object-Oriented Methodologies and Systems. Proceedings, 1994. X, 386 pages. 1994.

Vol. 859: T. F. Melham, J. Camilleri (Eds.), Higher Order Logic Theorem Proving and Its Applications. Proceedings, 1994. IX, 470 pages. 1994.

Vol. 860: W. L. Zagler, G. Busby, R. R. Wagner (Eds.), Computers for Handicapped Persons. Proceedings, 1994. XX, 625 pages. 1994.

Vol: 861: B. Nebel, L. Dreschler-Fischer (Eds.), KI-94: Advances in Artificial Intelligence. Proceedings, 1994. IX, 401 pages. 1994. (Subseries LNAI).

Vol. 862: R. C. Carrasco, J. Oncina (Eds.), Grammatical Inference and Applications. Proceedings, 1994. VIII, 290 pages. 1994. (Subseries LNAI).

Vol. 863: H. Langmaack, W.-P. de Roever, J. Vytopil (Eds.), Formal Techniques in Real-Time and Fault-Tolerant Systems. Proceedings, 1994. XIV, 787 pages. 1994.

Vol. 864: B. Le Charlier (Ed.), Static Analysis. Proceedings, 1994. XII, 465 pages. 1994.

Vol. 865: T. C. Fogarty (Ed.), Evolutionary Computing. Proceedings, 1994. XII, 332 pages. 1994.

Vol. 866: Y. Davidor, H.-P. Schwefel, R. Männer (Eds.), Parallel Problem Solving from Nature - PPSN III. Proceedings, 1994. XV, 642 pages. 1994.

Vol 867: L. Steels, G. Schreiber, W. Van de Velde (Eds.), A Future for Knowledge Acquisition. Proceedings, 1994. XII, 414 pages. 1994. (Subseries LNAI).

Vol. 868: R. Steinmetz (Ed.), Multimedia: Advanced Teleservices and High-Speed Communication Architectures. Proceedings, 1994. IX, 451 pages. 1994.

Vol. 869: Z. W. Raś, Zemankova (Eds.), Methodologies for Intelligent Systems. Proceedings, 1994. X, 613 pages. 1994. (Subseries LNAI).

Vol. 870: J. S. Greenfield, Distributed Programming Paradigms with Cryptography Applications. XI, 182 pages. 1994.

Vol. 871: J. P. Lee, G. G. Grinstein (Eds.), Database Issues for Data Visualization. Proceedings, 1993. XIV, 229 pages. 1994.

Vol. 873: M. Naftalin, T. Denvir, M. Bertran (Eds.), FME '94: Industrial Benefit of Formal Methods. Proceedings, 1994. XI, 723 pages. 1994.

Vol. 874: A. Borning (Ed.), Principles and Practice of Constraint Programming. Proceedings, 1994. IX, 361 pages. 1994.

Vol. 875: D. Gollmann (Ed.), Computer Security – ESORICS 94. Proceedings, 1994. XI, 469 pages. 1994.

Vol. 876: B. Blumenthal, J. Gornostaev, C. Unger (Eds.), Human-Computer Interaction. Proceedings, 1994. IX, 239 pages. 1994.

Vol. 877: L. M. Adleman, M.-D. Huang (Eds.), Algorithmic Number Theory. Proceedings, 1994. IX, 323 pages. 1994.

Vol. 878: T. Ishida; Parallel, Distributed and Multiagent Production Systems. XVII, 166 pages. 1994. (Subseries LNAI).

Vol. 879: J. Dongarra, J. Waśniewski (Eds.), Parallel Scientific Computing. Proceedings, 1994. XI, 566 pages. 1994.

Vol. 880: P. S. Thiagarajan (Ed.), Foundations of Software Technology and Theoretical Computer Science. Proceedings, 1994. XI, 451 pages. 1994.

Vol. 882: D. Hutchison, A. Danthine, H. Leopold, G. Coulson (Eds.), Multimedia Transport and Teleservices. Proceedings, 1994. XI, 380 pages. 1994.